PROGRAMMED CELL DEATH IN HIGHER PLANTS

PROGRAMMED CELL DEATH IN HIGHER PLANTS

Edited by

ERIC LAM

Biotech Center, Rutgers University, New Brunswick, NJ, USA

HIROO FUKUDA

Department of Biological Sciences, University of Tokyo, Tokyo, Japan

and

JEAN GREENBERG

Department of Molecular Genetics and Cell Biology, University of Chicago, Chicago, IL, USA

Reprinted from *Plant Molecular Biology*, Volume 44 (3), 2000

SPRINGER SCIENCE+BUSINESS MEDIA, LLC

A C.I.P. Catalogue record for this book is available from the Library of Congress

ISBN 978-94-010-3797-6 ISBN 978-94-010-0934-8 (eBook)
DOI 10.1007/978-94-010-0934-8

Printed on acid-free paper

CONTENTS

Cover illustration

Spontaneous formation of lesions resembling hypersensitive response (HR) lesions in the absence of a pathogen, is a dramatic phenotype occasionally found to accompany the expression of different, mostly unrelated, transgenes in plants. Recent studies indicated that transgene-induced lesion formation is not a simple case of necrosis, but results from the activation of the pathogen-induced programmed cell death (PCD) pathway in transgenic plants. The study of lesion mimic transgenes is important to our understanding of PCD and the signals that control it in plants. Cell death-inducing transgenes may also be useful in biotechnology, and some lesion mimic transgenes were found to induced in plants a state of enhanced systemic resistance against pathogen attack. The cover shows the spontaneous formation of HR-like lesions on flowers of tobacco plants that constitutively express the bacterio-opsin (*bO*) gene, a lesion mimic transgene. This gene is thought to activate the HR in plants by mimicking one of the early signal transduction events that occur during pathogen infection, i.e., enhanced proton flow across the plasma membrane. (Picture contributed by Eric Lam and Ron Mittler). *Plant Molecular Biology* 44, pp. 91–100.

Plant Molecular Biology **44:** vii–viii, 2000.
E. Lam, H. Fukuda and J. Greenberg (Eds.), Programmed Cell Death in Higher Plants.

Preface

Cell death: the 'Yin' path in the balancing act of the life cycle

Eric Lam[1] and Jean Greenberg[2]
[1]*Biotech Center, Foran Hall, Cook College, Rutgers University, 59 Dudley Road, New Brunswick, NJ 08903, USA;* [2]*Molecular Genetics and Cell Biology Department, University of Chicago, 1103 E57 Street, Chicago, IL 60637, USA*

For the Taoists of China in the ancient world, the balance between Yin and Yang is considered the ultimate path of becoming one with Nature. Yin, the Chinese word for shade, symbolizes repose, darkness, as well as death. Yang represents the opposite of these qualities – vigor, light, and life. Thus, it seems that the ancients recognized a long time ago that life is continuously being balanced by opposing forces in order to achieve the optimal state. In an analogous fashion, biologists now also appreciate that organisms are under an equilibrium determined by opposite forces: signals for cellular destruction and others that drive proliferation and/or differentiation. The delicate balance and control of these processes determine the size, shape and well-being of the organism as a whole.

The number of cells contained by a multicellular organism at any specific point in time is the net result of two opposite pathways: cell division and cell death. Of the two processes, cell death has traditionally been more difficult to study since the cells under study are continuously being destroyed and removed. Thus, it is not surprising that our understanding of cell death is lagging behind that of cell division. However, like the study of mitosis, the virtual explosion in cell death research during the past 6 years was ignited by a convergence of results from studies with different model systems. For mitosis, the cloning of the *cdc2* kinase gene from yeasts, the finding that this gene is conserved in all eukaryotes and, finally, the demonstration that its product is the homologue of the kinase subunit of the maturation-promoting factor (MPF) defined in frog eggs united the field of cell division. Perhaps the most spectacular consequence from this unification is the cross-feeding of information, tools and expertise that have been gained from studies using divergent model systems. For cell death research, the breakthrough came about in 1993 with the cloning of the *ced-3* gene from the nematode *Caenorhabditis elegans*. *ced-3* is required for the programmed death of the 131 cells that are normally produced and subsequently removed by this worm. Its sequence revealed *ced-3* encodes a homologue of a mammalian cysteine protease, called ICE, previously identified by its prointerleukin 1-β processing activity. With the cloning of *ced-9*, whose product encodes a negative regulator of cell death in *C. elegans*, and its homology with the mammalian cell death suppressor gene *BCL2* realized in 1994, the field of programmed cell death (PCD) began its steep rise and it is still in its rapid phase of information expansion since these seminal findings.

For plants, it is perhaps ironic that the field of PCD has lagged behind that of the animal systems, since two terms closely associated with this field are botanical in origin. The descriptive noun 'cell' originated from the early microscopist Robert Hooke in the 1650s who examined cork slices under crude light microscopes and likened the regular arrangement of 'pores' to those of cells in a monastery. These are now recognized as plant cells that had underwent PCD in order to form a functional structure that is vital to the plant's survival. Apoptosis, the most well-characterized form of animal PCD, is a Greek word that describes the falling of leaves and petals during their natural senescence process. In spite of the late entry into the field of PCD, we believe that plants promise to hold many surprises and differences from their animal counterparts in the way that their cells may carry out their own demise. As many articles in this special issue of *Plant Molecular Biology* illustrate, plant PCD occurs in a wide variety of cell types with distinct purposes. Unlike animal apoptosis, where the dead cell's corpse is rapidly 'removed' through ingestion by its neighbors, with perhaps the exception of keratinocytes where dead skin cells are continuously shed from the body, most plant cells remain at least for a time after their death. In many instances, such as the tracheary elements, the dead cells serve an important function and are not actively removed during the life of the plant. Comparison of key players in plant PCD with those defined in animal systems should provide interesting clues to the evolution of PCD in multicellular organisms. Several interesting new pieces of information that may help to integrate plant research into that of the general field of PCD studies are: (1) cloning of plant homologues to the retinoblastoma (Rb) gene, a tumor-suppressor gene from animal systems that can also regulate PCD; the cooperation of its product with plant viral factors for infection may be one component that determines the ability of the virus to proliferate in the plant host; (2) flow cytometry studies with tobacco cells revealed that reactive oxygen species, an inducer of PCD in animals, can induce cell cycle arrest at the G_1/S boundary, and accelerate their entry into the death pathway; (3) detection of caspase-like protease activity in plant extracts and the evidence for their essential role in cell death during the hypersensitive response (HR). This result suggests the first biochemical component critical for PCD induction that may be conserved between plants and animals. Recent studies also suggest that plant mitochondria could be playing an important role during plant PCD, at least in the context of the HR. Together with the view that this organelle may also control different forms of cell death in animal systems as well as yeast, these findings are consistent with the view that a conserved pathway for cell death induction may eventually emerge.

With rapid advances being made in genomic sciences, molecular genetics, and cell biological technologies, we anticipate the ramp-up of PCD research in higher plants is merely beginning. Together with our increased understanding of related pathways such as cell cycle control, hormonal signaling, plant-pathogen interactions and cloning of marker genes that may allow us to differentiate between different death pathways, we are optimistic that plant PCD research will make a significant impact on general plant biology in addition to enriching our understanding of the central, conserved mechanisms of the 'death engine'. From cell death in reproductive tissues to xylem differentiation to plant disease responses, we especially hope that this special issue will serve as a timely guide to the young plant scientists who are looking for a new niche of plant biology the potential of which is waiting to be tapped.

SECTION 1

DEVELOPMENTAL CELL DEATH IN PLANTS

Plant Molecular Biology **44:** 245–253, 2000.
E. Lam, H. Fukuda and J. Greenberg (Eds.), Programmed Cell Death in Higher Plants.

Programmed cell death of tracheary elements as a paradigm in plants

Hiroo Fukuda
Department of Biological Sciences, Graduate School of Science, University of Tokyo, Hongo, Tokyo 113-0033, Japan (e-mail: fukuda@biol.s.u-tokyo.ac.jp)

Key words: nuclease, programmed cell death, protease, tracheary element, vacuole, *Zinnia*

Abstract

Plant development involves various programmed cell death (PCD) processes. Among them, cell death occurring during differentiation of procambium into tracheary elements (TEs), which are a major component of vessels or tracheids, has been studied extensively. Recent studies of PCD during TE differentiation mainly using an *in vitro* differentiation system of *Zinnia* have revealed that PCD of TEs is a plant-specific one in which the vacuole plays a central role. Furthermore, there are recent findings of several factors that may initiate PCD of TEs and that act at autonomous degradation of cell contents. Herein I summarize the present knowledge about cell death program during TE differentiation as an excellent example of PCD in plants.

Introduction

In multicellular organisms, programmed cell death (PCD) plays essential roles in the abortion or formation of specific cells and tissues during development to organize their body plan (Ellis *et al.*, 1991; Yamada and Hashimoto, 1998). PCD occurring in a variety of animal cells is controlled by a common mechanism called apoptosis (Kerr and Harmon, 1991). Recent studies revealed that even in plants apoptosis-like cell death occurs in some developmental processes and in response to environmental stimuli (Jones and Dangl, 1996; Greenberg, 1996, 1997; Pennell and Lamb, 1997). However, plants appear to have also a unique cell death program, in which the vacuole plays a critical role. A typical example of such plant-specific PCD is cell death occurring during tracheary element (TE) differentiation (Jones and Dangl, 1996; Fukuda, 1996, 1997a, b, 1998).

Vascular plants have evolved two distinct conductive tissues, vessels/tracheids and sieve tubes, for transport of nutrients, ions, and water. Sieve tubes are formed from sieve elements which are still living but often lose their nucleus and inside of which fluids move. In contrast, the conductive cells of vessels and tracheids are TEs emptied by the loss of all cell contents including the nucleus to form hollow tubes as a pathway for fluids. Although both sieve elements and TEs are differentiated terminally from the procambium or cambium, only TE differentiation includes a program toward cell death. TEs possess a characteristic secondary cell wall of annular, spiral, reticulate or pitted wall thickenings, which add strength and rigidity to the wall and prevent TEs from collapse by high pressure resulting from fluid transport. Therefore the formation of the cell wall and PCD seems to be coupled with each other in TE differentiation. In this article, I summarize the present knowledge of PCD during TE differentiation.

Developmental program toward cell death

TE differentiation *in situ* involves a series of developmental stages, from meristematic stage to TE maturating stage via procambial initial, procambia, immature xylem, and TE precursor stages (Fukuda, 1996). Cell death occurs in TE maturating stage. When is the program of cell death determined during TE differentiation? To answer this question, the analysis with an *in vitro* experimental system is more desirable than that with the *in situ* system. Various lines of work on PCD have been performed with an *in vitro Zinnia* system in which mesophyll cells can transdifferentiate into TEs at high frequency without cell division (Fukuda and Komamine, 1980a, b). A number of molecular and physiological markers allowed us to divide the *in vitro* differentiation process into three stages, I, II and III, and to consider that the *in vitro* differentiation mimics *in situ* differentiation of TEs from the meristem, except for an extra dedifferentiation process *in vitro* which occurs in Stage I (Demura and Fukuda, 1994; Fukuda, 1997a). Dedifferentiated cells *in vitro*, which may be compared to meristem cells *in situ*, differentiate to TE precursor cells via procambium cells and immature xylem cells in Stage II. Stage III involves TE-specific events, that is secondary wall thickening followed by lignification and cell death-related events. Demura and Fukuda (1993, 1994) isolated, as molecular markers of Stage II, three cDNAs of TED2, TED3 and TED4 whose mRNAs accumulated preferentially in developing vascular cells of *Zinnia*. Among them, TED3 was considered as a good marker of TE precursor cells both *in vitro* and *in situ* based on *in situ* hybridization analysis. Transfer of Stage II-cells into non-inductive medium inhibits further development of TEs including cell death (Church and Galston, 1988), demonstrating that the cell death is not determined during Stage II, even in TE precursor cells. Therefore, the transition from Stage II to Stage III will be a critical step toward cell death. Next, I will discuss on factors that may induce the transition.

Factors initiating PCD

Transgenic tobacco expressing a ubiquitin variant that acts as an inhibitor of ubiquitin-dependent protein degradation exhibited abnormal vascular tissues (Bachmair *et al.*, 1990). Immunohistology with an anti-ubiquitin antibody revealed preferential localization of ubiquitin in vascular tissues including xylem cells (Stephenson *et al.*, 1996). Moreover, inhibition of proteasome with clasto-lactacystin β-lactone around the onset of Stage III delayed the differentiation process by 24 h, indicating that proteasome function is required for progression of the TE program including PCD in committed cells (Woffenden *et al.*, 1998). These results suggest that the ubiquitin-proteasome system may be involved in PCD in TEs. The addition of a membrane-permeable E-64 analogue before the secondary wall formation also resulted in the inhibition of both the secondary wall formation and cell death, but the addition at or just after the secondary wall formation starts showed no inhibitory effect on secondary wall formation or the early process of cell death (Yoriko Watanabe and Hiroo Fukuda, unpublished). This suggests that a cysteine protease(s) may also be involved in the entry into Stage III.

Chlorotetracycline staining of cultured *Zinnia* cells indicated that the increase in sequestered Ca^{2+} in differentiating TEs or TE precursor cells, which was speculated to result from the increase in Ca^{2+} uptake (Roberts and Haigler, 1989). Roberts and Haigler (1990, 1992), with calcium channel blockers, have also revealed that the preceding uptake of Ca^{2+} into the cell is necessary for entry into Stage III. Calmodulin antagonists also prevent *Zinnia* cells from entering Stage III (Kobayashi and Fukuda, 1994). Before entering Stage III, calmodulin levels increase transiently, and then a few calmodulin-binding proteins start being expressed in a differentiation-specific manner (Kobayashi and Fukuda, 1994). Therefore, the calcium/calmodulin system may be a key factor initiating entry into Stage III.

Very recently, biosynthesis of a brassinosteroid, a steroid-type phytohormone, was found to be another factor that is necessary for entry into Stage III (Yamamoto *et al.*, 1997). Iwasaki and Shibaoka (1991) have indicated that uniconazole, an inhibitor of gibberellin and brassinosteroid biosynthesis, suppresses TE differentiation. Exogenously supplied brassinolide, an active brassinosteroid, but not gibberellic acid counteracts its suppression. Yamamoto *et al.* (1997) revealed, with a variety of genes that are expressed in different stages of TE differentiation in *Zinnia*, that uniconazole specifically suppresses the accumulation of transcripts for genes that were induced in Stage III but did not inhibit the accumulation of transcripts that appear in Stage I and Stage II (Figure 1). This suppression was overcome with the addition of brassinolide. Indeed, some intermediates for the biosynthesis of brassinolide and an active brassinosteroid, castasterone rapidly increased in the late Stage II (Ryo Yamamoto, Shozo Fujioka, Shigeo Yoshida and Hiroo Fukuda, unpublished). Therefore endogenous brassinosteroids which are synthesized in the late Stage II may induce the transition from Stage II into Stage III toward cell death.

Although some factors mentioned above may initiate the transition from Stage II to Stage III, an interrelationship between these factors is still unknown.

The burst of reactive oxygen species including O_2^- and H_2O_2 can trigger a signal pathway leading to

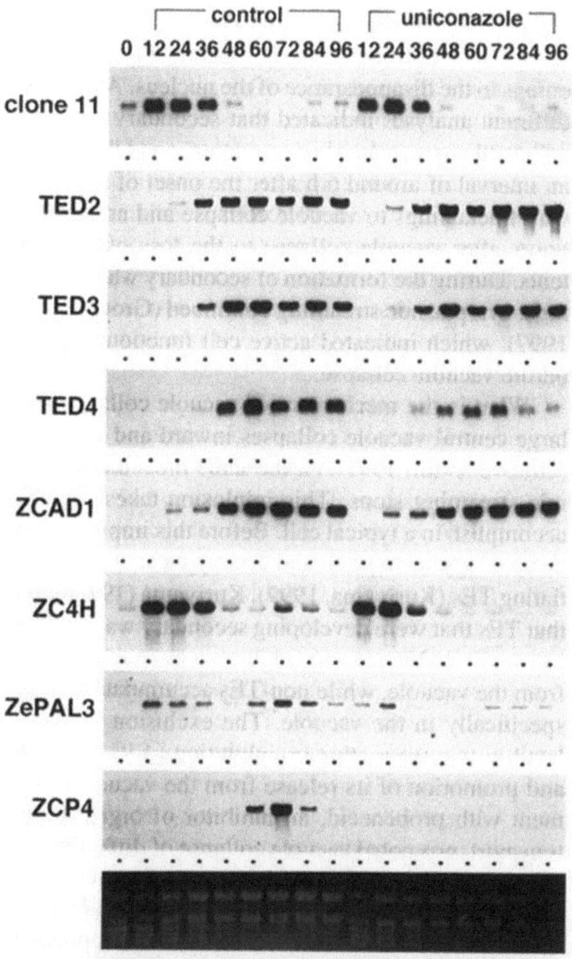

Figure 1. Changes in the accumulation of mRNAs for various genes during TE differentiation in the presence or absence of uniconazole. Total RNA was isolated from *Zinnia* mesophyll cells that had been cultured for indicated periods in a TE differentiation-inducing medium. mRNA for clone11 accumulates during Stage I. mRNAs for *TED2, TED3, TED4* and *ZCAD1* (a gene encoding cinnamyl alcohol dehydrogenase) begin to accumulate during Stage II. mRNAs for *ZePAL3* (a gene encoding phenylalanine ammonia-lyase) and *ZC4H* (a gene encoding cinnamate 4-hydroxylase) accumulate during both Stage I and III. mRNA for *ZCP4*, a gene encoding cysteine protease, accumulates specifically during Stage III. Note that stage III-specific accumulation of genes is preferentially inhibited by uniconazole. (After Yamamoto *et al.*, 1997.)

PCD not only in animals (Jacobson, 1996) but also in plants (Lamb and Dixon, 1997). To assay the production of an oxidative burst during PCD of TEs, hydrogen peroxide levels were monitored for cells cultured in TE-inductive or non-inductive medium using the fluorescent dye 2,7-dichloroflurorescein diacetate (DCF) (Groover *et al.*, 1997). DCF revealed that cells cultured in TE-inductive medium produced a modest increase in hydrogen peroxide, but the increase was much less than that in cells cultured in non-inductive medium. Hydrogen peroxide challenge induced necrotic cell death but not PCD of TEs (Groover *et al.*, 1997) or the expression of *ZRNaseI*, a TE cell death-specific gene (Ye and Droste, 1996). Diphenyleneiodonium, which is an inhibitor of NADPH oxidase, failed to inhibit TE PCD (Groover *et al.*, 1997). These results indicate that the oxidative burst is not involved in PCD of TEs.

Process of cell death

There have been a number of ultrastructural observations of cell death processes during TE differentiation (O'Brien and Thimann, 1967; Srivastava and Singh, 1972; Lai and Srivastava, 1976; Esau and Charvat, 1978; Burgess and Linstead, 1984; Groover *et al.*, 1997; Figure 2). These observations have indicated, during PCD of TEs, rapid and progressive degeneration of organelles, and finally the removal of protoplasts and parts of unlignified primary walls. Epifluorescent microscopic observations revealed that differentiating *Zinnia* TEs have all organelles such as the nucleus, vacuole, and many active mitochondria and chloroplasts, and that loss of such cell contents occurs abruptly several hours after the formation of visible secondary wall thickenings (Groover *et al.*, 1997). The earliest sign of organelle degradation is vacuole collapse (Fukuda, 1996; Groover *et al.*, 1997). After vacuole collapse, single-membrane organelles, endoplasmic reticulum (ER) and Golgi bodies first swell at the ends and then over their entire length, become balloon-like structures, and disappear. A little later than the sign of degeneration of single-membrane organelles, the degeneration of double-membrane organelles such as chloroplasts and mitochondria occurs first in their matrix and then in their membranes.

Nuclear condensation and fragmentation, which are typical features of apoptotic cell death in animals, do not occur during this process. A TdT-mediated dUTP nick- end labeling (TUNEL) assay (Gavrieli *et al.*, 1992) was performed with the aim of understanding specific degradation of nuclear DNA prior to cytological changes in the nucleus and cell morphology. Although the nucleus of TEs close to their maturation labeled with the TUNEL assay (Mittler and Lam 1995; Groover *et al.*, 1997), the nucleus of most of the developing TEs with pronounced secondary cell wall did not label (Groover *et al.*, 1997), indicating that

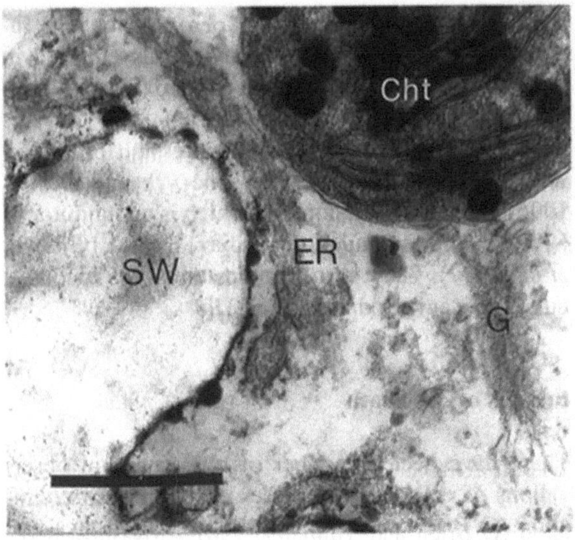

Figure 2. An electron microscopical observation of degrading organelles in a TE. The bar shows 0.5 μm. SW, secondary wall; ER, endoplasmic reticulum; G, Golgi apparatus; Cht, chloroplast.

DNA degradation does not precede cellular changes. TUNEL-positive signal in differentiating TEs may result from function of nucleases released by vacuole collapse, because a nuclease(s) with DNase activity is expressed specifically in differentiating TEs (Thelen and Northcote, 1989; Aoyagi *et al.*, 1998) and may accumulate in the vacuole (see below). In contrast, non-TEs did not label with TUNEL assay. DNA ladder has not been detected in differentiating TEs either. Thus, although death of TEs is a developmentally programmed and active process, its cytological feature differs from apoptosis.

Vacuole: a key organelle for performing cell death

The most striking feature of TE PCD is the rapid collapse of the large central vacuole, which leads to the final degradation of cell contents. Observation of cultured *Zinnia* cells by a time-lapse video microscopy indicated that the onset of secondary wall thickenings precedes vacuole collapse by 6 h, after which most cell contents are cleaned up in several hours (Groover *et al.*, 1997). In contrast, protoplast of cells killed during the isolation process or by exogenous azide persisted for many days. The time, 6 h, was consistent with the longevity of living TEs after the onset of secondary wall thickenings, when measured with fluorescein diacetate (Kuriyama, 1999). Aoyagi *et al.* (1998) also examined the timing of the disappearance of the nucleus in differentiating *Zinnia* TEs and found that it took 6 h, from the start of secondary wall thickenings to the disappearance of the nucleus. All of these different analyses indicated that secondary walls and cell death are regulated concomitantly and there exists an interval of around 6 h after the onset of secondary wall thickenings to vacuole collapse and another few hours after vacuole collapse to the loss of cell contents. During the formation of secondary wall thickenings, cytoplasmic streaming continued (Groover *et al.*, 1997), which indicated active cell function even just before vacuole collapse.

What is the mechanism of vacuole collapse? The large central vacuole collapses inward and fragments (Groover *et al.*, 1997). At the same time, all cytoplasmic streaming stops. This implosion takes 3 min to accomplish in a typical cell. Before this implosion, the vacuole appears to swell and occupy most of differentiating TEs (Kuriyama, 1999). Kuriyama (1999) found that TEs that were developing secondary wall thickenings excluded fluorescein, a fluorescent organic anion, from the vacuole, while non-TEs accumulated the dye specifically in the vacuole. The exclusion was regulated in two ways, that is, inhibition of its entry into and promotion of its release from the vacuole. Treatment with probenecid, an inhibitor of organic anion transport, promoted vacuole collapse of differentiating TEs and led to the ultimate disruption of the vacuole of non-TEs in a process similar to that observed in TEs. These changes in vacuole properties were suppressed by cycloheximide. Therefore a change in the organic anion permeability of the tonoplast may initiate vacuole collapse in TEs *in vivo*. The mechanism by which the inhibition of the transport of organic anion through the tonoplast leads to vacuole collapse is unknown, although Kuriyama (1999) speculates the possibility that osmoregulation by the vacuole is disturbed by the inhibition.

Vacuole collapse causes the release of insulated hydrolytic enzymes and allows them to attack the organelles. In addition, autophagy may be involved in degradation of some of the cell contents during autolysis in TEs, because Groover *et al.* (1997) reported that the cytoplasm becomes less dense in appearance before vacuole collapse. Groover and Jones (1999) examined effects of various factors on nuclear DNA degradation with TUNEL in *Zinnia* TEs and found that mastoparan, an activator of heterotrimeric G-proteins, induced significant numbers of cells to DNA degradation. Other agents, some of which induced cell death such as NaN_3, okadaic acid, hydrogen peroxide and Triton X-100, did not cause DNA degradation in TEs. These results indicate that mastoparan specifically activates an endogenous process required for rapid collapse of the vacuole, leading to autolysis and fragmentation of DNA. Mastoparan-induced cell death and DNA fragmentation were suppressed by the addition of antagonists of Ca^{2+} influx, and Ca^{2+} ionophore itself induced DNA fragmentation. This result strongly suggests that an influx of Ca^{2+} into the cell can cause rapid collapse of large hydrolytic vacuole. Groover and Jones (1999) also reported an interesting result that exogenously supplied trypsin induced DNA degradation, and that chelators and channel blockers of calcium suppressed its inhibition. They also found a 40 kDa protease secreted during PCD of TEs, activity of which can be prevented by a trypsin inhibitor. From these results, they proposed a model of cell death induction that a secreted protease digests some protein component of cell wall, the resultant protein fragments promote an influx of Ca^{2+} through the plasma membrane, and then the increase in intracellular Ca^{2+} causes the vacuole collapse. Although this is an attractive model for the mechanism of vacuole collapse in TEs, there are still many problems to be solved. For example, the timing of secretion of the protease is so late that this protease may not initiate the vacuole collapse. Furthermore, experiments with calcium channel blockers by Roberts and Haigler (1990, 1992) have revealed that the preceding entry of Ca^{2+} into the cell is necessary for secondary wall formation as well. Therefore the entry of Ca^{2+} does not seem to be a direct inducer of vacuole collapse *in situ*, but rather it may be involved in the progression of Stage III including cell death and secondary wall thickenings. In any case, it remains to be shown if an influx of Ca^{2+} indeed increases rapidly before the vacuole of developing TEs collapses *in situ*.

Autolytic process

It has been demonstrated that single cells can differentiate without direct association with neighboring cells by serial observation of cells (Fukuda and Komamine, 1980b) or using a time-lapse videomicroscope (Groover *et al.*, 1997). Using *Zinnia* cells immobilized on gelatin-coated slides, Beers and Freeman (1997) indicated the increased protease activity is specific to differentiating TEs but not to non-differentiating cells. These results strongly suggest that individual cells actively and autonomously degrade their contents without the assistance of neighboring cells. This autonomous degradation of cell contents is performed by a group of hydrolytic enzymes that are newly expressed in differentiating TEs. I will discuss the involvement of hydrolytic enzymes, in particular proteases and nucleases, in the autolytic process of PCD of TEs.

Involvement of proteases

Ye and Varner (1996) isolated a protease gene, p48h-17, whose mRNA accumulates specifically in Stage III of TE differentiation. This gene encodes a papain-like cysteine protease with a molecular mass of 38 kDa containing a putative signal peptide. The protease was predicted to have a mature form with a molecular mass of 23 kDa by protein processing, based on the comparison of the sequences between cysteine proteases, and on the result of transient expression of this gene in tobacco protoplasts. DNA gel blot analysis suggested the presence of another similar gene (Ye and Varner, 1996). Indeed, a similar gene, *ZCP4*, has been isolated from differentiating *Zinnia* cells and the two cysteine protease genes appear to be expressed in a similar pattern (Ye and Varner, 1996; Yamamoto *et al.*, 1997). In the extracts from *Zinnia* cells cultured in TE-inductive medium, E-64-sensitive 28 and 24 kDa protease activities were detected (Ye and Varner, 1996; Beers and Freeman, 1997). Minami and Fukuda (1995) also found about 30 kDa cysteine protease activity, which rapidly increases just before the start of autolysis in *Zinnia*. Thus a few 23 to 30 kDa cysteine proteases are induced prior to autolysis and function in autolysis during PCD of TEs. Where are these proteases located? Because the 30 kDa cysteine protease is prepared from cells and has a pH optimum of pH 5.5, the protease is predicted to accumulate in the vacuole. The inhibition of cysteine protease activities with carbobenzoxy-leucinyl-leucinyl-leucinal resulted in incomplete autolysis in *Zinnia* TEs (Woffenden *et al.*, 1998). The addition of a membrane-permeable E-64 analogue also caused conspicuous inhibition of complete disruption of the nucleus, although it did not inhibit the vacuole collapse (Yoriko Watanabe and Hiroo Fukuda, unpublished). Therefore cysteine proteases may play a role in degrading cell contents and cleaning up the cell after the vacuole collapse. In addition, serine proteases of 145 kDa and 60 kDa were detected specifically in differentiating TEs on substrate-impregnated gels (Ye and Varner, 1996; Beers and Freeman, 1997). The activity of the

60 kDa protease is higher above pH 7, which is unlike the pH optimum of the autolysis-specific cysteine protease, suggesting that the protease may function in the cytoplasm before the vacuole collapse. These results indicate that a complex set of proteases is involved in the autolytic process.

Involvement of nucleases

Elevation of nuclease activities is coupled with PCD of TEs. In *Zinnia* cultured cells, at least 7 RNase active bands of two 43 kDa , a 25 kDa, a 24 kDa, two 22 kDa, and a 17 kDa appeared, six of which were induced specifically in TEs (Thelen and Northcote, 1989). These activities are obtained from cell extract, indicating that differentiating cells have these activities inside. Recently, *ZRNaseI* cDNA which is anticipated to encode the differentiation-specific 22 kDa RNase was isolated (Ye and Droste, 1996). Its corresponding transcripts accumulated specifically in differentiating TEs just before or during secondary wall formation. Tissue print hybridization showed that the expression of the gene was associated preferentially with differentiating TEs in *Zinnia* stem.

The 43 kDa nuclease(s) exhibited activity hydrolyzing both single- and double-stranded DNA in addition to RNA. Because the 43 kDa nuclease(s), among *Zinnia* nucleases that are expressed in association with TE differentiation, is the only one that can hydrolyze DNA, the nuclease(s) is thought to be involved in nuclear degradation during PCD of TEs. The nuclease(s) requires Zn^{2+} for activation (Thelen and Northcote, 1989). cDNA for the 43 kDa nuclease was recently isolated and designated *ZEN1*. *ZEN1* is similar to S1 nuclease of *Aspergillus* and this type of nucleases is also found to be expressed specifically in other PCD in higher plants (see Sugiyama *et al.*, 2000). Therefore the S1 type of nucleases plays a common role in nuclear degradation during different PCD in plants. *ZEN1* nuclease possesses a putative signal peptide for targeting specific organelles or for secretion. Localization analysis of *ZEN1* nuclease with transgenic tobacco cells in which genomic DNA of *ZEN1* had been introduced revealed that this nuclease is transported specifically into the vacuole (Jun Ito, Munetaka Sugiyama and Hiroo Fukuda, unpublished). The fact that the pH optimum of ZEN1 nuclease is around 5 also supports that the nuclease functions in the vacuole or in the cytoplasm after the vacuole collapse. Sugiyama *et al.* (2000) proposed a hypothesis on the general involvement of both Zn^{2+}- and Ca^{2+}-activated DNase activities in PCD of plants. However, Ca^{2+}-activated DNase activity has not been reported for PCD of TEs. Transcripts for most of the cell death-related genes such as *ZEN1, ZCP4* (p48h-17 gene), and *ZRNase I* (Aoyagi *et al.*, 1989, Ye and Varner, 1996; Yamamoto *et al.*, 1997; Ye and Droste, 1996) accumulate transiently in a very similar pattern just before autolysis starts. Therefore, the expression of the genes encoding these enzymes may be regulated by a common mechanism, for example, by the same *cis* and *trans* activation factors.

Comparison of TE PCD with other PCD in plants

Plant PCD may be categorized by cytological features into three types: (1) apoptosis-like cell death, (2) cell death occurring during leaf senescence, (3) PCD in which the vacuole plays a central role, as seen in differentiating TEs.

Apoptosis-like cell death in plants as well as animals exhibits characteristic features such as nuclear shrinkage, chromatin condensation, nuclear fragmentation, and DNA laddering. Salt stress (Katsuhara, 1997), infection (Ryerson and Heath, 1996), or toxin (Wang *et al.*, 1996) can induce apoptosis-like cell death. Apoptotic cell death is also observed in some developmental processes such as the degeneration process of barley aleurone cells (Wang *et al.*, 1996b) and of onion root cap cells at the periphery (Wang *et al.*, 1996a). The first target of degradation in apoptosis-like cell death is the nucleus. Quick disruption of the nucleus allows cells to lose cellular organization system rapidly. Because plants do not have specific cells for cleaning up like macrophages, cell contents must be digested autonomously. Therefore quick disruption of the cellular organization system in apoptotic cell death in plants seems to result in incomplete degradation of cell organelles and then in low recovery of cell contents.

In contrast, cell death during leaf senescence proceeds very slowly (Smart, 1994). Thorough degradation of chloroplasts during a slow cell death process brings about high recovery of cell contents, which allows dying cells to supply a large amount of digested cell contents into new developing organs. In this cell death, disruption of the nucleus and the vacuole occurs at the end of cell death after thorough degradation of chloroplasts. Thorough degradation of cell contents before disruption of the nucleus and the vacuole seems to occur in senescence of various organs.

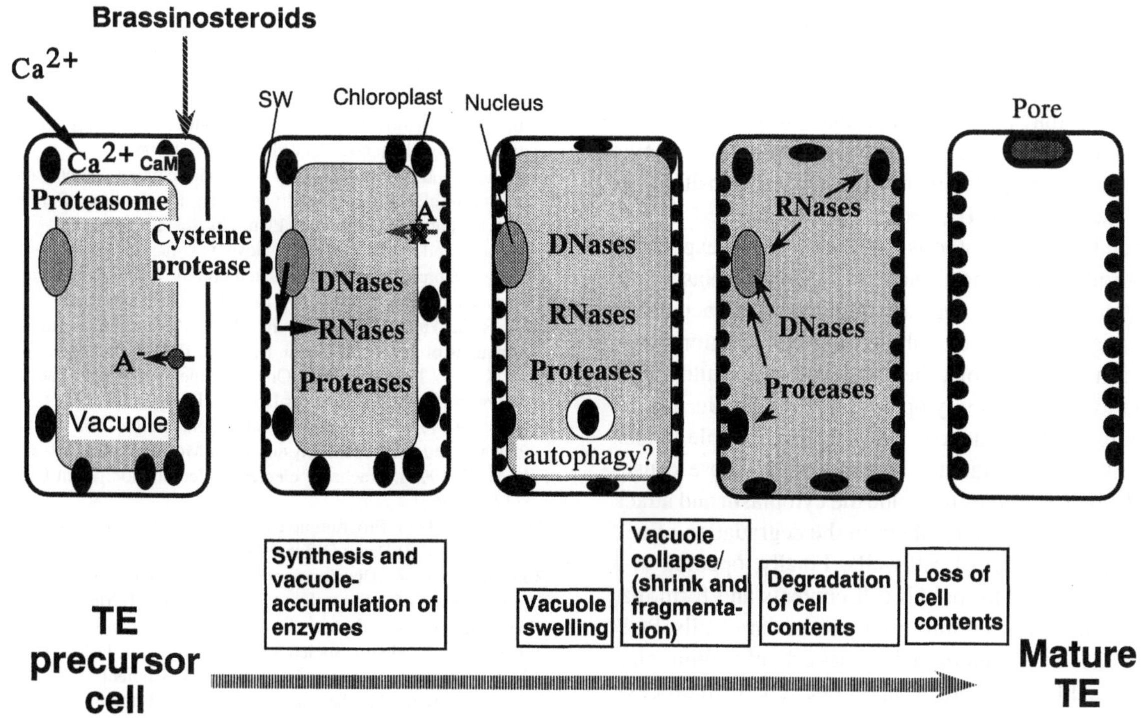

Figure 3. The process of PCD during TE differentiation. See details in text. CaM, calmodulin; SW, secondary wall.

PCD in which disruption of the vacuole plays an essential role is just between apoptotic cell death and leaf senescence-type cell death both in speed of death and in the degree of recovery of cell contents. This type of PCD is also seen in the death of endocarp cells during ovary senescence of pea (Vercher *et al.*, 1987). The aleurone cell death process involves vacuolar enlargement and then the disruption of the tonoplast and plasma membrane, although it is unknown whether the vacuolar collapse precedes that of the plasma membrane (Kuo *et al.*, 1996). Thus, during the development of the plants, there is one common type of cell death process in which a key event is active in vacuolar collapse. Interestingly, similar hydrolytic enzymes are commonly expressed during this type of cell death. During pea ovary senescence, a cysteine protease gene, *tpp*, is expressed preferentially in endocarp and ovule as well as vascular bundles in which *tpp*, is expressed even when the ovary is non-senescent (Granell *et al.* 1992). This gene is highly homologous to p48h-17 gene and *ZCP4*. Genes for a similar type of cysteine proteases are also expressed in aleurone of barley (Rogers *et al.*, 1985) and rice (Watanabe *et al.*, 1991). The S1 type nucleases and their mRNAs are expressed commonly in barley aleurone (Brown and Ho, 1986, 1987) and *Zinnia* TEs (Thelen and Northcote, 1989; Aoyagi *et al.*, 1998). Thus, there may be some common cell death-performing mechanism among the different developmental processes in which the vacuolar collapse is a critical event of PCD.

Is there a common mechanism between the vacuole collapse-dependent PCD and the other two types of PCD in plants? At present we do not have any evidence showing a common mechanism among them. Rather the vacuole collapse-dependent PCD may involve a unique cell death program, as typically shown by the fact that genes essential to PCD of TEs (*ZR-NaseI* and p48h-17 gene) and a TE-specific protease activity were not induced during leaf senescence or by peroxygen challenge (Ye and Droste, 1996; Ye and Varner, 1996; Beers and Freeman, 1997).

Conclusions

TE differentiation includes a unique type of PCD in plants, in which the vacuole plays a central role. Figure 3 summarizes what is known about PCD of TEs at present. The commitment to cell death during TE differentiation may occur in TE precursor cells. Brassinosteroids synthesized in late Stage II may induce TE precursor cells to enter Stage III during which events in the cell death program as well as in the formation of secondary walls are performed. The entry of calcium into cells may also be a factor necessary for transition from Stage II to Stage III. Proteasome and a cysteine protease(s) may also be involved in this transition. At the start of Stage III, genes that are involved in both secondary wall synthesis and autolysis are expressed. Hydrolytic enzymes, such as DNases, RNases and cysteine proteases, may accumulate in the vacuole. In association with the inhibition of organic anion transport into the vacuole, the vacuole swells. Autophagic degradation of the cytoplasm may occur during the swelling of the vacuole. The swelled vacuole bursts, shrinks and fragments. The vacuole collapse causes hydrolytic enzymes to invade the cytoplasm and attack various organelles, resulting in the degradation of cell contents and part of cell walls. Finally, opening of a pore leads TEs to lose all cell contents and form mature hollow tubes reinforced by secondary walls. PCD of TEs is autonomous and occurs actively, although it is quite different from the apoptotic cell death process.

Acknowledgements

The author thanks Dr M. Sugiyama for critical reading of the manuscript. This work was supported in part by Grants-in-Aid from the Ministry of Education, Science, Sports and Culture of Japan (No. 10304063, No. 10219201, No. 10182101), and from the Japan Society for the Promotion of Science (JSPS-RFTF96L00605).

References

Aoyagi, S., Sugiyama, M. and Fukuda, H. 1998. *BEN1* and *ZEN1* cDNAs encoding S1-type DNases that are involved in programmed cell death in plants. FEBS Lett. 429: 134–138.

Bachmair, A., Becker, F., Masterson, R.V. and Schell, J. 1990. Perturbation of the ubiquitin system causes leaf curling, vascular tissue alterations and necrotic lesions in a higher plant. EMBO J. 9: 4543–4550.

Beers, E.P. and Freeman, T.B. 1997. Protease activity during tracheary element differentiation in *Zinnia* mesophyll cultures. Plant Physiol. 113: 873–880.

Brown, P.H. and Ho, T.H.D. 1986. Barley (*Hordeum vulgare* cultivar Himalaya) aleurone layers secrete a nuclease in response to gibberellic acid: purification and partial characterization of the associated RNase, DNase, and 3′-nucleotidase activities. Plant Physiol. 82: 801–806.

Brown, P.H. and Ho, T.H.D. 1987. Biochemical properties and hormonal regulation of barley nuclease. Eur. J. Biochem. 168: 357–364.

Burgess, J. and Linstead, P. 1984. In vitro tracheary element formation: structural studies and the effect of triiodobenzoic acid. Planta 160: 481–489.

Church, D.L. and Galston, A.W. 1988. Kinetics of determination in the differentiation of isolated mesophyll cells of *Zinnia elegans* to tracheary elements. Plant Physiol. 88: 92–96.

Demura, T. and Fukuda, H. 1993. Molecular cloning and characterization of cDNAs associated with tracheary element differentiation in cultured *Zinnia* cells. Plant Physiol. 103: 815–821.

Demura, T. and Fukuda, H. 1994. Novel vascular cell-specific genes whose expression is regulated temporally and spatially during vascular system development. Plant Cell 6: 967–981.

Ellis, R.E., Yuan, J. and Horvitz, H.R. 1991. Mechanisms and functions of cell death. Annu. Rev. Cell Biol. 7: 663–698.

Esau, K. and Charvat, I. 1978. On vessel member differentiation in the bean (*Phaseolus vulgaris* L.). Ann Bot. 42: 665–677.

Fukuda, H. 1996. Xylogenesis: initiation, progression and cell death. Annu. Rev. Plant Physiol. Plant Mol. Biol. 47: 299–325.

Fukuda, H. 1997a. Tracheary element differentiation. Plant Cell 9: 1147–1156.

Fukuda, H. 1997b. Programmed cell death during vascular system formation. Cell Death Differ. 4: 684–688.

Fukuda, H. 1998. Developmentally programmed cell death in plants: death of xylem cells as a paradigm. In: T. Yamada and Y. Hashimoto (Eds.), Apoptosis: Its role and mechanism, Business Center for Academic Societies Japan, Tokyo, pp. 113–127.

Fukuda, H. and Komamine, A. 1980a. Establishment of an experimental system for the tracheary element differentiation from single cells isolated from the mesophyll of *Zinnia elegans*. Plant Physiol. 65: 57–60.

Fukuda, H. and Komamine, A. 1980b. Direct evidence for cytodifferentiation to tracheary elements without intervening mitosis in a culture of single cells isolated from the mesophyll of *Zinnia elegans*. Plant Physiol. 65: 61–64.

Gavrieli, Y., Sherman, Y. and Ben-Sasson, S.A. 1992. Identification of programmed cell death *in situ* via specific labeling of nuclear DNA fragmentation. J. Cell Biol. 119: 493–501.

Granell, A., Harris, N., Pisabarro, A.G. and Carbonell, J. 1992. Temporal and spatial expression of a thiolprotease gene during pea ovary senescence, and its regulation by gibberellin. Plant J. 2: 907–915.

Greenberg, J.T. 1996. Programmed cell death: a way of life for plants. Proc. Natl. Acad. Sci. USA 93: 12094–12097.

Greenberg, J.T. 1997. Programmed cell death in plant-pathogen interactions. Annu. Rev. Plant Physiol. Plant Mol. Biol. 48: 525–545.

Groover, A., DeWitt, N., Heidel, A. and Jones, A. 1997. Programmed cell death of plant tracheary elements differentiating *in vitro*. Protoplasma 196: 197–211.

Groover, A. and Jones, A. 1999. Tracheary element differentiation uses a novel mechanism coordinating programmed cell death and secondary cell wall synthesis. Plant Physiol. 119: 375–384.

Iwasaki, T. and Shibaoka, H. 1991. Brassinosteroids act as regulators of tracheary-element differentiation in isolated *Zinnia* mesophyll cells. Plant Cell Physiol. 32: 1007–1014.

Jacobson, M.D. 1996. Reactive oxygen species and programmed cell death. Trends Biochem. Sci. 21: 83–86.

Jones, A.M. and Dangl, J. 1996. Logjam at the Styx: programmed cell death in plants. Trends Plant Sci. 1: 114–119.

Katsuhara, K. 1997. Apoptosis-like cell death in barley roots under salt stress. Plant Cell Physiol. 38: 1087–1090.

Kerr, J. F. R. and Harmon, B. V. 1991. Definition and incidence of apoptosis: a historical perspective. In: L.D. Tomei and F.O.

Cope (Eds.), Apoptosis: The Molecular Basis of Cell Death, Cold Spring Harbor Lab. Press, Plainview, NY, pp. 5–29.

Kobayashi, H. and Fukuda, H. 1994. Involvement of calmodulin and calmodulin-binding proteins in the differentiation of tracheary elements in *Zinnia* cells. Planta 194: 388–394.

Kuo, A., Cappelluti, S., Cervantes-Cervantes, M., Rodriguez, M. and Bush, D.S. 1996. Okadaic acid, a protein phosphatase inhibitor, blocks calcium changes, gene expression, and cell death induced by gibberellin in wheat aleurone cells. Plant Cell 8: 259–269.

Kuriyama, H. 1999. Loss of tonoplast integrity programmed in tracheary element differentiation. Plant Physiol. 121, in press.

Lai, V. and Srivastava, L.M. 1976. Nuclear changes during differentiation of xylem vessel elements. Cytobiologie 12: 220–243.

Lamb, C. and Dixon, R.A. 1997. The oxidative burst in plant disease resistance. Annu. Rev. Plant Physiol. Plant Mol. Biol. 48: 251–275.

Minami, A. and Fukuda, H. 1995. Transient and specific expression of a cysteine endopeptidase during autolysis in differentiating tracheary elements from *Zinnia* mesophyll cells. Plant Cell Physiol. 36: 1599–1606.

Mittler, R. and Lam, E. 1995. *In situ* detection of nDNA fragmentation during the differentiation of tracheary elements in higher plants. Plant Physiol. 108: 489–493.

O'Brien, T.P. and Thimann, K.V. 1967. Observation on the fine structure of the oat coleoptile III. Correlated light and electron microscopy of the vascular tissues. Protoplasma 63: 443–478.

Pennell, R.I. and Lamb, C. 1997. Programmed cell death in plants. Plant Cell 9: 1157–1168.

Roberts, A.W. and Haigler, C.H. 1989. Rise in chlortetracycline accompanies tracheary element differentiation in suspension cultures of *Zinnia*. Protoplasma 152: 37–45.

Roberts, A.W. and Haigler, C.H. 1990. Tracheary-element differentiation in suspension-cultured cells of *Zinnia* requires uptake of extracellular calcium ion: experiments with calcium-channel blockers and calmodulin inhibitors. Planta 180: 502–509.

Roberts, A.W. and Haigler, C.H. 1992. Methylxanthines reversibly inhibit tracheary-element differentiation in suspension cultures of *Zinnia elegans* L. Planta 186: 586–592.

Rogers, J.C., Dean, D. and Heck, G.R. 1985. Aleurain: a barley thiol protease closely related to mammalian cathepsin H. Proc. Natl. Acad. Sci. USA 82: 6512–6516.

Ryerson, D.E. and Heath, M.C. 1996. Cleavage of nuclear DNA into oligonucleosomal fragments during cell death induced by fungal infection or by abiotic treatments. Plant Cell 8: 393–402.

Smart, C.M. 1994. Gene expression during leaf senescence. New Phytol. 126: 419–448.

Srivastava, L.M. and Singh, A.P. 1972. Certain aspects of xylem differentiation in corn. Can J. Bot. 50: 1795–1804.

Stephenson, P., Collins, B.A., Reid, P.D. and Rubinstein, B. 1996. Localization of ubiquitin to differentiating vascular tissues. Am. J. Bot. 83: 140–147.

Sugiyama, M., Ito, J., Aoyagi, S. and Fukuda, H. 2000. Endonuclease. Plant Mol. Biol. this issue.

Thelen, M.P. and Northcote, D.H. 1989. Identification and purification of a nuclease from *Zinnia elegans* L.: a potential molecular marker for xylogenesis. Planta 179: 181–195.

Vercher, Y., Molowny, A. and Carbonell, J. 1987. Gibberellic acid effects on the ultrastructure of endocarp cells of unpollinated ovaries of *Pisum sativum*. Physiol. Plant. 71: 302–308.

Wang, H., Li, J., Bostock, R.M. and Gilchrist, D.G. 1996a. Apoptosis: a functional paradigm for programmed plant cell death induced by a host-selective phytotoxin and invoked during development. Plant Cell 8: 375–391.

Wang, M., Oppedijk, B.J., Lu, X., Duijn, B.V. and Schilperoort, R.A. 1996b. Apoptosis in barley aleurone during germination and its inhibition by abscisic acid. Plant Mol. Biol. 32: 1125–1134.

Watanabe, H., Abe, K., Emori, Y., Hosoyama, H. and Arai, S. 1991. Molecular cloning and gibberellin-induced expression of multiple cysteine proteases of rice seeds (Oryzains). J. Biol. Chem. 266: 16897–16907.

Woffenden, B.J., Freeman, T.B. and Beers, E.P. 1998. Proteasome inhibitors prevent tracheary element differentiation in *Zinnia* mesophyll cell cultures. Plant Physiol. 118: 419–430.

Yamada, T. and Hashimoto Y. 1998. Apoptosis its roles and mechanism. Business Center for Academic Societies, Japan, Tokyo.

Yamamoto, R., Demura, T. and Fukuda, H. 1997. Brassinosteroids induce entry into the final stage of tracheary element differentiation in cultured *Zinnia* cells. Plant Cell Physiol. 38: 980–983.

Ye, Z.-H. and Droste, D.L. 1996. Isolation and characterization of cDNAs encoding xylogenesis-associated and wound-induced ribonucleases in *Zinnia elegans*. Plant Mol. Biol. 30: 697–709.

Ye, Z.-H. and Varner, J.E. 1996. Induction of cysteine and serine proteases during xylogenesis in *Zinnia elegans*. Plant Mol. Biol. 30: 1233–1246.

Plant Molecular Biology **44:** 255–266, 2000.
E. Lam, H. Fukuda and J. Greenberg (Eds.), Programmed Cell Death in Higher Plants.

Programmed cell death in cereal aleurone

Angelika Fath, Paul Bethke, Jennifer Lonsdale, Roberto Meza-Romero and Russel Jones
Department of Plant and Microbial Biology, University of California, Berkeley, CA 94720-3102, USA

Key words: abscisic acid, apoptosis, gibberellic acid, nuclease, programmed cell death, protease

Abstract

Progress in understanding programmed cell death (PCD) in the cereal aleurone is described. Cereal aleurone cells are specialized endosperm cells that function to synthesize and secrete hydrolytic enzymes that break down reserves in the starchy endosperm. Unlike the cells of the starchy endosperm, aleurone cells are viable in mature grain but undergo PCD when germination is triggered or when isolated aleurone layers or protoplasts are incubated in gibberellic acid (GA). Abscisic acid (ABA) slows down the process of aleurone cell death and isolated aleurone protoplasts can be kept alive in media containing ABA for up to 6 months. Cell death in barley aleurone occurs only after cells become highly vacuolated and is manifested in an abrupt loss of plasma membrane integrity. Aleurone cell death does not follow the apoptotic pathway found in many animal cells. The hallmarks of apoptosis, including internucleosomal DNA cleavage, plasma membrane and nuclear blebbing and formation of apoptotic bodies, are not observed in dying aleurone cells. PCD in barley aleurone cells is accompanied by the accumulation of a spectrum of nuclease and protease activities and the loss of organelles as a result of cellular autolysis.

Introduction

The cereal endosperm consists of two specialized tissues, the aleurone layer and the starchy endosperm (Figure 1A). Programmed cell death (PCD) is an integral part of the development of these tissues, although they die at different times. Starchy endosperm cells die after grain filling is complete. Unlike PCD of most other plant cells, including aleurone cells, the death of starchy endosperm cells is not followed by rapid destruction of the corpse. Dead starchy endosperm cells function to store reserves of carbon and nitrogen that will be used by the embryo upon germination. These reserves, which account for most of the mass of dead starchy endosperm cells, are not broken down until germination has been triggered. Enzymes to break down reserves are synthesized in aleurone. Mutations have been identified, however, in which this pattern is altered. The *shrunken2* mutation in *Zea mays* results in premature endosperm cell death and degradation of the corpse. This causes the endosperm to become deformed and produces aberrant kernels (Young *et al.*, 1997).

In contrast, the cells of the aleurone layer are alive in the mature grain, and do not die until a few days after germination. Death of the cereal aleurone was first described more than one hundred years ago by Haberlandt in maize, oat, rye and wheat grains (Haberlandt, 1884). Haberlandt characterized the changes that occurred within the cereal aleurone cell upon imbibition, including the progressive vacuolation of the cytoplasm which preceded the death and collapse of the protoplast. He described the cell remnants that he saw within the aleurone cell wall after endosperm depletion as ‘refractive globules’.

In this review, we describe recent advances in our understanding of aleurone cell death. After producing enzymes used for the mobilization of starchy endosperm reserves, aleurone cells undergo autolysis and die. Gibberellins (GA) and abscisic acid (ABA) tightly regulate this process. While GA stimulates the onset of PCD in the aleurone layer of barley (Wang *et al.*, 1996; Bethke *et al.*, 1999) and wheat (Kuo *et al.*, 1996), ABA postpones PCD (Figure 2). The effects of ABA on cell death in barley aleurone are dramatic. ABA-treated barley aleurone protoplasts can

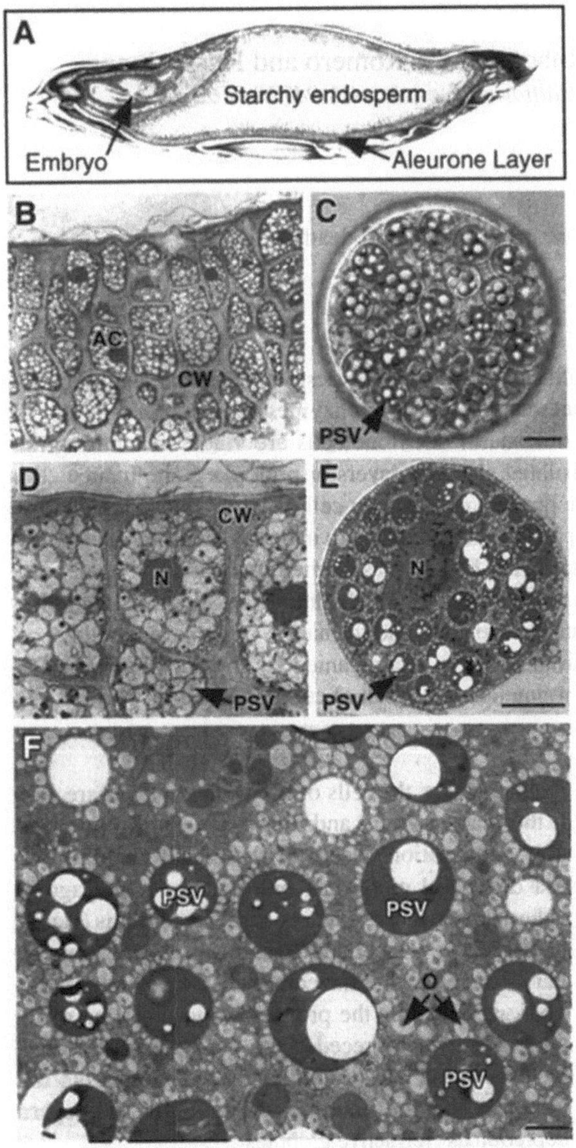

Figure 1. Anatomy of the barley grain and structure of the aleurone. A. Longitudinal section through a barley grain. B, D. Light microscopy of isolated aleurone layer after chemical fixation shown at low (B) and higher magnification (D). Individual aleurone cells (AC), cell wall (CW), nuclei (N) and protein storage vacuoles (PSV) are seen. C. Light microscopy of living aleurone protoplast. E, F. Transmission electron microscopy of aleurone protoplast after high-pressure freezing-freeze substitution shown at low (E) and higher magnification (F). Oleosomes (O) and PSV are prominent. Bars in C, E = 5 μm, bar in F = 1 μm.

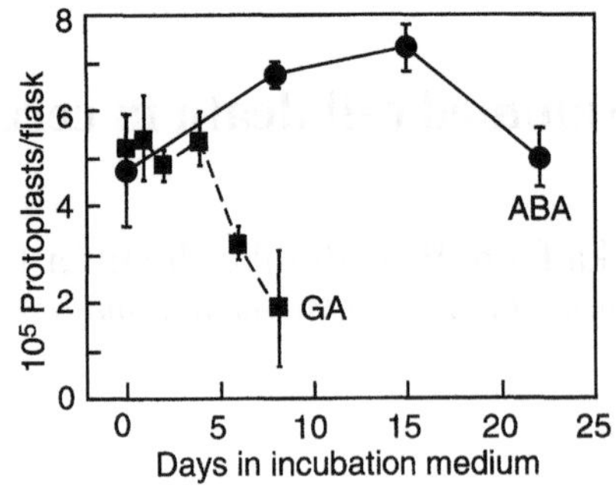

Figure 2. Cell death in barley aleurone is tightly regulated by GA and ABA. Freshly isolated barley aleurone protoplasts were either incubated in 25 μM GA (■) or 5 μM ABA (●) for the times indicated. Each point represents the mean ± SD of three or four flasks of protoplasts, and each flask was used for only one time point.

be kept alive for more than 6 months (J.E. Lonsdale, P.C. Bethke and R.L. Jones, unpublished observations), whereas GA treatment of protoplasts brings about death of most cells in 5–8 days.

Structure and function of the cereal aleurone

Aleurone cell structure

The aleurone layer of most cereals is one cell thick, the notable exceptions being barley and rice where the aleurone layer can be up to four cells thick (Figure 1B). Differentiation of the triploid, coenocytic endosperm into aleurone and starchy endosperm begins about 8 to 10 days after pollination in barley (Bosnes *et al.*, 1992). Except for a small region near the micropyle, the aleurone layer envelops the entire embryo and starchy endosperm (Bethke *et al.*, 2000). Aleurone cells appear distinctly different from the rest of the endosperm and at least three types can be identified: crease aleurone cells located in the ventral furrow of the grain, germ aleurone cells that surround the embryo; and the cells in the aleurone layer that envelops the starchy endosperm. Aleurone cells characteristically have a thick cell wall (up to 5 μm thick) composed almost entirely of non-cellulosic polysaccharides (Figure 1B, D; Fincher, 1992). This wall is made up of a thin inner wall, referred to as the resistant wall, and a much thicker cell wall matrix. The wall matrix consists of arabinoxylan and (1-3, 1-4)-β-glucan and is rapidly broken down during germination by glucanases and xylanases secreted from the aleurone cell (Taiz and Jones, 1970; Fincher, 1992). The resistant wall lies adjacent to the plasma membrane and persists during the early stages of germination (Taiz and Jones, 1970).

The cytoplasm of the mature aleurone cell is filled with organelles, the most prominent of which are the protein storage vacuoles (PSV) (Figures 1C–F and 3). To quickly produce enzymes to mobilize endosperm, amino acid building blocks are stored in PSV. These vacuoles originate from the endoplasmic reticulum, and in addition to containing storage proteins, they have lipid-containing oleosomes embedded in their limiting membranes (Figure 1F; Bethke *et al.*, 1998). PSV store a small amount of non-starch carbohydrate (Jacobsen *et al.*, 1971) and abundant mineral reserves that are chelated into phytin (Ca, Mg, and K salt of inositol hexaphosphate) (Stewart *et al.*, 1988). Mitochondria, ER, glyoxysomes and Golgi are also prominent in aleurone cells (Jones, 1969; Bechtel and Pomeranz, 1977; Lonsdale *et al.*, 1999). Protoplasts isolated from barley aleurone layers are indistinguishable in structure from cells in the intact layer except that protoplasts are spherical because they lack a cell wall (Figures 1C, 1E and 3C).

The responses of cereal aleurone cells to hormones have been studied in detail in de-embryonated half-grains, isolated aleurone layers and protoplasts of barley, wheat and wild oat. Although there are quantitative differences in the responses of these different experimental materials to GA and ABA, the qualitative nature of these responses are similar and most likely reflect conserved regulatory mechanisms. Because protoplasts are often the experimental system of choice for investigations into aleurone cell function, it is important to emphasize that the responses of aleurone protoplasts treated with GA or ABA, although slower, are nearly the same as those of aleurone layers

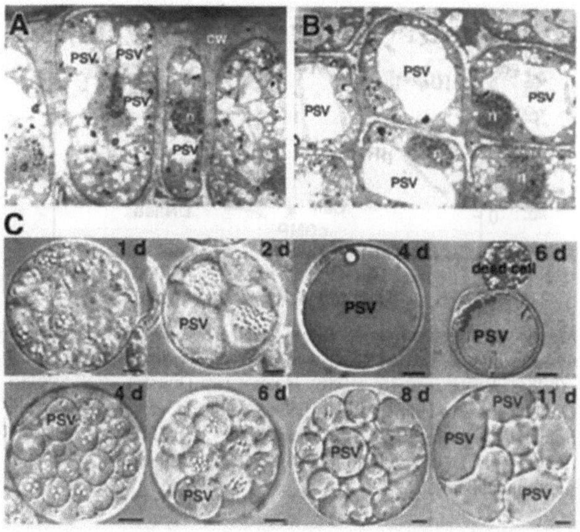

Figure 3. Vacuolation of aleurone cells in response to GA treatment. A, B. Light microscopy of chemically fixed aleurone layers incubated in GA for 12 h (A) and 30 h (B). C. Light microscopy of aleurone protoplasts incubated in GA (top row) or ABA (bottom row) for up to 11 days. Bars in C = 5 μm. PSV, protein storage vacuole; n, nuclei; cw, cell wall.

Changes in the aleurone layer in response to hormones

The embryo of the germinating cereal grain produces GAs that trigger a sequence of signaling events (Figure 4) and morphological changes (Figure 3A, B for layer and 3C for protoplasts) in the aleurone layer that culminate in cell death. When stimulated by GA, the aleurone functions as a secretory tissue that synthesizes a spectrum of hydrolases and secretes them into the dead starchy endosperm (Fincher, 1989; Jones and Jacobsen, 1991). Signal transduction pathways leading to hydrolase synthesis utilize changes in cytosolic Ca^{2+} (Gilroy and Jones, 1992; Gilroy, 1996), calmodulin (Schuurink *et al.*, 1996), cGMP (Penson *et al.*, 1996) and the *trans*-activating protein GAMyb (Gubler *et al.*, 1995). These signaling molecules all change in response to GA within the first 2 h of hormone treatment in aleurone layers (Figure 4). The synthesis of α-amylase, which represents as much as 60% of the newly synthesized protein in GA-stimulated aleurone layers, begins ca. 3–4 h after GA perception (Higgins *et al.*, 1982). Glucanases, xylanases, proteases and nucleases are other classes of enzymes that are secreted into the endosperm at this time (Jones and Jacobsen, 1991). Amylases break down starch reserves in the starchy endosperm while glucanases and xylanases participate in digestion of the aleurone cell wall. Cell wall breakdown can be observed within a few hours of GA treatment and wall hydrolysis is extensive by 12 h after exposure to GA. Proteases and nucleases break down starchy endosperm storage proteins and residual nucleic acids.

One feature that is shared by the secreted hydrolases is that they are synthesized *de novo* from amino acids that arise from the breakdown of storage proteins in PSV (Jones and Jacobsen, 1991; Bethke *et al.*, 1998). A consequence of the hydrolysis of storage proteins in the PSV lumen is that these organelles swell and coalesce to form one large vacuole (Figure 3). Vacuolation is characteristic of GA-treated cells and the size and number of PSV is strongly correlated with duration of GA treatment (Figure 3). Aleurone cells

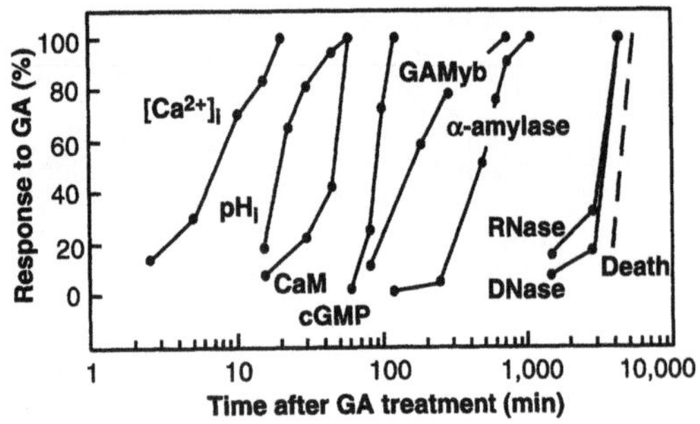

Figure 4. Time-course of GA-induced responses of barley and wheat aleurone layers. Data were extracted from Chandler *et al.* (1984), Brown and Ho (1986), Heimovaara-Dijkstra *et al.* (1994), Gubler *et al.* (1995), Bush (1996), Penson *et al.* (1996) and Schuurink *et al.* (1996). Cytosolic free Ca^{2+} ($[Ca^{2+}]_i$), cytosolic pH (pH_i), calmodulin (CaM), cGMP (3′5′-cyclic guanosine monophosphate).

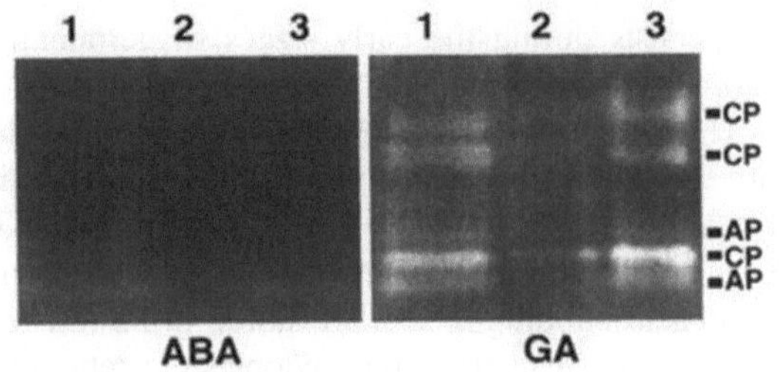

Figure 5. GA treatment of barley aleurone protoplasts increases protease activities in PSV. Proteolytic activity gels of ABA- (left panel) and GA-treated (right panel) protoplasts (lanes 1), protoplast lysate (lanes 2) and purified PSV (lanes 3). Samples were prepared from barley aleurone protoplasts incubated with 5 μM ABA or 5 μM GA for 17.5 h. Each lane contains 4 μg protein. The positions of cysteine protease (CP) and aspartic protease (AP) activities are indicated.

or protoplasts that have not been exposed to GA have PSV with a diameter of about 1–2 μm (Figure 1B-F). With increasing duration of exposure of cells to GA, the size of the PSV increases to a diameter of 40–50 μm and the number of PSV decreases.

The synthesis of secretory proteins by GA-treated aleurone cells is also associated with proliferation of the endoplasmic reticulum, its organization into stacks of rough-surfaced membranes (Jones and Price, 1970), and an increase in the size and complexity of the Golgi apparatus (Cornejo *et al.*, 1988). Other changes in the endomembranes of the aleurone cell that coincide with the synthesis and secretion of hydrolases include a decrease in the number of oleosomes and an increase in the number of glyoxysomes and mitochondria (Jones, 1972, 1980).

Some of the enzymes whose synthesis is stimulated by GA are not secreted into the endosperm but are retained within aleurone cells. Many of these enzymes are targeted to the PSV, oleosomes or glyoxysomes where they play a role in the metabolism of stored materials. Among these enzymes are proteases, nucleases, phytase, acid phosphatase, lipase, malate synthase and isocitrate lyase (reviewed by Jones and Jacobsen, 1991).

Protein storage vacuoles become lytic organelles after GA treatment

Dramatic changes occur in PSV upon GA treatment of aleurone cells as they are transformed from nutrient storing compartments to lytic organelles. In freshly isolated barley and oat aleurone cells, the pH of the PSV lumen was close to neutrality (Davies *et al.*, 1996; Swanson and Jones, 1996). In this respect, aleurone PSV differ from vacuoles found in most plant cells whose contents are acidic (Kurkdjian and Guern, 1989). Within an hour of GA perception, however, the PSV within barley aleurone cells became acidic and after about 6 h vacuolar pH was less than 5.5 (Swanson and Jones, 1996). ABA, on the other hand, greatly slowed the acidification of the PSV lumen. A pH of 6.8 was maintained in the PSV of vacuoles in barley aleurone cells treated with ABA for 20 h (Swanson and Jones, 1996). After prolonged incubation in ABA, the pH of PSV became acidic, but was still higher than that of PSV in GA-treated protoplasts (Swanson *et al.*, 1998). Acidification of aleurone PSV can be brought about by tonoplast V-type H^+-ATPases or H^+-pyrophosphatases, but the relative contributions of these proton pumps to PSV acidification *in vivo* are not known.

The PSV of GA-treated aleurone cells accumulate a spectrum of acidic hydrolases, including proteases, phosophatases and phytase. Most of these are inactive or absent in vacuoles from ABA-treated cells (Figure 5; Bethke *et al.*, 1996). There are two phases of protease accumulation in aleurone PSV. One occurs relatively early in the GA response, and is probably associated with the mobilization of storage proteins in the PSV lumen. The other phase occurs later, and is correlated with autolysis and cell death. PSV from aleurone protoplasts incubated in GA for 17.5 h contain three cysteine protease and two aspartic protease activities (Figure 5). One of the aspartic proteases was identified as phytepsin, formally known as *Hordeum vulgare* aspartic protease (HvAP, Figure 5; Törmäkangas *et al.*, 1994; Bethke *et al.*, 1996). All of the protease activities that accumulate in PSV during the first 17.5 h after incubation in GA have acidic pH optima. None of these enzymes show activity above pH 6.5, and most have maximal activity at pH 4.5 or below (Bethke *et al.*, 1998).

It is not known whether these PSV proteases exist as inactive precursors that become proteolytically processed to their active forms when the vacuolar lumen is acidified, or whether they are synthesized *de novo*. Phytepsin, however, could play a role in activating other PSV proteases. Immunolocalization showed that phytepsin was present in PSV isolated from dry, unimbibed grain (Törmäkangas *et al.*, 1994; Bethke *et al.*, 1996) as well as in PSV from cells treated with GA. Phytepsin has also been shown to proteolytically processed probarley lectin in barley root vacuoles (Runeberg-Roos *et al.*, 1994). We speculate that acidification of the PSV lumen upon GA treatment activates phytepsin which in turn proteolytically processes other PSV proteases.

Hormonal regulation of autolysis and death in barley aleurone cells

Formation of the lytic vacuole may allow for autolysis of the aleurone cell. In an extensive TEM survey of barley aleurone protoplasts, autophagic vacuoles similar to those seen in animal cells undergoing apoptosis have not been observed (J.E. Lonsdale and R.L. Jones, unpublished data). We hypothesize that degradation of the aleurone cytoplasm occurs in the lytic PSV, the cytoplasm itself, or most likely in both.

Aleurone cell death culminates in a loss of plasma membrane integrity

GA-treated aleurone cells do not die before they reach a highly vacuolate stage (Figure 3). Highly vacuolate aleurone cells have mobilized most of their stored protein and have used the resulting amino acids to synthesize secreted hydrolases. We used video time-lapse microscopy of barley aleurone protoplasts to characterize death of these highly vacuolate cells (Bethke *et al.*, 1999). Death of the aleurone cell was manifested in a loss of plasma membrane integrity. Morphological changes foreshadowing death were not apparent until just before death. At that time, the plasma membrane (PM) appeared ruffled (Figure 6; Bethke *et al.*, 1999) and the cells became permeable to vital stains, indicating that the permeability of the plasma membrane had changed. Protoplasts then lost turgor and collapsed rapidly, producing a compact mass of cellular debris (Figure 6). These corpses do indeed look like 'refractive globules' as described by Haberlandt more than 100 years ago.

Signal transduction elements leading to PCD in cereal aleurone

The cellular signals proposed to lead to programmed cell death in plants are quite varied. To date, a common signaling pathway for plant PCD has not been identified, and it may well be that different signal transduction cascades are utilized in response to different death-inducing stimuli. Changes in the cytosolic concentration of Ca^{2+} and calmodulin (CaM), however, appear to be important in the death of many plant tissues. Ca^{2+} and CaM are players in the programmed cell death of *Zinnia* tracheary elements (Fukuda, 1996; Groover and Jones, 1999) and cytosolic Ca^{2+} levels increase during hypersensitive cell death in soybean cells exposed to the pathogen *Pseudomonas syringae* pv. *glycinea* (Levine *et al.*, 1996). Cell death during hypoxia-induced aerenchyma formation in maize roots was promoted by an increase in cytosolic Ca^{2+} and inhibited by a Ca^{2+}-CaM antagonist (He *et al.*, 1996). Aerenchyma formation was also stimulated by an inhibitor of protein phosphatases and blocked by a protein kinase inhibitor and an antagonist of inositol phospholipids (He *et al.*, 1996). Reactive oxygen species may also play a significant role in plant PCD. Reactive oxygen species are produced during the hypersensitive response in senescing pea leaves, and an NADPH oxidoreductase is up-regulated in mesophyll cells of zinnia during tracheary element differentiation (Demura and Fukuda, 1994; Levine *et al.*, 1994; Pastori and Del Rio, 1997).

The signal transduction cascades leading to the GA-induced synthesis and secretion of hydrolases in cereal aleurone have been studied extensively (for reviews, see Bethke *et al.*, 1997, 1998; Ritchie and Gilroy, 1998b). Cytosolic free Ca^{2+}, cytosolic pH, cGMP, CaM and protein kinases and protein phosphatases have been identified as elements of these signaling pathways (Figure 4; Bush, 1996; Heimovaara-Dijkstra *et al.*, 1996; Kuo *et al.*, 1996; Penson *et al.*, 1996; Schuurink *et al.*, 1996; Ritchie and Gilroy, 1998a). Much less is known about the signaling components leading to GA-induced death in cereal aleurone cells. Kuo *et al.* (1996) showed that the protein phosphatase inhibitor okadaic acid blocked GA-induced increases in cytosolic Ca^{2+} and

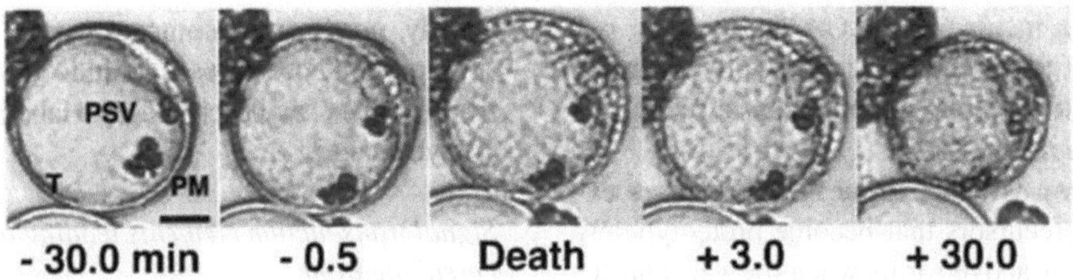

Figure 6. Aleurone cell death is accompanied by a rapid loss of plasma membrane integrity and turgor. PCD in barley aleurone was investigated by time-lapse photomicrography. Images were captured every 30 s and those 30 and 0.5 min before death (–30.0 and –0.5), at the time of death, and +3 and +30 min after death are shown. Time of death was defined as the time when the cell lost its plasma membrane integrity. Bar = 20 μm. Protein storage vacuole (PSV), tonoplast (T), plasma membrane (PM).

prevented cell death. It therefore seems likely that a protein phosphatase is part of the signaling cascade leading to PCD in wheat aleurone.

Microinjection into barley aleurone protoplasts of syntide-2, a synthetic substrate for Ca^{2+}- and CaM-dependent protein kinase, inhibited GA-induced gene expression and α-amylase secretion but did not prevent the GA-induced increase in cytosolic Ca^{2+} (Ritchie and Gilroy, 1998a). Interestingly, the development of large vacuoles typical of a GA response was inhibited by the presence of syntide-2 (Ritchie and Gilroy, 1998a). We infer from this experiment that protein phosphorylation is involved in the signaling cascade leading to barley aleurone PCD, since only protoplasts that reach the highly vacuolated stage are destined to die.

Blocking a GA-signaling pathway that contains cGMP with the guanylyl cyclase inhibitor LY83583 resulted in reduced expression of genes encoding the transactivating protein GAMyb and α-amylase, and prevented α-amylase secretion (Penson *et al.*, 1996). LY83583 also prevented GA-induced increases in intracellular nuclease activities, DNA degradation in living cells, and cell death (Bethke *et al.*, 1999; Fath *et al.*, 1999). This makes cGMP a likely candidate for the GA signal transduction pathway that leads to cell death in barley aleurone cells. These data, however, do not allow us to conclude that any of these signal transduction components are uniquely involved in a signaling pathway leading to PCD. It is equally likely that they are involved in a more global response to GA.

GA-induced cell death can be uncoupled from enzyme secretion

Death of the aleurone cell occurs after the cell has reached the highly vacuolate stage (Figure 3; Bethke *et al.*, 1999) and hydrolase secretion is essentially complete. It could be argued that aleurone cells die as a consequence of the depletion of resources that occurs during the phase of hydrolase synthesis. Experiments with ABA show that cell death can be uncoupled from synthesis of secreted hydrolases (Figure 7). Not only does ABA postpone the death of cells that have not been triggered by GA (Figure 2), but ABA also retards death in cells that have been pretreated with GA for 24 h before ABA addition (Figure 7B). Thus, when ABA was added to cells 24 h after GA was added, α-amylase synthesis was not inhibited (Figure 7A) but cell death was significantly delayed (Figure 7B). These results show two important facets of the response of barley aleurone cells to GA. First, hydrolase synthesis and secretion and PCD are separate parts of the response to GA. Second, ABA can significantly alter the temporal relationship between these two distinct parts.

DNA degradation

DNA degradation is an important diagnostic factor for PCD. By using methods for DNA isolation that avoid DNA degradation during the purification procedure, DNA was extracted from living aleurone cells at various times after incubation in GA or ABA, and the amount of DNA was quantified. Degradation of nuclear DNA in GA-treated protoplasts began around the time that aleurone cells reached the highly vacuolate stage (Figure 8A). DNA degradation was rapid and extensive. Living barley aleurone protoplasts lost more than half of their DNA between the fourth and fifth day after GA treatment (Figure 8). ABA treatment, on the other hand, prevented degradation of barley aleurone DNA for at least 5 days (Figure 8A).

A commonly used method for investigating DNA degradation when studying PCD is the TUNEL (terminal dUTP nick-end labeling) assay, which allows for *in situ* detection of nicked or broken DNA at the single-cell level (Gavrieli *et al.*, 1992). This method is

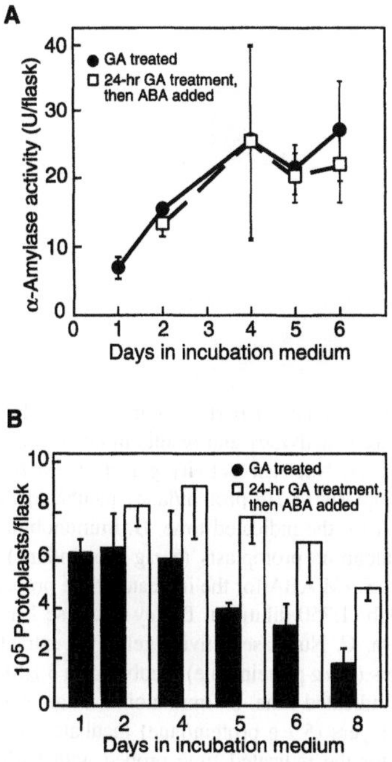

Figure 7. ABA delays PCD in GA-treated barely aleurone protoplasts. Freshly isolated barley aleurone protoplasts were treated with 25 μM GA. ABA (125 μM) was added to the incubation medium of some flasks 24 h later. A. α-Amylase activity in the incubation media of barley aleurone protoplasts incubated in GA for the number of days indicated with or without the addition of ABA at 24 h. U, units. B. The number of living cells per flask after incubation in GA for the number of days indicated with (open bars) or without (filled bars) the addition of ABA at 24 h. Each data point in A or bar in B represents the mean ± SD of four flasks of protoplasts.

based on the labeling of free 3′-OH ends of DNA with fluorescein-labeled deoxyuridine using terminal deoxynucleotidyl transferase. Populations of chemically fixed GA-treated aleurone protoplasts that contained cells at all morphological stages of barley aleurone development (Bush *et al.*, 1986) were examined by the TUNEL assay (Figure 8B; Fath *et al.*, 1999). Cells were counter-stained with DAPI to visualize nuclei. TUNEL-positive staining of the nucleus was most intense in protoplasts that had reached the highly vacuolate stage and in cells that had undergone PCD suggesting that these cells contain degraded DNA. Protoplasts that were at earlier stages of development were much less intensely labeled (Fath *et al.*, 1999), as were 4-day old ABA-treated protoplasts (Figure 8B) indicating that the DNA in these cells is largely intact. TUNEL-positive fluorescence was also observed from nuclei in aleurone layers 2–3 days after germination (Wang *et al.*, 1998). It is important to point out that the onset of extensive DNA degradation occurs in cells

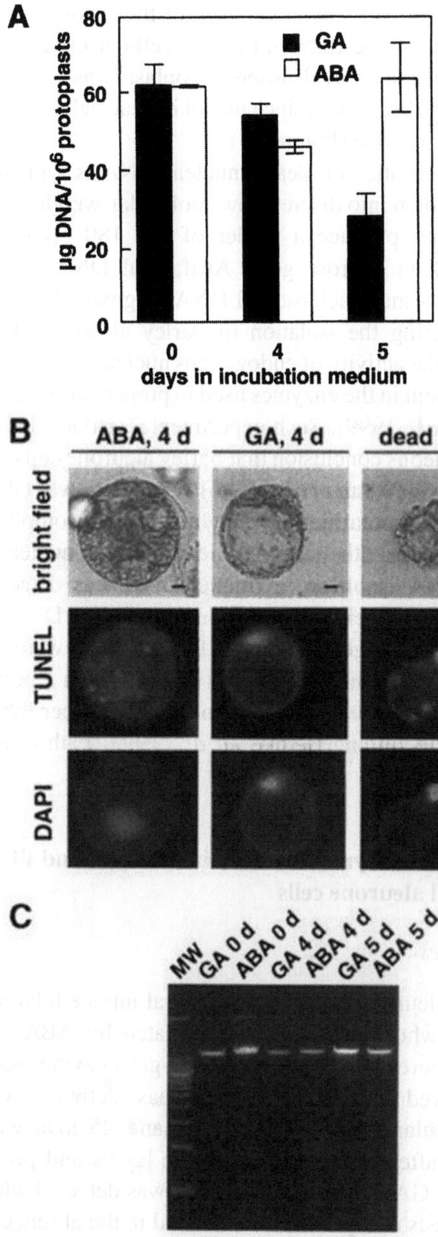

Figure 8. DNA degradation is induced by GA and inhibited by ABA. A. Amount of DNA in 10^6 living protoplasts isolated 0, 4 or 5 days after treatment with 25 μM GA (filled bars) or 5 μM ABA (open bars). DNA was quantified spectrophotometrically. Values represent the mean ± SD of 3 flasks of protoplasts. B. *In situ* detection of DNA cleavage in barley aleurone protoplasts. Chemically fixed barley aleurone protoplasts at different stages of development were observed by bright-field (upper panels) or fluorescence microscopy after TUNEL (middle panels) or DAPI (lower panels) staining. Cells were either treated with 5 μM ABA (left panels) or 25 μM GA for 4 days (middle and right panels). The protoplast in the right panels died prior to fixation. Bar = 5 μm. C. DNA is present as high-molecular-weight DNA. DNA (1.75 μg/lane) from cells incubated in GA or ABA for 0, 4 and 5 days was loaded on a 1.5% agarose gel, electrophoresed and stained with ethidium bromide. DNA cleaved into fragments differing by 500 bp was used as a marker.

that have already lost a majority of their storage proteins and a large fraction of their cellular organelles. DNA breakdown in aleurone protoplasts was not complete when cells died, since nuclei in dead cells stained TUNEL-positive (Figure 8B).

In apoptotic cell death, nuclear DNA is typically broken down into discrete low-molecular-weight fragments that produce a ladder of ca. 180 bp when analyzed on agarose gels. Artifactual DNA breakdown into internucleosomal DNA fragmentation can occur during the isolation of barley aleurone DNA through the activity of endogenous nucleases or nucleases present in the enzymes used to prepare protoplasts (Fath *et al.*, 1999). Such experimental artifacts led to the erroneous conclusion that barley aleurone cells die by apoptosis (Wang *et al.*, 1996). When care was taken to remove contaminating enzymes from protoplasts and to reduce the action of endogenous nucleases during DNA isolation, extracted DNA was detected almost exclusively as high-molecular-weight DNA on a 1.5% agarose gel (Figure 8C). DNA ladders were not observed even when DNA was isolated from a population of cells in which the amount of DNA per living cell was declining (Bethke *et al.*, 1999; Fath *et al.*, 1999).

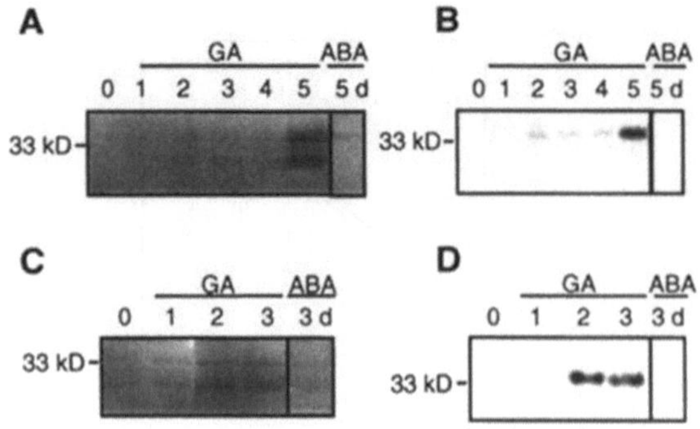

Figure 9. GA treatment of barley aleurone protoplasts and layers increases nuclease activities and results in an accumulation of barley nuclease I. A. Nuclease activity gel of an extract from barley aleurone protoplasts (5 μg protein/lane) incubated in 25 μM GA or 5 μM ABA for the indicated time. B. Immunoblot of an extract from barley aleurone protoplasts (5 μg protein/lane) incubated in 25 μM GA or 5 μM ABA for the indicated time probed with barley nuclease I Mab (1:750 dilution). Barley nuclease I has a predicted size of 33 kDa. C. Nuclease activity gel of an extract from barley aleurone layers (1 μg protein/lane) incubated in 5 μM GA or 5 μM ABA for the indicated time. D. Immunoblot of an extract from barley aleurone layers (5 μg protein/lane) incubated in 5 μM GA or 5 μM ABA for the indicated time probed with barley nuclease I Mab (1:750 dilution).

Candidate enzymes involved in autolysis and PCD of cereal aleurone cells

Nucleases

Barley aleurone cells contain several intracellular nucleases whose activities are regulated by ABA and GA (Figure 9A, C). Using an in-gel enzyme assay we showed that at least three nuclease activities with a molecular mass of ca. 33, 27 and 25 kDa were present after exposure of aleurone layers and protoplasts to GA, but only low activity was detected when protoplasts or layers were incubated in the absence of GA or in ABA (Figure 9A, C). Intracellular accumulation of nucleases in response to GA did not begin until after synthesis and secretion of α-amylase and other hydrolases is essentially complete. The increase in nuclease activities after incubation of protoplasts in GA was most dramatic after 5 days of incubation, corresponding to the time when death is most frequent. In aleurone layers, an increase in nuclease activities was detected after 2 days of incubation with GA, and reached a maximum after 3 days (Figures 4 and 9C). The nuclease activity at 33 kDa was identified as barley nuclease I by immunoblotting using a monoclonal antibody against this protein (Brown and Ho, 1986), whereas the nuclease activities at 27 and 25 kDa did not cross-react with the nuclease I antibody (Figure 9B, D).

All three of these nucleases can cleave single-stranded DNA and RNA, as well as double-stranded DNA (Fath *et al.*, 1999). They are activated by $MnCl_2$ and are strongly inhibited by EDTA and EGTA. Barley nuclease I has endonuclease activity and the resulting free 3′-OH ends can be detected by the TUNEL assay. A cDNA encoding an S1-type DNase, BEN1, was cloned from barley aleurone (Aoyagi *et al.*, 1998), and the deduced amino acid sequence of this cDNA showed high homology to the partial amino acid sequence of purified nuclease I (Brown and Ho, 1987). This suggests that *BEN1* encodes the 33 kDa barley nuclease I. Northern blotting shows that *BEN1* mRNA accumulates to high levels in barley aleurone layers incubated in GA for 72 h, but *BEN1* transcripts were not detected in freshly isolated aleurone layers (Aoyagi *et al.*, 1998).

The striking increases in barley nuclease activities and *BEN1* mRNA late in the response of this cell to GA correlate with the onset of DNA degradation and TUNEL-positive labeling. This suggests that these enzymes may be players in aleurone PCD and may function to degrade nuclear DNA in living barley aleurone cells prior to death. DNA breakdown represents an irreversible step in the cell death program and may indicate the time at which aleurone cells are committed to die.

Proteases

The role of proteases in PCD has been well documented for animals and proteases have often been suggested to play a role in plant PCD (Del Pozo and Lam, 1998; Fukuda *et al.*, 1998; Runeberg-Roos and Saarma, 1998; Groover and Jones, 1999; Solomon *et al.*, 1999). The data for proteases and PCD of zinnia tracheary elements are particularly interesting. Incubation of differentiating tracheary elements with trypsin, but not with a variety of other hydrolytic enzymes including macerozyme, pectinase, cellulase and proteinase K, caused a significant increase in the rate of death (Groover and Jones, 1999). Furthermore, addition of trypsin inhibitor to media in which zinnia mesophyll cells were induced to differentiate brought about a dramatic decrease in cell death (Groover and Jones, 1999). These data were taken to indicate that PCD of *Zinnia* tracheary elements is triggered by an extracellular serine protease. A candidate for this death-inducing protease was identified (Groover and Jones, 1999).

Germinating barley grain and isolated aleurone layers secrete a broad range of hydrolases, including aspartic, cysteine and serine proteases (Koehler and Ho, 1988; Wrobel and Jones, 1992; Törmäkangas *et al.*, 1994; Bethke *et al.*, 1996). We explored the possibility that secreted hydrolases bring about cell death in barley aleurone cells. Aleurone protoplasts were incubated for 8 days in a medium containing GA, during which time almost all cells died. The resulting hydrolase-containing medium (conditioned medium) was filtered to remove cellular debris and incubated with freshly prepared aleurone protoplasts. We compared PCD of cells in conditioned medium with a parallel set of protoplasts incubated in fresh medium and found that the presence of hydrolases in conditioned medium did not influence the rate of death (Bethke *et al.*, 1999). From these results we concluded that cell death in barley aleurone cells is not brought about by a lethal accumulation of hydrolases in the incubation medium.

Only a few proteases are known to be retained within the aleurone. The best characterized of these is the aspartic protease phytepsin, a cathepsin D homologue (Sarkkinen *et al.*, 1992). Phytepsin was localized to the PSVs of barley aleurone cells and its activity was slightly up-regulated after treatment with GA for 17.5 h (Bethke *et al.*, 1996). Sarkkinen *et al.* (1992) found that phytepsin was capable of removing 13 to 15 amino acids from the propeptide of barley lectin and they suggested that phytepsin may be a processing enzyme rather than being involved in digesting storage proteins. In a more recent study, phytepsin was found to be highly expressed during autolysis of developing tracheary elements and sieve cells in barley roots (Runeberg-Roos and Saarma, 1998). Thus, phytepsin may play a role in the active autolysis of plant cells, perhaps by activating other proteases.

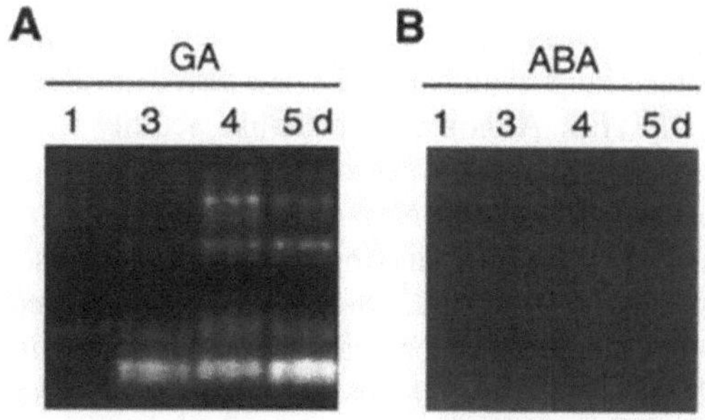

Figure 10. GA treatment of barley aleurone protoplasts increases cysteine protease activities. Aleurone protoplasts were incubated either in 25 μM GA or 5 μM ABA for the time indicated and homogenates (3 μg protein/lane) were analyzed on protease activity gels in the presence of the aspartic protease inhibitor pepstatin (20 μM). (A) Protoplast incubated in GA or ABA (B).

Using protease activity gels, we found at least five intracellular cysteine proteases in extracts from protoplasts incubated in GA for 5 days (R. Meza-Romero and R.L. Jones, unpublished; Figure 10). These cysteine protease activities were absent in ABA-treated and freshly isolated protoplasts and in gel activity they were maximal 4–5 days after GA treatment (Figure 10). Using antibodies to the mature form of the cysteine protease aleurain, we have shown that this cathepsin H homologue accumulates within aleurone protoplasts beginning 2–3 days after incubation in GA and reaches a maximum shortly before cells die (A. Fath and R.L. Jones, unpublished observations). Although the timing of the accumulation of cysteine proteases suggests that these activities may play a role in PCD of the aleurone cell, a specific role for these GA-induced proteases remains to be elucidated.

The best studied cysteine proteases involved in PCD in animal cells are caspases (Cohen, 1997). Caspases are cysteine proteases that cleave their substrate after an aspartic acid residue, and all caspases share unique catalytic properties. Fluorogenic substrates for caspase 1 and caspase 3 as well as their inhibitors are commercially available and were tested in aleurone cells *in vivo*. Although caspase-like activity was detected in living barley aleurone cells using fluorogenic substrates, the involvement of these activities in aleurone PCD is unlikely since no difference in caspase activty could be detected between ABA- and GA-treated protoplasts. Furthermore, caspase activity in aleurone cells was localized to the vacuole, whereas caspases in animal cells are cytoplasmic proteins (Cohen, 1997).

Aleurone cell death: apoptosis or autolysis?

Although arguments have been brought forward in favor of cell death by apoptosis in the cereal aleurone cell (Wang *et al.*, 1996, 1998), several lines of evidence indicate that PCD in this tissue is better described as a form of autolysis. The hallmarks of apoptosis are not observed in barley aleurone cells (Bethke *et al.*, 1999; Fath *et al.*, 1999). Changes in nuclear morphology, formation of internucleosomal DNA ladders, and the formation of apoptotic bodies do not accompany death of the aleurone cell. Rather, what has been observed is that these cells mobilize almost all of their cellular contents before death. Cellular sources of nitrogen, phosphorus and carbon are catabolized and exported to the growing seedling leaving a highly lytic compartment within the cell. Electron microscopy and fluorescence microscopy using fluorophores that stain mitochondria and antibodies that label glyoxysomes and western blotting of aleurone extracts all show that organelles decline in number as cells approach death (J.E. Lonsdale, P.C. Bethke and R.L. Jones, unpublished observations). In this regard, aleurone cell death resembles autolyis as found in senescing leaves and other plant organs (Thomas, 1994; Bleecker and Patterson, 1997). We speculate that autolysis reduces organellar function to a level that is insufficient to support cellular homeostasis and that reduced ATP levels or an inability of the cell to repair or replace essential proteins leads to cell death.

Acknowledgement

We thank Eleanor Crump for editorial assistance. This work was supported by grants from the National Science Foundation and Novartis Agricultural Discovery Institute to R.L.J.

References

Aoyagi, S., Sugiyama, M. and Fukuda, H. 1998. *BEN1* and *ZEN1* cDNAs encoding S1-type DNases that are associated with cell death in plants. FEBS Lett. 429: 134–138.

Bechtel, D.B. and Pomeranz, Y. 1977. Ultrastructure of the mature rice (*Oryza sativa*) caryopsis. The caryopsis coat and the aleurone cells. Am. J. Bot. 64: 966–973.

Bethke, P.C., Hillmer, S. and Jones, R.L. 1996. Isolation of intact protein storage vacuoles from barley aleurone. Plant Physiol. 110: 521–529.

Bethke, P.C., Schuurink, R.C. and Jones, R.L. 1997. Hormonal signaling in cereal aleurone. J. Exp. Bot. 48: 1337–1356.

Bethke, P.C., Swanson, S.J., Hillmer, S. and Jones, R.L. 1998. From storage compartment to lytic organelle: the metamorphosis of the aleurone protein storage vacuole. Ann. Bot. 82: 399–412.

Bethke, P.C., Lonsdale, J.E., Fath, A. and Jones, R.L. 1999. Hormonally regulated programmed cell death in barley aleurone cells. Plant Cell 11: 1033–1046.

Bethke, P.C., Jacobsen, J.V. and Jones, R.L. 2000. Barley biotechnology: progress, problems and prospects. In: M. Black and D. Bewley (Eds.), Seed Technology and its Biological Basis, Sheffield Academic Press, Sheffield, pp. 184–225.

Bleecker, A.B. and Patterson, S.E. 1997. Last exit: senescence, abscission, and meristem arrest in arabidopsis. Plant Cell 9: 1169–1179.

Bosnes, M., Weideman, F. and Olsen, O.-A. 1992. Endosperm differentiation in barley wild-type and sex mutants. Plant J 2: 661–674.

Brown, P.H. and Ho, T.-H.D. 1986. Barley aleurone layers secrete a nuclease in response to gibberellic acid. Plant Physiol. 82: 801–806.

Brown, P.H. and Ho, T.-H.D. 1987. Biochemical properties and hormonal regulation of barley nuclease. Eur. J. Biochem. 52: 1–8.

Bush, D.S. 1996. Effects of gibberellic acid and environmental factors on cytosolic calcium in wheat aleurone cells. Planta 199: 89–99.

Bush, D.S., Cornejo, M.-J., Huang, C.-N. and Jones, R.L. 1986. Ca^{2+}-stimulated secretion of α-amylase during development in barley aleurone protoplasts. Plant Physiol. 82: 566–574.

Chandler, P.M., Zwar, J.A., Jacobsen, J.V., Higgins, T.J.V. and Inglis, A.S. 1984. The effect of gibberellic acid and abscisic acid on α-amylase mRNA levels in barley aleurone layers: studies using an α-amylase cDNA clone. Plant Mol. Biol. 3: 407–418.

Cohen, G.M. 1997. Caspases: the executioners of apoptosis. Biochem J. 326: 1–16.

Cornejo, M.J., Platt-Aloia, K.A., Thomson, W.W. and Jones, R.L. 1988. Effects of GA_3 and Ca^{2+} on barley aleurone protoplasts: a freeze-fracture study. Protoplasma 146: 157–165.

Davies, T.G.E., Steele, S.H., Walker, D.J. and Leigh, R.A. 1996. An analysis of vacuole development in oat aleurone protoplasts. Planta 198: 356–364.

Del Pozo, O. and Lam, E: 1998. Caspases and programmed cell death in the hypersensitive response of plants to pathogens. Curr. Biol 8: 1129–1132.

Demura, T. and Fukuda, H. 1994. Novel vascular cell-specific genes whose expression is regulated temporally and spatially during vascular system development. Plant Cell 6: 967–981.

Fath, A., Bethke, P.C. and Jones, R.L. 1999. Barley aleurone cell death is not apoptotic: characterization of nuclease activities and DNA degradation. Plant J., 20: 305.

Fincher, G.B. 1989. Molecular and cellular biology associated with endosperm mobilization in germinating cereal grains. Annu. Rev. Plant Physiol. Plant Mol. Biol. 40: 305–346.

Fincher, G.B. 1992. Cell wall metabolism in barley. In: P.R. Shewry (Ed.), Barley: Genetics, Biochemistry, Molecular Biology and Biotechnology, CAB, Wallingford, UK, pp. 413–437.

Fukuda, H. 1996. Xylogenesis: initiation, progression and cell death. Annu. Rev. Plant Physiol. Plant Mol. Biol. 47: 299–325.

Fukuda, H., Watanabe, Y., Kuriyama, H., Aoyagi, S., Sugiyama, M., Yamamoto, R., Demura, T. and Minami, A. 1998. Programming of cell death during xylogenesis. J. Plant Res. 111: 253–256.

Gavrieli, Y., Sherman, Y. and Ben-Sasson, S.A. 1992. Identification of programmed cell death in situ via specific labeling of nuclear DNA fragmentation. J. Cell. Biol. 119: 493–501.

Gilroy, S. 1996. Signal transduction in barley aleurone protoplasts is calcium dependent and independent. Plant Cell 8: 2193–2209.

Gilroy, S. and Jones, R.L. 1992. Gibberellic acid and abscisic acid coordinately regulate cytoplasmic calcium and secretory activity in barley aleurone protoplasts. Proc. Natl. Acad. Sci. USA 89: 3591–3595.

Groover, A. and Jones, A.M. 1999. Tracheary element differentiation uses a novel mechanism coordinating programmed cell death and secondary cell wall synthesis. Plant Physiol. 119: 375–384.

Gubler, F., Kalla, R., Roberts, J.K. and Jacobsen, J.V. 1995. Gibberellin-regulated expression of a *myb* gene in barley aleurone cells: evidence for Myb transactivation of a high-pI α-amylase gene promoter. Plant Cell 7: 1879–1891.

Haberlandt, G. 1884. Physiologische Pflanzenanatomie, W. Engelman, Leipzig, Germany.

He, C.-J., Morgan, P.W. and Drew, M.C. 1996. Transduction of an ethylene signal is required for cell death and lysis in the root cortex of maize during aerenchyma formation induced by hypoxia. Plant Physiol. 112: 463–472.

Heimovaara-Dijkstra, S., Heistek, J.C. and Wang, M. 1994. Counteractive effects of ABA and GA3 on extracellular and intracellular pH and malate in barley aleurone. Plant Physiol. 106: 359–365.

Heimovaara-Dijkstra, S., Nieland, T.J.F., van der Meulen, R.M. and Wang, M. 1996. Abscisic acid-induced gene-expression requires the activity of protein(s) sensitive to the protein-tyrosine phosphatase inhibitor phenylarsine oxide. Plant Growth Regul. 18: 115–123.

Higgins, T.J.V., Jacobsen, J.V. and Zwar, J.A. 1982. Changes in protein synthesis and mRNA levels in barley aleurone layers treated with gibberellic acid. Plant Mol. Biol. 1: 191–215.

Jacobsen, J.V., Knox, R.B. and Pyliotis, N.A. 1971. The structure and composition of aleurone grains in the barley aleurone layer. Planta 101: 189–209.

Jones, R.L. 1969. The fine structure of barley aleurone cells. Planta 85: 359–374.

Jones, R.L. 1972. Fractionation of the enzymes of the barley aleurone layer: evidence for a soluble mode of enzyme release. Planta 103: 95–109.

Jones, R.L. 1980. Quantitative and qualitative changes in the endoplasmic reticulum of barley aleurone layers. Planta 150: 70–81.

Jones, R.L. and Jacobsen, J.V. 1991. Regulation of synthesis and transport of secreted proteins in cereal aleurone. Int. Rev. Cytol. 126: 49–88.

Jones, R.L. and Price, J.M. 1970. Gibberellic acid and the fine structure of barley aleurone cells. III. Vacuolation of the aleurone cell during the phase of ribonuclease release. Planta 94: 191–202.

Koehler, S. and Ho, T.-H.D. 1988. Purification and characterization of gibberellic acid- induced cysteine endoprotease in barley aleurone layers. Plant Physiol .87: 95–103.

Kuo, A., Cappelluti, S., Cervantes-Cervantes, M., Rodriguez, M. and Bush, D.S. 1996. Okadaic acid, a protein phosphatase inhibitor, blocks calcium changes, gene expression, and cell death induced by gibberellin in wheat aleurone cells. Plant Cell 8: 259–269.

Kurkdjian, A. and Guern, J. 1989. Intracellular pH: measurement and importance in cell activity. Annu. Rev. Plant Physiol. Plant Mol. Biol. 40: 271–303.

Levine, A., Tenhaken, R., Dixon, R. and Lamb, C. 1994. H_2O_2 from the oxidative burst orchestrates the plant hypersensitive disease resistance response. Cell 79: 583–593.

Levine, A., Pennell, R.I., Alvarez, M.E., Palme, R. and Lamb, C. 1996. Calcium-mediated apoptosis in a plant hypersensitive disease resistance response. Curr. Biol 6: 427–437.

Lonsdale, J.A., McDonald, K.L. and Jones, R.L. 1999. High pressure freezing and freeze substitution reveal new aspects of fine structure and maintain protein antigenicity in barley aleurone cells. Plant J. 17: 221–229.

Pastori, G.M. and Del Rio, L.A. 1997. Natural senescence of pea leaves: an activated oxygen-mediated function for peroxisomes. Plant Physiol. 113: 411–418.

Penson, S.P., Schuurink, R.C., Fath, A., Gubler, F., Jacobsen, J.V. and Jones, R.L. 1996. Cyclic GMP is required for gibberellic acid-induced gene expression in barley aleurone. Plant Cell 8: 2325–2333.

Ritchie, S. and Gilroy, S. 1998a. Calcium-dependent protein phosphorylation may mediate the gibberellic acid response in barley aleurone. Plant Physiol. 116: 765–776.

Ritchie, S. and Gilroy, S. 1998b. Tansley Review No. 100: Gibberellins: regulating genes and germination. New Phytol. 140: 363–383.

Runeberg-Roos, P. and Saarma, M. 1998. Phytepsin, a barley vacuolar aspartic proteinase, is highly expressed during autolysis of developing tracheary elements and sieve cells. Plant J. 15: 139–145.

Runeberg-Roos, P., Kervinen, J., Kovaleva, V., Raikhel, N.V. and Gal, S. 1994. The aspartic protease of barley is a vacuolar enzyme that processes probarley lectin. Plant Physiol. 105: 321–329.

Sarkkinen, P., Kalkkinen, P., Tilgmann, C., Siuro, J., Kervinen, J. and Mikola, L. 1992. Aspartic protease from barley grains is related to mammalian lysosomal cathepsin D. Planta 186: 317–323.

Schuurink, R.C., Chan, P.V. and Jones, R.L. 1996. Modulation of calmodulin mRNA and protein levels in barley aleurone. Plant Physiol. 111: 371–380.

Solomon, M., Belenghi, B., Delledonne, M., Menachem, E. and Levine, A. 1999. The involvement of cysteine proteases and protease inhibitor genes in the regulation of programmed cell death in plants. Plant Cell 11: 431–443.

Stewart, A., Nield, H. and Lott, J.N.A. 1988. An investigation of the mineral content of barley grains and seedlings. Plant Physiol. 86: 93–97.

Swanson, S.J. and Jones, R.L. 1996. Gibberellic acid induces vacuolar acidification in barley aleurone. Plant Cell 8: 2211–2221.

Swanson, S., Bethke, P.C. and Jones, R.L. 1998. Barley aleurone cells contain two types of vacuoles: characterization of lytic compartments using fluorescent probes. Plant Cell 13: 685–698.

Taiz, L. and Jones, R.L. 1970. Gibberellic acid, β-1,3-glucanase and the cell walls of barley aleurone layers. Planta 92: 73–84.

Thomas, H. 1994. Aging in the plant and animal kingdoms-the role of cell death. Rev. Clin. Gerontol. 4: 5–20.

Törmäkangas, K., Kervinen, J., Östman, A. and Teeri, T. 1994 Tissue-specific localization of aspartic proteinase in developing and germinating barley grains. Planta 195: 116–125.

Wang, M., Oppedijk, B., Lu, X., van Duijn, B. and Schilperoort, R.A. 1996. Apoptosis in barley aleurone during germination and its inhibition by abscisic acid. Plant Mol. Biol. 32: 1125–1134.

Wang, M., Oppedijk, B.J., Caspers, M.P.M., Lamers, G.E.M., Boot, M.J., Geerlings, D.N.G., Bakhuizen, B., Meijer, A.H. and van Duijn, B. 1998. Spatial and temporal regulation of DNA fragmentation in aleurone of germinating barley. J. Exp. Bot. 49: 1293–1301.

Wrobel, R. and Jones, B.L. 1992. Appearance of endoproteolytic enzymes during the germination of barley. Plant Physiol. 100: 1508–1516.

Young, T.E., Gallie, D.R. and DeMason, D.A. 1997. Ethylene-mediated programmed cell death during maiz endosperm development of wild-type and shrunken2 genotypes. Plant Physiol. 115: 737–751.

Plant Molecular Biology **44:** 267–281, 2000.
E. Lam, H. Fukuda and J. Greenberg (Eds.), Programmed Cell Death in Higher Plants.

Programmed cell death in plant reproduction

Hen-ming Wu[1,2] and Alice Y. Cheung[1,2,3,*]
[1]*Department of Biochemistry and Molecular Biology,* [2]*Graduate Program in Molecular and Cellular Biology, and* [3]*Plant Biology Graduate Program, University of Massachusetts, Amherst, MA 01003, USA (*author for correspondence; e-mail: acheung@biochem.umass.edu)*

Key words: fertilization, male sterility, pollination, sex determination, tapetum

Abstract

Reproductive development is a rich arena to showcase programmed cell death in plants. After floral induction, the first act of reproductive development in some plants is the selective killing of cells destined to differentiate into an unwanted sexual organ. Production of functional pollen grains relies significantly on deterioration and death of the anther tapetum, a tissue whose main function appears to nurture and decorate the pollen grains with critical surface molecules. Degeneration and death in a number of anther tissues result ultimately in anther rupture and dispersal of pollen grains. Female sporogenesis frequently begins with the death of all but one of the meiotic derivatives, with surrounding nucellar cells degenerating in concert with embryo sac expansion. Female tissues that interact with pollen undergo dramatic degeneration, including death, to ensure the encounter of compatible male and female gametes. Pollen and pistil interact to kill invading pollen from an incompatible source. Most observations on cell death in reproductive tissues have been on the histological and cytological levels. We discuss various cell death phenomena in reproductive development with a view towards understanding the biochemical and molecular mechanisms that underlie these processes.

Introduction

With an origin that may be rooted in unicellular organisms that emerged between one and two billion years ago (Ameisen, 1996; Yao, 1996), programmed cell death (PCD) is a fact of life in multicellular organisms (Vaux and Korsmeyer, 1999). From embryogenesis to fertilization, cell and tissue death is an integral part of plant development and morphogenesis and its response to the environment (Barlow, 1982; Greenberg, 1996; Mittler and Lam, 1996; Pennell and Lamb, 1997; Buckner *et al.*, 1998; this issue). In contrast to senescence and cell death during vegetative plant development that may be accompanied by a vivid display of colors, PCD processes that are key to sexual plant reproduction are hidden from view. Cell death occurs early in sexual development in the selective abortion of primordia for one or the other sexual organ in some unisexual plants and during the development of the male and female reproductive organs and gametes. Late in the reproductive phase, cell death occurs in both pollen and the pistil. Failure to properly enter the various cell death programs (Figure 1) compromises reproductive success, sometimes even resulting in sterility. The cellular deterioration patterns that have been described for these reproductive cells are, in many respects, similar to those observed in PCD in other plant and animal developmental systems (Pennell and Lamb, 1997; Allen *et al.*, 1998; Vaux and Korsmeyer, 1999; this issue). Relatively unusual in plant reproductive cell death is that it often involves cell rupture, releasing cellular contents that are incorporated into useful functional components of other differentiating cell types. The cell death processes during reproductive development have been described extensively at the histological and the cytological levels. These provide an excellent background for the emerging genetic, molecular and biochemical studies to understand the mechanisms underlying PCD in plant reproductive development.

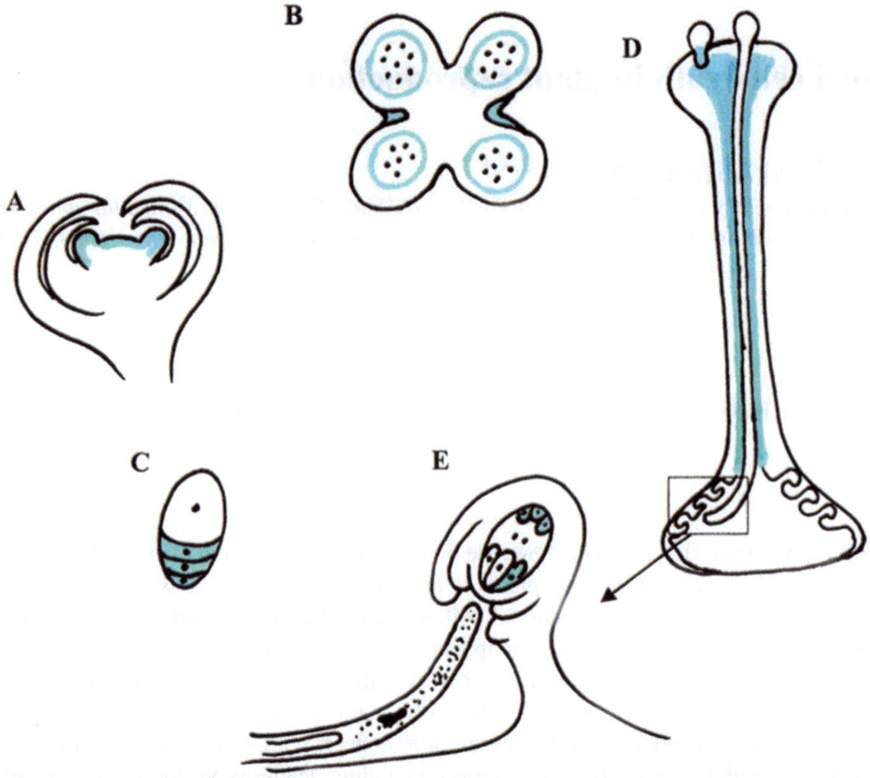

Figure 1. Tissues and cells that most prominently undergo PCD (in blue) during reproductive development. A. Floral apical meristem. The central zone gives rise to stamen (outer blue zones) and carpel (central blue zone) primordia. In some unisexual plants, such as maize, cells in one or the other primordia die, giving rise to either male or female flowers. B. A cross-section of an anther showing the four locules. The tapetum (blue circles) and stomium (blue area in the mid-line) degenerate and die to complete pollen development and dispersal. C. One functional megaspore develops into the female gametophyte, the three (blue cells) degenerate. D. Pollination and pollen tube growth are accompanied with cell degeneration and death in the secretory and transmitting tissues (blue). Arrested incompatible pollen tubes (blue) dies. E. Pollen tube entrance into an ovule. In the seven-celled female gametophyte, one of the two synergid cells (blue, towards the pollen tube) that surround the egg cell degenerates and bursts, and the pollen tube enters via this cell. The antipodal cells (blue, opposite to the egg cell/synergid cells) degenerate after fertilization.

Sex determination and PCD

Most angiosperms are hermaphrodites, producing male (stamen) and female (carpel or pistil) organs in a spatially and temporally regulated manner at the floral meristem (Ma, 1994). About 10% of flowering species develop unisexual flowers (Irish and Nelson, 1989; Lebel-Hardenack and Grant, 1994; Wu and Cheung, 1998). For example, *Zea mays* (maize) is a monoecious species in which male and female flowers develop on separate inflorescences, tassel and ears, respectively, on one plant; *Silene latifolia* (white campion) and *Rumex acetosa* (sorrel) are dioecious, bearing flowers of a single sex on one plant. In between dioecy and monoecy exists a spectrum of intermediate species in which unisexual and bisexual flowers are produced on the same plant. It is generally believed that sex determination occurs subsequent to the initiation of sexual organogenesis (Hardenack *et al.*, 1994; Juarez and Banks, 1998). During early development, the flowers of some unisexual plants (e.g. *Z. mays* and *Silene*) are hermaphroditic, producing promordia for both stamens and carpels at all the floral meristems. Unisexuality is accomplished either by abolition of the unwanted sexual organ initials or by inhibition of their further differentiation, resulting in the differentiation of male or female flowers.

In maize, sex determination involves abolition of the gynoecial initials in the male flowers and a relatively late arrest of staminate development in the female flowers when anther lobes are already visible. In both cases, cellular vacuolization and the loss of organelle and cytoplasmic integrity (Cheng *et al.*, 1983) precede primordial abortion. Genetic dissection of sex determination and the characterization of a few sex determination genes indicate that phytohormones play critical roles in this process. For example, the masculinized ear phenotypes in maize *dwarf* and anther *ear 1* mutants could be mitigated or reversed by gibberellic acid treatment (see Irish and Nelson, 1989; Lebel-Hardenack and Grant, 1997). The *ANTHER EAR 1* gene and one of the *dwarf* genes (dwarf 3) both encode enzymes involved in the biosynthesis of gibberellic acid (Bensen *et al.*, 1995; Winkler and Helentjaris, 1995). A series of *tasselseed (ts)* mutants display various degrees of feminization of the tassel (Neuffer *et al.*, 1997), including a complete transition to pistillate tassels in *ts1* and *ts2* plants. TS2 encodes a protein with structural features similar to short-chain alcohol dehydrogenases and is most similar to steroid dehydrogenases (Delong *et al.*, 1993). In wild-type plants, TS2 mRNA is located subepidermally in the gynoecial primordium before its abortion in developing florets in the tassel, but is not detected in florets in the tassel of the *ts2* mutant. The cell death process appears to progress from the subepidermal region of the gynoecium, ultimately aborting the entire organ in staminate flowers (Calderon-Urrea and Dellaporta, 1999). The patterns of TS2 expression and cell death process are consistent with a function for TS2 to arrest female development in a hermaphrodite meristem by causing, directly or indirectly, cell death in the gynoecial primordia. A steroid hormone, brassinosteroid (Mandava, 1988; Clouse, 1996), is important for plant development and affects pollen fertility in a number of plants (e.g. Li *et al.*, 1996). Whether the *TS2* gene product functions via its probable ability to modify a steroid-like substance to arrest carpel development in maize remains to be explored. A cell death mechanism similar to that in maize appears to be conserved in its wild relative *Tripsacum dactyloides* (Li *et al.* 1997), but is apparently not universal in unisexual plants (see Lebel-Hardenack *et al.*, 1977; Ainsworth *et al.*, 1995).

PCD in anther development and male fertility

Microsporogenesis and microgametogenesis, the production of male gametophytes (pollen grains), occurs in the anther locules of the stamen and is initiated from a linear group of sporogenous mother cells (Mascarenhas, 1990; Bedinger, 1992; McCormick, 1993). Anther development is viewed to proceed in two phases (Koltunow *et al.*, 1990; Bedinger, 1992; Goldberg *et al.*, 1993; McCormick, 1993). In phase I, histo-specification takes place in which specific anther tissues differentiate and meiotic divisions of the microspore mother cell give rise to four microspores enveloped in a callose (β-1,3-glucan)-walled tetrad. In phase II, microspores are released from the tetrad, subsequently undergoing mitosis to produce the male gametophytes (pollen grains), each with a generative and a vegetative cell. In some species, the generative cell divides again to produce two sperm cells prior to anthesis. In others, this step occurs after pollen germination. Cell degeneration programs in various anther tissues, which are critical for pollen maturation and dispersal, are initiated during this time. Mutants that are perturbed in this degenerative phase of anther development are often male-sterile (e.g. Mariani *et al.*, 1990; Dawson *et al.*, 1993; Loukides *et al.*, 1995; Beals and Goldberg, 1997).

Tapetal cell death

To appreciate the importance of PCD in male fertility, one has to examine the mature pollen grains and the role of the tapetum in their formation (Chapman, 1987; Bedinger, 1992; Loukides *et al.*, 1995). Pollen grains are covered with a coat of proteinaceous and lipoidal materials and further decorated by an elaborately sculptured layer called the exine. The exine and components of the pollen coat are derived largely from the tapetum. The tapetal cell layer lines the anther locules (Figure 1) where microsporogenesis and pollen maturation take place. Mature tapetal cells lack a well-developed primary cell wall and are packed with secretory organelles on surfaces exposed toward the locular region, suggesting highly polar secretory activities into the locule. During microspore development, the tapetum is believed to play a nutritive role. The tapetum also secretes enzymes that degrade callose (β-1,3-glucanase) and pectic molecules (Rhee and Somerville, 1998) to release individual microspores from the tetrad. The temporal regulation of β-1,3-glucanase secretion is critical for the production of viable pollen. Microspore abortion and defective pollen coatings have both been observed when β-1,3-glucanase is prematurely secreted (Izhar and Frankel, 1971; Worral *et al.*, 1992).

In normal pollen development, shortly after microspore release from the tetrad and before mitosis, tapetal cells begin to degenerate. Tapetal cell deterioration is marked by cell shrinkage, polarization of cytoplasmic materials, vacuolation, and thinning of cell walls, which become less distinct between adjacent cells. Ultimately the cells rupture and degeneration of the entire tapetum occurs (Chapman, 1987; Bedinger, 1992). Through their death, the tapetal cells further contribute to completing the extracellular sculpting of the pollen grains as well as providing them with adhesive and signaling molecules of proteinaceous and lipoidal nature that are critical for interacting with the pistil during pollination. It appears that tapetal cells, prior to their death, have already secreted proteinaceous molecules that form a scaffold on the maturing pollen upon which other decorating molecules are deposited (Wu *et al.*, 1997; Piffanelli and Murphy, 1998; Murphy and Ross, 1998). Absence of key molecules from the pollen coat, defects in tapetal-produced enzymes, arrest in tapetal development and precocious tapetal deterioration can all result in male sterility (e.g. Mariani *et al.*, 1990; Preuss *et al.*, 1993; Hölskamp *et al.*, 1995a; Loukides *et al.*, 1995; Aarts *et al.*, 1997). Presumably, in the absence of a differentiated tapetum, critical nutrient resources are not available for proper microspore development and/or for their differentiation into functional pollen grains. If tapetal cell death is precocious and occurs before microspores are released from the tetrad, important tapetal-derived pollen coat components could be dislodged into the anther locule before pollen surfaces are exposed for their deposition. Alternatively, precocious tapetal cell death and rupture may release molecules that are deleterious to pollen maturation.

Anther dehiscence and PCD of specific cell types

Release of mature pollen requires the breakdown of anther cells at the stomium (Figure 1), a specialized structure that is contiguous with the epidermis and runs along the lateral side of each anther half (Bonner and Dickinson, 1989; Koltunow *et al.*, 1990; Goldberg *et al.*, 1993; Beals and Goldberg, 1997). Failure to dehisce leads to male sterility (Dawson *et al.*, 1993; Beals and Goldberg, 1997). Prior to anther dehiscence, the endothecium, which lies between the tapetum and the epidermis, and surrounding connective tissues degenerate. A specialized cell cluster that is adjacent to the stomium, known as the circular cell cluster, also undergoes cell death, releasing cell debris and calcium crystals (Horner and Wagner, 1992; Beals and Goldberg, 1997). Selective cytotoxin ablation of the stomium, circular cell clusters and their surrounding connective tissues blocks dehiscence, although ablation of the stomium alone adequately blocks this process as well (Beals and Goldberg, 1997). This suggests that the stomium structure is important for dehiscence, probably acting as a fault line where rupture can occur.

Cytoplasmic male sterility (CMS), mitochondria and PCD

CMS is a maternally inherited but male-expressed trait in which the plant fails to produce viable pollen grains (Levings, 1993; Schnabe and Wise, 1998). CMS has been found in more than 150 species. The best understood systems have resulted from aberrations in their mitochondrial genome, giving rise to novel polypeptides. The most extensively analyzed CMS system is probably the *cms-T* maize because of its significant agronomic value as well as its role in the Southern corn leaf blight epidemic in the late 1960s (see Levings, 1993). In *cms-T*, mitochondrial DNA rearrangement creates a novel gene, *T-urf13*, which encodes a 13 kDa polypeptide (URF13) uniquely found in the mitochondria of these plants (Dewey *et al.*, 1986, 1987). In *cms-T* plants, the tapetum undergoes precocious vacuolation and degeneration and the pollen developmental pathway is blocked soon after microspore mother cell meiosis.

The URF13 polypeptide is located in the inner mitochondrial membrane of *cms-T* plants, allowing it to interfere with the critical processes of electron transport and ATP synthesis. Sensitivity of the *cms-T* line to a toxin produced by the fungal pathogen that causes the leaf blight epidemic was, ironically, instrumental in advancing an understanding of how URF13 mediates its effect on mitochondrial function. Mitochondria in the pathotoxin-treated *cms-T* tissue show massive ion leakage and thus a dissipation of membrane potential (Holden and Sze, 1987). *Escherichia coli* expressing *T-urf13* accumulate URF13 in the plasma membrane (Dewey *et al.*, 1988). They become sensitive to the toxin, and glucose-driven respiration and growth is inhibited. These *E. coli* also show massive ion leakage and spheroplast swelling, not unlike phenomena associated with other PCD systems (Vaux and Korsmeyer, 1999).

URF13 apparently induces physiological defects that herald organelle and cell death via the ability of this protein to form membrane pores in the presence of the fungal toxin. Normally, however, the only major defect in *cms-T* plant is male sterility. The specific deleterious effect of *T-urf13* on pollen development has been attributed to the extraordinary metabolic demand on mitochondrial function in anther cells (Levings, 1993). In the maize tapetal and sporogenous cells, mitochondrial numbers increase 40- and 20-fold, respectively, reflecting a significant demand for energy during early pollen development. In *cms-T*, tapetal degeneration occurs after the rapid mitochondrial divisions at a window of time that is subsequent to pollen mother cell meiosis but before microspores are released from the tetrads, resulting in abortion of pollen development. Mitochondria in the *cms-T* anther tapetum may be uniquely sensitive to the presence of a foreign inner membrane protein (URF13) that has the potential to interfere with normal mitochondria physiology and biochemistry. Therefore, in CMS plants that have rearranged mitochondrial genomes resulting in the expression of novel polypeptides, mitochondrial malfunctions underlie precocious tapetal degeneration and male sterility.

Given our current knowledge of the role of mitochondria in PCD (Green and Reed, 1998; Susin *et al.*, 1998; Earnshaw, 1999; Susin *et al.*, 1999), it is possible to hypothesize that during normal anther development, tapetal PCD may be triggered by mitochondrial fatigue in maintaining the metabolic needs of developing pollen. Investigating the role of the mitochondria in signaling tapetal PCD may extend into territories thus far not explored in plant PCD.

PCD in pistil development and pollen-pistil interactions

The female organ experiences PCD in multiple tissue types throughout its development, from inception to fertilization (Figure 1). Postgenital fusion of carpels in some multi-carpelloid flowers, though not involving cell death per se, may be regarded a process in which adnate epidermal cells are triggered by certain signals to rejuvenate from an already destined developmental fate to de-differentiate and re-differentiate (Walker, 1975a, b, c; Seigel and Verbeke, 1989) to create a conduit for the transportation of the male gametes. Development of the embryo sac, where the egg cell resides, begins with and is facilitated by PCD. During pollination, cells in the stigma, the stylar transmitting tissue and ovules undergo PCD in preparation for the incoming pollen grains and tubes, either facilitating or preventing the passage of the male gametophytes from a compatible or an incompatible source, respectively (see Cheung, 1996b). In incompatible pollination, i.e. not destined to complete fertilization (see Matton *et al.*, 1994; Nasrallah, 1997), pollen-pistil interactions set off a signaling cascade that leads to the arrest of pollen tube growth activities. Prior to fertilization, PCD in a special cell pair in the ovules, the synergids, occurs, allowing the entrance of a pollen tube to the embryo sac.

Cellular rejuvenation in early pistil development

The gynoecium of many flowering plants is multi-carpelloid in which the individual carpel fuses to form the internal transmitting tissue through which pollen tubes elongate. After the emergence of carpel primordia in the floral meristem, an early event is the adherence of cells along the opposing carpels as they grow into contact with one another (Walker, 1975a, b, c). Epidermal cells in developing organs are destined to a developmental fate that rarely involves periclinal cell divisions. Instead, they are characterized by anticlinal divisions, which are required to fulfill the spatial needs for an expanding epidermis as an organ enlarges. The epidermal cells in developing carpels, upon contact with the epidermis of an opposite carpel, are unique in their ability to reverse this developmental fate and de-differentiate. This reversal is marked by a change from rectangularly shaped and densely cytoplasmic cells that strictly divide anticlinally from isodiametrically shaped secretory cells that predominantly divide periclinally. Walls of the adnate cells adhere and the fusion line is ultimately blurred by the copious amount of cell wall material deposited there. The periclinal divisions initiated from these epidermal cells ultimately produce the stigmatic secretory zone and the stylar transmitting tissue. These regions, in turn, provide a highly enriched conduit for pollen penetration of the female tissues. Although the cytological changes that have been observed during postgenital fusion are far from being marked by phenomena that are characteristics of PCD, it nonetheless involves a halt to a cell fate destined to quiescence and the cellular de-differentiation may be considered a rejuvenation event. Postgenital fusion can also be viewed as a case in which the differentiation of a tissue is diverted to give rise to the production of other essential cell types that are better suited for reproductive success. This is not unlike the altruistic purpose of PCD observed in developmental systems and in defense (Vaux and Kosmeyer, 1999; Pennell and Lamb, 1997; this issue). Understanding the mechanism that underlies a cell fate reversal process may lead to insights into how a cell death program may be diverted.

PCD in stigmatic and stylar transmitting tissue and their role in pollen germination and pollen tube growth

Pollination and pollen tube growth occurs in multiple pistil tissue types (Figure 1). The most proximal structure on a pistil, the stigma, is composed of an epidermal papilla, two to three layers of secretory cells, and a zone of non-glandular cells. Pollen germinate on the stigmatic papilla and their tubes elongate through the extracellular matrix secreted by the secretory cells. The tubes continue to elongate into the style in the extracellular matrix secreted by the transmitting tissue (Cheung, 1996a, b). Prior to pollination, cells in the mature stigmatic papilla and the underlying glandular regions of plants with wet stigmas degenerate and release large amounts of cellular materials to support pollination on the stigmatic surface (Cresti et al., 1986; Kandasamy and Kristen, 1987; Goldman *et al.*, 1994; Wolters-Art *et al.*, 1996, 1998).

Pollen tube growth in the transmitting tissue is also accompanied by severe transmitting tissue cell degeneration, including cell rupture (Bell and Hicks, 1976; Herrero and Dickinson, 1979; Wang *et al.*, 1996). In plants with a solid style, such as tobacco, the stylar transmitting tissue is biochemically enriched but spatially restricted for the incoming pollen tubes. The well organized files of transmitting tissue cells vacuolate even before the penetration of large numbers of pollen tubes. By the time most pollen tubes reach half way down the styles, the transmitting tissue becomes highly disorganized, often devoid of cytoplasm and grossly misshapen. The intercellular space is filled with cytosolic materials that are released from the transmitting tissue cells and with the penetrating pollen tubes, which occupy a significant volume. The pollination-induced cell death in transmitting tissue appears specific since adjoining cortical tissue remains intact. This contrasts with a general degeneration of stylar tissues by the action of ethylene, which is known to induce senescence. However, the transmitting tissue in these ethylene-treated but unpollinated styles does not reach the same level of deterioration as in pollinated styles, suggesting pollen tube penetration is necessary to elicit the full extent of cell deterioration processes (Wang *et al.*, 1996).

Cytoplasmic constituents are mobilized from the transmitting tissue cells to the extracellular matrix and to the pollen tubes (Kroh *et al.*, 1970; Labraca and Loewus, 1973; Sassen, 1974; Cheung *et al.*, 1995; Wu *et al.*, 1995, 2000; Lind *et al.*, 1996; de Graaf *et al.*, 1998). The pollination-induced cell death in the transmitting tissue could be interpreted as a means to provide a biochemically more enriched and physically more accommodating medium for pollen tube penetration. It has also been postulated that death of the transmitting tissue after pollination prevents the invasion of the ovary by pathogens, thus protecting fertilization (Herrero, 1992).

PCD in the female gametophyte and fertilization

The female gametophyte, or embryo sac, develops within the ovules inside the ovary (Figure 1). A typically diploid megaspore mother cell undergoes meiosis, giving rise to four megaspores containing half of the chromosome complement of the vegetative plant. Typically, three of these megaspores undergo PCD, and are with the remaining 'functional' megaspore undergoing three rounds of mitosis, subsequent nuclear migrations, nuclear fusion and cellularization and produce the typically seven cell/eight nucleate embryo sac (Russell, 1979, 1993; Reiser and Fischer, 1993; Bell, 1996; Christensen *et al.*, 1998). Degenerating megaspores show cell shrinkage, cytoplasmic disorganization and a pinotic nucleus (Bell, 1996), similar to events in other cells that undergo PCD. They are compressed, partly, by the expansion of the embryo sac which is accompanied by the degeneration of the surrounding nucellar cells (Russell, 1979).

PCD in early female gametophytic development ensures the production of an egg cell that is protected deep inside the ovule within the embyro sac. To ensure that sperm cells are provided with access and guidance to the embryo sac for fertilization, an equally important PCD process occurs within this chamber (Huang and Russell, 1992a; Russell, 1996). The embryo sac is oriented such that the egg cell adjoins a pair of synergid cells that occupy the micropylar end of the ovule where the pollen tube enters (Figure 1). Synergid degeneration appears to be a PCD process that prepares this synergid to receive the pollen tubes. In some species, such as cotton, one of the two synergids completely degenerates during pollen tube elongation in the style (Jensen and Fisher, 1968; Fisher and Jensen, 1969). In other species, such as tobacco, synergid degeneration occurs along with the arrival of the pollen tube (Huang and Russell, 1993, 1994). In the degenerating synergid cell, cytoplasmic contents increase in electron density, the nucleus becomes distorted, the vacuole collapses, autolysis of cellular content occurs, organelles degenerate and plasma membrane disintegrates. Synergid cells also show a high level of Ca^{2+} before pollen tube penetration, especially in the degenerative synergid (Huang and Russell, 1992b; Chaubal and Reger, 1992a, b; Tian and Russell, 1997).

In some plants, such as *Arabidopsis*, barley and sunflower, the degenerate synergid is usually located on the side close to the placenta from where the pollen tubes turn into the ovule (Mogensen, 1984; Yan *et al.*, 1991; Murgia *et al.*, 1993), suggesting that synergid degeneration is not random in these plants. Although degeneration of the synergids is not absolute in all plant species and may precede pollination or occur without pollination at all (Sumner, 1992), this degeneration is believed to serve a vital function during fertilization in species in which a synergid does undergo PCD.

Pollen tubes emerge from the style elongating in a basally oriented direction. However, the micropyle of the ovule (Figure 1), the conduit created by the integuments that protect the embryo sac, is oriented such that pollen tubes need to turn, sometimes as much as 90°, to gain access into the ovules (Cheung, 1996a; Wu and Cheung, 1998). Guidance cues are believed to emanate from within the ovules, and a viable embryo sac is essential for pollen tube entrance (Hölskamp *et al.*, 1995b; Ray *et al.*, 1997; Hagashiyama *et al.*, 1998; Cheung *et al.*, 2000). A burst of cytosolic contents from the degenerating synergid, e.g. Ca^{2+}, may set up a gradient of chemoattractant to guide pollen tubes (Chaubal and Reger, 1992a, b; Russell, 1996).

The antipodal cells (located on the chalazal end of the embryo sac, opposite from the micropyle) in some plants degenerate after fertilization (e.g. Webb and Gunning, 1990; Murgia *et al.*, 1993), their cytoplasm becomes more electron-dense and their organelle structure deteriorates. Ultimately, these cells collapse and almost disappear with only remnants of cytoplasm remaining within the cell wall.

Cell death in pollen tubes in self-incompatible pollination

In self-incompatible pollination systems, pollen grains are either arrested on the stigmatic surface or their tubes are arrested after some distance of penetration into the stylar transmitting tract. In these plants, self-male gametes are prevented from fusing with the egg cell by a variety of mechanisms (see Matton *et al.*, 1994; Nasrallah, 1997; Franklin-Tong, 1999) to preclude self-fertilization. Pollen tubes elongating in incompatible transmitting tissues develop thick cell walls, their tips swell, some tubes burst and others are enveloped by callose at their tips (Nettancourt *et al.*, 1973). It has been reported that within 24 h of pollination, most cytoplasmic inclusions in incompatible *Petunia* pollen tubes in the style have lost their identifiable forms (Herrero and Dickinson, 1981). Cytological features observed in various self-incompatible pollen tubes suggest substantial degradative activities resulting in the blurring of structural features, consistent with their being consequences of PCD processes (Gietmann, 1999). Induced pollen cell death in self-incompatible pollination systems ensures out-crossing as the predominant mechanism for breeding, maintaining hybrid vigor.

Towards molecular and biochemical characterization and understanding of PCD in reproduction

Cell and tissue degeneration and death in plant-reproductive tissues have not been examined with the full array of tools and information of PCD accumulated during the past decade. However, the physiology of cell death may apply to all systems in that cellular homeostasis must by perturbed through, for example, membrane degeneration and massive degradation of cytoplasmic contents. Some of the predictable effector molecules that carry out cell death-associated processes, including proteases, nucleases, and reactive oxygen species, may be involved and perform functions analogous to those in animal and plant systems that have been studied more extensively with a perspective specifically on PCD. Only a few studies in reproductive tissues have provided a glimpse of linkage between some molecular and biochemical events and PCD (e.g. Holden and Sze, 1987; Li *et al.*, 1995; Wang *et al.*, 1996; Orzaez and Granell, 1997; Cercos *et al.*, 1999). On the other hand, many studies on senescence physiology, which we have not covered here, have identified genes and proteins that potentially qualify for functional roles in a degenerating cellular environment (e.g. Grannell *et al.*, 1998). Closer inspection of the relationship between these genes and their products in reproductive development may reveal functional roles in reproductive cell death processes.

DNA and RNA degradation

DNA fragmentation, a phenomenon that accompanies PCD in a variety of systems (Wyllie, 1980), serves as a useful marker for PCD. *In situ* DNA end labeling revealed DNA fragmentation in poppy pollen tubes treated with incompatible S-proteins (V. Franklin-Tong and F.C.H. Franklin, personal communication). A DNA laddering pattern on gel electrophoretic analysis, a prominent feature in some animal and plant PCD processes that reflects DNA cleavage into internucleosomal-sized fragments, has so far not been reported in the reproductive tissues discussed, although it has been observed in senescing pea fruits (Orzaez and Granell, 1997).

The gynoecial tissues have an abundance of RNase activities (Green, 1994). Specific RNA metabolic activities appear to have defined roles in some reproductive cells that are undergoing cell death. In many gametophytic self-incompatible plants (see Matton *et al.*, 1994), the incompatibility response is dependent on a pistil-produced ribonuclease, the S-RNase, that enters the incompatible pollen tubes, degrades their RNA and arrest growth (McClure *et al.*, 1990).

In the tobacco stylar transmitting tissue, pollination stimulates RNA degradative activities that trim the poly(A) tails of mRNAs, subjecting them to further nucleolytic degradation. Wang *et al.* (1996) showed that poly(A)-tail trimming results in a decline in the level of several transmitting tissue-specific mRNAs. Among these is a specific class of mRNAs (TTS, for transmitting tissue-specific mRNAs) that encode a pollen tube growth-promoting glycoprotein (Wang *et al.*, 1993; Cheung *et al.*, 1995; Wu *et al.*, 1995). Contrary to the rapid decline in the levels of other poly(A)-tail trimmed mRNA, the level of 3′-end shortened TTS mRNAs is enhanced after pollination. The post-pollination pool of TTS mRNAs still has substantial poly(A) tails and is highly translatable as evidenced by the increased TTS protein levels after pollination (Wang *et al.*, 1993). It has been suggested that the degenerating transmitting tissue has an overall RNA degradative cellular environment. Unlike other 3′-end shortened transmitting tissue mRNAs which are subjected to further hydrolysis, special mechanisms must have evolved to maintain the stability of the poly(A)-tail-trimmed TTS mRNAs and ensure an adequate level of TTS proteins in the transmitting tissue for successful pollination (Wang *et al.*, 1996).

Like the pollination-induced cell death in the transmitting tissue described above, the mRNA 3′-end trimming activities are dependent on pollen tube penetration, since similar RNA degradation is not observed upon pollination by pollen grains from incompatible sources whose tubes do not enter the tobacco style. The dependence of mRNA poly(A) tail trimming on compatible pollination and the presence of mechanisms protecting specific classes of mRNAs in an overall degradative environment argue strongly that the cell deterioration and death process induced by pollination is a programmed process which includes considerations for maximum pollination success.

mRNA polyadenylation also appears to be an important aspect in a CMS system in sunflower (Gagliardi and Leaver, 1999). In the male-sterile lines, a novel gene (*orf522*) is fused to the 3′ end of the mitochondrial *atpA* gene and they are expressed as a dicistronic transcript, along with the monocistronic *atpA* transcripts. When restored to fertility by a nuclear gene, the dicistronic transcript becomes polyadenylated and is preferentially subjected to degradation. Apparently, the presence of the *atpA-orf522* mRNAs, with *orf522* encoding a 18 kDa polypeptide, interferes with mitochondria functions, resulting in male sterility. In plants restored to fertility, RNA polyadenylation and degradation of an apparently deleterious RNA is the counter-death measure adopted by sunflower anther tissue.

Proteolysis and other lytic processes

Caspases are cysteine proteases that play key roles in mediating PCD in animal systems (Allen *et al.*, 1998; Cryns and Yuan, 1998; Thornberry and Lazebnik, 1998), triggering other cellular degenerative events, including activation of other hydrolytic activities (e.g. Liu *et al.*, 1997) or regulatory proteins (e.g. Pai *et al.*, 1996). The collective name caspase denotes the cysteine protease mechanism of action and their activity of cleavage after an aspartic acid residue (Alnemri *et al.*, 1996). In plants, caspase activity has been detected in tobacco mosaic virus-infected tobacco plants when a short peptide substrate was used (del Pozo and Lam, 1998). In poppy, a caspase inhibitor appears to reduce the level of DNA fragmentation observed in pollen tubes elongating in the presence of incompatible S-proteins (V. Franklin-Tong and F.H.C. Franklin, personal communication).

Participation of other cysteine proteases in plant PCD and their inhibition by specific protease inhibitors (Minami and Fukuda, 1995; Solomon *et al.*, 1999; this issue) has been better documented. In reproductive organs, cysteine proteases have been associated with a number of senescing tissues. For instance, in pea, carpel development is associated with the induction of a thiol-protease (Cercos and Carbonell,

1993; Cercos *et al.*, 1999). In tobacco anthers, a thiol-endopeptidase mRNA accumulates in the connective tissue, circular cell cluster and stomium right before dehiscence, suggesting correlation with PCD in these tissues (Koltunow *et al.*, 1990). A mRNA for a flower-expressed cysteine protease is detected in anther and ovule cells of brinjal, suggesting a participation of these enzymes in cell degenerative processes in these organs (Xu and Chye, 1999).

Another class of proteases, the aspartic proteases, are found in the seeds of a variety of plants (Mutu and Gal, 1999), including the pistil of *Cynara cardunculus* (Vieira *et al.*, 1999; Faro *et al.*, 1999). The exact role of these pistil aspartic proteases in the cellular degeneration that goes on in reproductive tissues is not known. They may be involved in restructuring the extracellular matrix to provide a physically more permissive medium for pollen tubes or for resource mobilization to nurture the developing pollen grains, the elongating pollen tubes in the pistil, the developing embyro sac and the zygote in the seed. Alternatively, they may activate other proteins that have a more direct role in pistil cell death processes. Whatever the role individual proteases play in reproductive cells, they, as a whole, most likely contribute in a vital way to the completion of a developmental program that is critical for the perpetuation of a species.

Ubiquitination plays a central role in tagging proteins for proteolysis (Vierstra, 1996). The role of ubiquitination in PCD in plant reproductive tissue has not been examined in systematically. Promoters of *Arabidopsis* ubiquitin-extension proteins are highly active in transgenic pollen, ovules and cells lining the ovary locules (Callis *et al.*, 1990). Immunolabeling has indicated the presence of ubiquitinated structures in *N. alata* anthers and pistils (Li *et al.*, 1995). In maize, a ten-fold and fifty-fold decline, respectively, of the levels of free ubiquitin and ubiquitinated proteins has been observed from young microspores to mature pollen, a situation that is apparently distinct from other developmental accumulation patterns observed in other plant or animal systems (Callis and Bedinger, 1994). The 26S proteasomes are multicatalytic protease complexes, which degrade proteins that have been targeted for destruction by ubiquitination. A cDNA corresponding to a non-catalytic α-type subunit of 26S proteasomes has been isolated from the tobacco style (Bahrami and Gray, 1999); however, its putative nuclear location has led to suggestions that it has a role as a regulatory molecule rather than to degrade and recycle proteins during senescence. Given the central role of proteolysis in cellular degeneration, a more systematic analysis of protein unbiquitination and PCD in reproductive cells is warranted.

In addition to degenerative events in the protoplasts, plant cell death most likely involves lytic processes that degrade or restructure cell wall components. For example, expression of a pectate lyase has been shown to associate with the differentiation of tracheary elements in *Zinnia* (Domingo *et al.*, 1998). A battery of cell wall digestive and modifying enzymes are known to be produced by pollen tubes and the secretory cells along the pollen tube growth pathway in the pistil. An even greater number of cDNAs corresponding to pollen-produced cell wall hydrolytic proteins have been reported (Mascarenhas, 1990). Other than their hydrolytic activities, the functional role for these enzymes in reproductive development remains to be determined. One intriguing possibility could be in hydrolyzing cell wall polymers, releasing low-molecular-weight oligosaccharrides capable of signaling developmental events (e.g. Tran Thanh Van *et al.*, 1985; Eberband *et al.*, 1989), including PCD.

Signals and signal transduction in reproductive tissue PCD

The signaling of PCD in reproductive tissue is virtually unknown, so much of the discussions here will draw upon what is known about signal transduction in other systems that involve elements that are also prominent components of reproductive tissues.

Cell death in most reproductive cells is associated with the maturation and senescence phase of floral organs. It is not unlikely that signaling of cell death has some features common to signal transduction pathways in other senescing system (this issue). Ethylene and metabolites in the ethylene biosynthetic pathways, such as auxin (Zhang and O'Neill, 1993), are credible candidates for signal molecules. In cell death that occurs during pollen tube growth in the style and penetration into the ovary, cell death is often initiated before the arrival of pollen tubes (Jensen and Fisher, 1968; Fisher and Jensen, 1969; Wang *et al.*, 1996). This suggests a signaling mechanism involving diffusible molecules. Inhibitors of ethylene biosynthesis and reception inhibit pollination-induced transmitting tissue mRNA 3′ poly(A) tail trimming (Wang *et al.*, 1996). Application of ethylene, its precursor aminocyclopropane-1-carboxylic acid or the phosphatase inhibitor okadaic acid induces poly(A) tail trimming in unpollinating transmitting tissue. These observations are consistent with ethylene being involved in some way in the signaling of this RNA degradative process in the deteriorating transmitting tissue. However, the full spectrum of cell death responses revealed on the cytological level requires both ethylene and pollen tube penetration (Wang *et al.*, 1996).

In solid styles where pollen tubes elongate in a spatially restricted transmitting tissue extracellular matrix, mechanical stress in pollinated tissue is significant and may be conceived as signals for death by pistil cells. That this is plausible is suggested by the observation that one of the *Arabidopsis* touch-inducible genes, TCH4, encodes xyloglucan endotransglycosylase, an enzyme that acts on hemicellulose in plant cell walls (Xu *et al.*, 1995). TCH4 and a related gene from soybean (BRU1) are both induced by brassinosteroid (Zurek and Clouse, 1994; Xu *et al.*, 1995). Pollen is enriched in brassinosteroid and the penetration of pollen tubes in the style must involve some kind of cell wall restructuring. PCD in pistil tissues could be triggered by pollen-borne brassinosteroid and/or mechanical stress on the cell walls. Brassinosteroid has been shown to be involved in inducing the final stage of tracheary element differentiation in *Zinnia* (Yamamoto *et al.*, 1997).

Arabinogalactan proteins (AGPs) are a family of extracellular matrix hydroxyproline-rich glycoproteins ubiquitously present in plants (Nothnagel, 1997; Cheung and Wu, 1999). In carrot cell cultures, a specific AGP epitope is associated with the surface of cells that undergoes a functionally asymmetric division, giving rise to a cell that is competent to undergo embryogenesis, and another that is destined to die (McCabe *et al.*, 1997). In maize, AGPs are present on the secondary wall thickenings of cells destined to undergo PCD to become xylem (Schindler *et al.*, 1995). These observations suggest that some AGPs mark cells that are destined to die, although apparently some AGPs are critical for cell proliferation (Serpe and Nothnagel, 1994).

AGPs are abundantly found along the pollen tube growth pathway in pistil tissues and in the pollen tubes as well (see Cheung and Wu, 1999). Some of these appear to be at least transiently membrane-anchored (Youl *et al.*, 1998), others are secreted and associate with the extracellular matrix with different affinity (Wang *et al.*, 1993; Cheung *et al.*, 1999). Whether any of these AGPs are involved functionally in cell death processes is not known at this point. However, sugar molecules on AGPs in the transmitting tissue are known to be cleaved by pollen-produced enzymes (Wu *et al.*, 1995), providing a source of low-molecular-weight carbohydrate molecules for nutrients and/or signaling purposes.

A class of glucanases, chitinase, is abundant in pistil tissues, including the stylar transmitting tissue (Leung, 1992; Wemmer *et al.*, 1994; Harikrishna *et al.*, 1996). Plant cells are usually not known to produce chitins, β-(1,4)-linked polymers of N-acetylglucosamine, and chitinases are in general attributed roles in defense against fungal or insect pathogens. However, recent studies on the role of a chitinase important in embryo somatic embryogenesis led to the finding of embryo-produced AGPs as a substrate for this chitinase (de Jong *et al.*, 1992, 1993; van Hengel, 1998). Whether there are AGPs or other glycosylated molecules in pistil tissues or on the invading pollen tube surface that could serve as substrates for chitinases remains to be determined. If chitinase-mediated hydrolysis does exist in reproductive tissues, this may be another way a hydrolytic protein and glycoproteins may be involved in the signaling of PCD or in resource mobilization that is accomplished through cell death.

Other low-molecular-weight moieties that may serve as signal molecules for PCD in reproductive tissues may include phenolic compounds and Ca^{2+}. A mutation in an aromatic ring-hydroxylating enzyme, which degrades phenolic compounds, underlies the *lethal leaf spot1* mutation in maize which no longer shows a contained hypersensitive response (Gray *et al.*, 1997), suggesting that phenolic compounds may be the mediator of cell death. Style and ovary tissues have high levels of phenolic compounds as well as a variety of dioxygenases (Milligan and Gasser, 1995; Lantin *et al.*, 1999). Cytotoxic hydroperoxides and free radicals may be generated by high levels of dioxygenases in carpel tissues that are not destined for further development, such as in the senescing carpel of unpollinated pea (Rodriguez-Concepcion and Beltran, 1995).

Ca^{2+} has been shown to mediate apoptosis in the plant hypersensitive response (Levine *et al.*, 1996). In *Papaver rhoeus* (poppy), the S-protein from an incompatible pistil induces a rapid burst of Ca^{2+} influx in poppy pollen tubes and could well serve a signaling role in the pollen tube death that follows (Franklin-Tong, 1999). Rupture of the synergids as a consequence of cell death and release of Ca^{2+} to the extracellular matrix has been viewed as providing guidance cues for pollen tube approach (Chaubal and Reger, 1992a, b; Huang and Russell, 1992b). However, more subtle redistribution of internal Ca^{2+} concentration within a synergid cell prior to visible cell degeneration (Tian and Russell, 1997) could signal and initiate its own demise.

It has been reported that the unpollinated petunia transmitting tissue has high levels of unesterified pectins which may serve as a Ca^{2+} store (Lenartowska, personal communication). Pollination induces their progressive esterification, presumably releasing large amounts of Ca^{2+} to the transmitting tissue extracellular matrix. It was speculated the released Ca^{2+} may play a role in supporting pollen tube growth. It is equally possible that the Ca^{2+} released from pectin esterification could signal pistil tissue degeneration.

The brief list of studies discussed above underscore the paucity of our understanding of key biochemical processes that one expects to occur in reproductive tissues that undergo PCD. They however indicate that the basic elements for signaling PCD and for the degradation of cellular components are abundantly present. The challenge is to dissect the roles that these molecules play in each reproductive tissue and cell type as they undergo cell death at a specific developmental time. PCD in plant reproductive tissues occurs in limited cell numbers, in tissues closely associated with multiple non-degenerating tissues, and in locations that are not easily accessible. These together render the biochemical characterization of PCD in reproductive development difficult. Mutant analysis will likely be instrumental in providing insight into the initial signaling of PCD in reproductive cells. *In situ* studies will be the necessary first steps to obtain indications that particular molecules, identified by genetic or molecular means, are potential players in PCD events in these cells. Experiments can then be realistically designed to understand the biochemical and molecular mechanisms of how these molecules act to signal and mediate PCD in reproductive development.

Acknowledgements

We thank Drs Bernard Rubinstein and D. Scott Russell for their helpful comments on the manuscript. Research in our laboratory cited here was supported by grants from USDA, NIH and DOE. We also thank Drs V. Franklin-Tong, C. Franklin and M.G.R. Lenartowska for communicating unpublished results.

References

Aarts, M.G.M., Hodge, R., Kalantidis, K., Florack, D., Wilson, Z.A., Mulligan, B.J., Stiekema, W.J., Scott, R. and Pereira, A. 1997. The *Arabidopsis* MALE STERILITY 2 protein shares similarity with reductases in elongation/condensation complexes. Plant J. 12: 615–623.

Ainsworth, C., Crossley, S., Buchanan-Wollaston, V., Thangavalu, M. and Parker, J. 1995. Male and female flowers of the dioecious plant sorrel show different patterns of MADs box gene expression. Plant Cell 7: 1593–1598.

Alnemri, E.S., Livingston, D.J., Nicholson, D.W., Salvesen, G., Thornberry, N.A., Wong, W.W. and Yuan, J. 1996. Human ICE/CED-3 protease nomenclature. Cell 87: 171.

Ameisen, J.D. 1996. The origin of programmed cell death. Science 272: 1278–1279.

Allen, R.T., Cluck, M.W. and Agrawal, D.K. 1998. Mechanisms controlling cellular suicide: role of Bcl1 and caspases. Cell. Mol. Life Sci. 54: 427–445.

Bahrami, A.R. and Gray, J.E. 1999. Expression of a proteasome α-type subunit gene during tobacco development and senescence. Plant. Mol. Biol. 39: 325–333.

Barlow, P.W. 1982. Cell death: an integral part of plant development. In: M.B. Jackson, B. Grout and I.A. Mackenzie (Eds.), Growth Regulators in Plant Senescence, Wantage, Oxon British Plant Growth Regulator Group, pp. 27–45.

Beals, T.P. and Goldberg, R.B. 1997. A novel cell ablation strategy blocks tobacco anther dehiscence. Plant Cell 9: 1527–1545.

Bedinger, P. 1992. The remarkable biology of pollen. Plant Cell 4: 879–887.

Bell, P.R. 1996. Megaspore abortion: a consequence of selective apoptosis? Int. J. Plant Sci. 157: 1–7.

Bell, J. and Hicks, G. 1976. Transmitting tissue in the pistil of tobacco: light and electron microscopic observations. Planta 131: 187–200.

Bensen, R.J., Johal, G.S., Crane, V.D., Tossberg, J.T., Schnable, P.S., Meeley, R.B. and Briggs, S.P. 1995. Cloning and characterization of the maize An1 gene. Plant Cell 7: 75–84.

Bonner, L.J. and Dickinson, H.G. 1989. Anther dehiscence in *Lycopersicon esculentum*. I. Structural aspects. New Phytol. 113: 97–115.

Buckner, B., Janick-Buckner, D., Gray, J. and Johal, G.S. 1998. Cell-death mechanisms in maize. Trend Plant Sci. 3: 218–223.

Calderon-Urrea, A. and Dellaporta, S.L. 1999. Cell death and cell protection genes determine the fate of pistils in maize. Development 126: 435–441.

Callis, J. and Bedinger, P. 1994. Developmentally regulated loss of ubiquitin and ubiquitinated proteins during pollen maturation in maize. Proc. Natl. Acad. Sci. USA. 91: 6074–6077.

Callis, J., Raasch, J.A. and Vierstra, R.D. 1990. Ubiquitin extension proteins of *Arabidopsis thaliana*. J. Biol. Chem. 265: 12486–12493.

Cercos, M. and Carbonnell, J. 1993. Purification and characterization of a thiol-protease induced during the senescence of unpollinated ovaries of *Pisum sativum*. Physiol. Plant. 88: 267–274.

Cercos, M., Santamaria, S. and Carbonnell, J. 1999. Cloning and characterization of TPE4A, a thiol-protease gene induced during ovary senescence and seed germination in pea. Plant Physiol. 119: 1341–1348.

Chapman, G.P. 1987. The tapetum. Int. Rev. Cytol. 107: 111–125.

Chaubal, R. and Reger, B.J. 1992a. Calcium in the synergid cells and other regions of pearl millet ovaries. Sex. Plant Reprod. 5: 34–46.

Chaubal, R. and Reger, B.J. 1992b. The dynamics of calcium distribution in the synergid cells of wheat after pollination. Sex. Plant Reprod. 5: 206–213.

Cheng, P.C., Greyson, R.I. and Walden, D.B. 1983. Organ initiation and the development of unisexual flowers in the tassel and ear of *Zea mays*. Am. J. Bot. 70: 450–462.

Cheung, A.Y. 1996a. Pollen-pistil interactions during pollen tube growth. Trends Plant Sci. 1: 45–51.

Cheung, A.Y. 1996b. The pollen tube growth pathway: its molecular and biochemical contributions and responses to pollination. Sex. Plant Reprod. 9: 330–336.

Cheung, A.Y. and Wu, H-M. 1998. Arabinogalactan proteins in plant sexual reproduction. Protoplasma 208: 87–98.

Cheung, A.Y., Wang, H. and Wu, H-M. 1995. A floral transmitting tissue-specific glycoprotein attracts pollen tubes and stimulates their growth. Cell 82: 383–393.

Cheung, A.Y., Zhan, X.-y., Wong, E., Wang, H. and Wu, H.M. 2000. Transcriptional, post-transcriptional and posttranslational regulation of a transmitting tissue-specific pollen tube growth-promoting arabinogalactan protein. In: E.A. Nothnagel, A. Bacic and A.E. Clarke (Eds.), Cell and Developmental Biology of Arabinogalactan-proteins, Kluwer Academic Publishers, Boston, pp. 133–148.

Cheung, A.Y, Wu, H.-M., Di Stilio, V., Glaven, R., Chen, C., Wong, E., Ogdahl, J. and Estavillo, A. 2000. Pollen-pistil interactions in *Nicotiana tabacum*. Ann. Bot. 85, 29–37.

Christensen, C.A., Subramanian, S. and Drews, G.N. 1998. Identification of gametophytic mutations affecting female gametophyte development in *Arabidopsis*. Dev. Biol. 202: 136–151.

Clouse, S.D. 1996. Molecular genetic studies confirm the role of brassinosterioids in plant growth and development. Plant J. 10: 1–8.

Cresti, M., Keijzer, C.J., Tiezzi, A., Ciampolini, F. and Focardi, S. 1986. Stigma of *Nicotiana*: ultrastructural and biochemical studies. Am. J. Bot. 73: 1713–1722.

Cryns, V. and Yuan, J. 1998. Proteases to die for. Genes Dev. 12: 1551–1570.

Domingo, D., Roberts, K., Stacey, N.J., Connerton, I., Ruiz-Teran, F. and McCann, M.C. 1998. A pectate lyase from *Zinnia elegans* is auxin inducible. Plant J. 13: 17–28.

Dawson, J., Wilson, Z.A., Aarts, M.G.M., Braithwaite, A.F., Briarty, L.G. and Milligan, B.J. 1993. Microspore and pollen development in six male-sterile mutants of *Arabidopsis thaliana*. Can. J. Bot. 71: 629–638.

de Graaf, B.H.J., Knuiman, B. and Mariani, C. 1998. The PELPs in the transmitting tissue of *Nicotiana tabacum* are translocated through the pollen walls in vivo. XVth International Congress on Sexual Plant Reproduction, p. 29.

de Jong, A.J., Cordewener, J., Lo Schiavo, F., Terzi, M., Vandekerckhove, J., van Kammen, A. and de Vries, S.C. 1992. A carrot somatic embryo mutation is rescued by chitinase. Plant Cell 4: 425–433.

de Jong, A.J., Heidstra, R., Spaink, H.P., Hartog, V., Meijer, E.A., Hendriks, T., Lo Schiavo, F., Terzi, M., Bisseling, T., van Kammen, A. and de Vries, S.C. 1993. *Rhizobium* liposaccharides rescue a carrot somatic embryo mutant. Plant Cell 5: 615–620.

Del Pozo, O. and Lam, E. 1998. Caspases and programmed cell death in the hypersensitive response of plants to pathogens. Curr. Biol. 8: 1129–1132.

DeLong, A., Calderon-Urrea, A. and Dellaporta, S.L. 1993. Sex determination gene TASSELSEED2 of maize encodes a short-chain alcohol dehydrogenase required for stage-specific floral organ abortion. Cell 74: 757–768.

Dewey, R.E., Levings, C.S. III and Timothy, D.H. 1986. Novel recombinations in the maize mitochondrial genome produce a unique transcriptional unit in the Texas male-sterile cytoplasm. Cell 44: 439–449.

Dewey, R.E., Timothy, D.H. and Levings, C.S. III. 1987. A mitochondrial protein associated with cytoplasmic male sterility in the T-cytoplasm of maize. Proc. Natl. Acad. Sci. USA 84: 5374-5378.

Dewey, R.E., Siedow, J.N., Timothy, D.H. and Levings, C.S. III. 1988. A 13 kilodalton maize mitochondrial protein in *E. coli* confers sensitivity to *Bipolaris maydis* toxin. Science 239: 293–295.

Domingo, C., Roberts, K., Stacey, N., Connerton, I., Tuiz-Teran, F. and McCann, M.C. 1998. A pectate lyase from *Zinnia elegans* is auxin inducible. Plant J. 13: 17–28.

Earnshaw, W.C. 1999. A cellular poison cupboard. Nature 397: 387–389.

Eberband, S., Doubrava, N., Marfa, V., Mohnen, D., Southwick, A., Darvill, A. and Albersheim, P. 1989. Pectic cell wall fragments regulate tobacco thin-cell-layer explant morphogenesis. Plant Cell 1: 747–755.

Faro, C., Ramalho-Santos, M., Viera, M., Mendes, A., Simoes, I., Andrade, R., Verissimo, P., Lin, X., Tang, J. and Pires, E. 1999. Cloning and characterization of a cDNA encoding cardosin A, an RGD-containing plant aspartic proteinase. J. Biol. Chem. 274: 28724–28729.

Fisher, D.B. and Jensen, W.A. 1969. Cotton embryogenesis: the identification of the X-bodies in the degenerated synergid. Planta 84: 122–133.

Franklin-Tong, V. 1999. Signaling and the modulation of pollen tube growth. Plant Cell 11: 727–738.

Gagliardi, D. and Leaver, C.J. 1999. Polyadenylation accelerates the degradation of the mitochondrial mRNA associated with cytoplasmic male sterility in sunflower. EMBO J. 18: 3757–3766.

Geitmann, A. 1999. Cell death of self-incompatible pollen tubes: necrosis or apoptosis? In: M. Cresti, G. Cai and A. Moscatelli (Eds.), Fertilization in Higher Plants: Molecular and Cytological Aspects, Springer-Verlag, Berlin/Heidelberg, pp. 113–137.

Goldberg, R.B., Beals, T.P. and Sanders, P.M. 1993. Anther development: basic principles and practical applications. Plant Cell 5: 1217–1229.

Goldman, M.H.S., Goldberg, R.B. and Mariani, C. 1994. Female sterile tobacco plants are produced by stigma-specific cell ablation. EMBO J. 13: 2976–2984.

Granell, A., Cercos, M. and Carbonell, J. 1998. Plant cysteine proteases in germination and senescence. In: A.J. Barret, N.D. Rawlings and J.F. Woessner (Eds.), Handbook of Proteolytic Enzymes, Academic Press, London, pp. 578–583.

Gray, J., Close, P.S., Briggs, S.P. and Johal, G.S. 1997. A novel suppressor of cell death in plants encoded by the Lls1 gene of maize. Cell 89: 25–31.

Green, P.J. 1994. The ribonuclease of higher plants. Annu. Rev. Plant Physiol. Plant Mol. Biol. 45: 421–445.

Green, D.R. and Reed, J.C. 1998. Mitochondria and apoptosis. Science 281: 1309–1312.

Greenberg, J.T. 1996. Programmed cell death: a way of life for plants. Proc. Natl. Acad. Sci. USA. 93: 12094–12097.

Hagashiyama, T., Kuroiwa, H., Kawano, S. and Kuroiwa, R. 1998. Guidance in vitro of the pollen tube to the naked embryo sac of *Torenia fournieri*. Plant Cell 10: 2019–2031.

Hardenack, S., Ye, D., Saedler, H. and Grant, S. 1994. Comparison of MADS box gene expression in developing male and female flowers of the dioecious plant White campion. Plant Cell 6: 1775–1787.

Harikrishna, K., Jampates-Beale, R., Milligan, S.B. and Gasser, C.S. 1996. An endochitinase gene expressed at high levels in the stylar transmitting tissue of tomatoes. Plant Mol. Biol. 30: 899–911.

Herrero, M. 1992. From pollination to fertilization in fruit trees. Plant Growth Regul. 11: 27–32.

Herrero, M. and Dickinson, H.G. 1979. Pollen-pistil incompatibility in *Petunia hybrida*: changes in the pistil following compatible and incompatible intraspecific crosses. J. Cell Sci. 36: 1–18.

Herrero, M. and Dickinson, H.G. 1981. Pollen tube development in *Petunia hybrida* following compatible and incompatible intraspecific matings. J. Cell Sci. 47: 365–383.

Holden, M.J. and Sze, H. 1987. Dissipation of the membrane potential in susceptible corn mitochondria by the toxin of *Helminthosporium maydis*, race T, and toxin analogs. Plant Physiol. 84: 670–676.

Hölskamp, M., Kopczak, S.D. Horejsi, T.F., Kihl, B.K. and Pruitt, R.E. 1995a. Identification of genes required for pollen-stigma recognition in *Arabidopsis thaliana*. Plant J. 8: 703–714.

Hölskamp, M, Schneitz, K. and Pruitt, R.E. 1995b. Genetic evidence for a long-range activity that directs pollen tube guidance in *Arabidopsis*. Plant Cell 7: 57–64.

Horner, H.T. and Wagner, B.L. 1992. Association of four different calcium crystals in the anther connective tissue and hypodermal stomium of *Capsicum annuum* (Solanaceae) during microsporogenesis. Am. J. Bot. 79: 531–541.

Huang, B.-Q. and Russell, S.D. 1992a. Female germ unit: organization, isolation and function. Int. Rev. Cytol. 140: 233–293.

Huang, B.-Q. and Russell, S.D. 1992b. Synergid degeneration in *Nicotiana*: a quantitiative, fluorochromatic and chlorotetracycline study. Sex. Plant Reprod. 5: 151–155.

Huang, B.-Q. and Russell, S.D. 1994. Fertilization in *Nicotiana tabacum*: cytoskeletal modifications in the embryo sac during synergid degeneration. Planta 194: 200–214.

Huang, B.-Q., Strout, G.W. and Russell, S.D. 1993. Fertilization in *Nicotiana tabacum*: ultrastructural organization of propane-jet-frozen embryo sacs in vivo. Planta 191: 256–264.

Irish, E. and Nelson, T. 1989. Sex determination in monoecious and dioecious plants. Plant Cell 1: 737–744.

Izhar, S. and Frankel, R. 1971. Mechanism of male sterility in *Petunia*: the relationship between pH, callase activity in the anthers, and the breakdown of the microsporogenesis. Theor. Appl. Genet. 41: 104–108.

Jensen, W.A. and Fisher, D.B. 1968. Cotton embryogenesis: the entrance and discharge of the pollen tube in the embryo sac. Planta 78: 158–183.

Juarez, C. and Banks, J.A. 1998. Sex determination in plants. Curr. Opin. Plant Biol. 1: 68–72.

Kandasamy, M.M. and Kristen, U. 1987. Developmental aspects of ultrastructure, histochemistry and receptivity of the stigma of *Nicotiana sylvestris*. Am. J. Bot. 60: 427–437.

Koltunow, A.M., Truettner, J., Cox, K.H., Wallroth, M. and Goldberg, R.B. 1990. Different temporal and spatial gene expression patterns occur during anther development. Plant Cell 2: 1201–1224.

Kroh, M., Miki-Hirosige, H., Rosen, W. and Loewus, F. 1970. Incorporation of label into pollen tube walls from myo-inositol labeled *Lilium longiflorum* pistils. Plant Physiol. Lancaster 45: 92–94.

Labraca, C. and Loewus, F. 1973. The nutritional role of pistil exudate in pollen tube wall formation in *Lilium longiflorum*. II. Production and utilization of exudate from the stigma and stylar canal. Plant Physiol. 52: 87–92.

Lantin, S., O'Brien, M. and Matton, D.P. 1999. Pollination and wounding of the style induce the expression of a developmentally regulated pistil dioxygenase at a distance in the ovary. Plant Mol. Biol. 41: 371–386.

Lebel-Hardenack, S., and Grant, S.R. 1997. Genetics of sex determination in flowering plants. Trends Plant Sci. 2: 130–136.

Lebel-Hardenack, S., Ye, D., Koutnikova, H., Saedler, H. and Grant, S.R. 1997. Conserved expression of a TASSELSEED2 homolog in the tapetum of the dioecious *Silene latifolia* and *Arabidopsis thaliana*. Plant J. 12: 515–526.

Leung, D.W.M. 1992. Involvement of plant chitinase in sexual reproduction of higher plants. Phytochemistry 31: 1899–1900.

Levine, A., Pennell, R.I., Alvarez, M.E., Palmer, R. and Lamb, C. 1996. Calcium-mediated apoptosis in a plant hypersensitive disease resistance response. Curr. Biol. 6: 427–437.

Levings, C.S. III. 1993. Thoughts on cytoplasmic male sterility in *cms-T* maize. Plant Cell 5: 1285–1290.

Li, D., Blakey, C.A., Dewald, C. and Dellaporta, S.L. 1997. Evidence for a common sex determination mechanism for pistil abortion in maize and in its wild relative *Tripsacum*. Proc. Natl. Acad. Sci. USA 94: 4217–4222.

Li, J., Nagpal, P., Cook, R.K., Elich, T., Lopez, E., Pepper, A., Poole, D. and Chory, J. 1996. Molecular characterization of DET2 mutant in *Arabidopsis*. Science 272: 398–401.

Li, Y-Q., Southworth, D., Linskens, H.F., Mulcahy, D.L. and Cresti, M. 1995. Localization of ubiquitin in anthers and pistils of *Nicotiana*. Sex. Plant Reprod. 8: 123–128.

Lind, J.L., Bonig, I., Clarke, A.E. and Anderson, M.A. 1996. A style-specific 120 kD glycoprotein enters pollen tubes of *Nicotiana alata* in vivo. Sex. Plant Reprod. 9: 75–86.

Liu, X., Zou, H., Slaughter, C. and Wang, X. 1997. DFF, a heterodimeric protein that functions downstream of caspase-3 to trigger DNA fragmentation during apoptosis. Cell 89: 175–184.

Loukides, C.A., Broadwater, A.H. and Bedinger, P.A. 1995. Two new male-sterile mutants of *Zea mays* (Poaceae) with abnormal tapetal cell morphology. Am. J. Bot. 82: 1017–1023.

Ma, H. 1994. The unfolding drama of flower development: recent results from genetic and molecular analyses. Genes Dev. 8: 745–756.

Mandava, N.B. 1988. Plant growth promoting brassinosteroids. Annu. Rev. Plant Physiol. Plant Mol. Biol. 39: 23–52.

Mariani, C., de Beuckeleer, M., Truettner, J., Leemans, J. and Goldberg, R.B. 1990. Induction of male sterility in plants by a chimaeric ribonuclease gene. Nature 347: 737–741.

Mascarenhas, J.P. 1990. Gene activity during pollen development. Annu. Rev. Plant Physiol. Plant Mol. Biol. 41: 317–338.

Matton, D.P., Nass, N., Clarke, A.E. and Newbingen, E. 1994. Self-incompatibility: how plants avoid illegitimate offspring. Proc. Natl. Acad. Sci. USA 91: 1992–1997.

McCabe, P.F., Valentine, T.A., Forsberg, L.S. and Pennell, R.I. 1997. Soluble signals from cells identified at the cell wall establish a developmental pathway in carrot. Plant Cell 9: 2225–2241.

McClure, B.A., Gray, J.E., Anderson, M.A. and Clarke, A.E. 1990. Self-incompatibility in *Nicotiana alata* involves degradation of pollen rRNA. Nature 347: 757–760.

McCormick, S. 1993. Male gametophyte development. Plant Cell 5: 1265–1275.

Milligan, S.R. and Gasser, C.S. 1995. Nature and regulation of pistil-expressed genes in tomato. Plant Mol. Biol. 28: 691–711.

Minami, A. and Fukuda, H. 1995. Transient and specific expression of a cysteine endopeptidase associated with autolysis during differentiation of *Zinnia* mesophyll cells into tracheary elements. Plant Cell Physiol. 36: 1599–1606.

Mittler, R. and Lam, E. 1996. Sacrifice in the face of foes: pathogen-induced programmed cell death in plants. Trends Microbiol. 4: 10–15.

Mogensen, H.L. 1984. Quantitative observations on the pattern of synergid degeneration in barley. Am. J. Bot. 71: 1448–1451.

Murgia, M., Huang, B-Q., Tucker, S.C. and Musgrave, M.E. 1993. Embryo sac lacking antipodal cells in *Arabidopsis thaliana* (Brassicaceae). Am. J. Bot. 80: 824–838.

Murphy, D.J. and Ross, J.H.E. 1998. Biosynthesis, targeting and processing of oleosin-like proteins, which are major pollen coat components in *Brassica napus*. Plant J. 13: 1–16.

Mutu, A. and Gal, S. 1999. Plant aspartic proteinases: enzymes on the way to a function. Physiol. Plant. 105: 569–576.

Nasrallah, J.B. 1997. Signal perception and response in the interactions of self-incompatiblity in *Brassica*. Essays Biochem. 32: 143–160.

Nettancourt, D., Devreux, M., Bozzini, A., Cresti, M. and Sarfatti, G. 1973. Ultrastructural aspects of the self-incompatibility mechanism in *Lycopersicum peruvianum* Mill. J. Cell Sci. 12: 403–419.

Neuffer, M.G., Coe, E.H. and Wessler, S.R. 1997. The Mutants of Maize, Cold Spring Harbor Laboratory Press, Plainview, NY.

Nothnagel, E.A. 1997. Proteoglycan and related components in plant cells. Int. Rev. Cytol. 174: 195–291.

Orzaez, D. and Granell, A. 1997. DNA fragmentation is regulated by ethylene during carpel senescence in *Pisum sativum*. Plant J. 11: 137–144.

Pai, J.-T., Brown, M.S. and Goldstein, J.L. 1996. Purification and cDNA cloning of a second apoptosis-related cysteine protease that cleaves and activates sterol regulatory element binding proteins. Proc. Natl. Acad. Sci. USA 93: 5437–5442.

Pennell, R.I. and Lamb, C. 1997. Programmed cell death in plants. Plant Cell 9: 1157–1168.

Piffanelli, P. and Murphy, D.J. 1998. Novel organelles and targeting mechanisms in the anther tapetum. Trends Plant Sci. 3: 250–253.

Preuss, D., Lemieux, B., Yen, G. and Davis, R.W. 1993. A conditional sterile mutation eliminates surface components from *Arabidopsis* pollen and disrupt cell signaling during fertilization. Genes Dev. 7: 974–985.

Ray, S., Park, S.-S. and Ray, A. 1997. Pollen tube guidance by female gametophyte. Development 124: 2489–2494.

Reiser, L. and Fischer, R.L. 1993. The ovule and the embryo sac. Plant Cell 5: 1291–1301.

Rhee, S.Y. and Somerville, C.R. 1998. Tetrad pollen formation in quartet mutants of *Arabidopsis thaliana* is associated with persistence of pectic polysaccharides of the pollen mother cell wall. Plant J. 15: 79–88.

Rodriguez-Concepcion, M. and Beltrain, J.P. 1995. Repression of the pea lipoxygenase gene *loxg* is associated with carpel development. Plant Mol. Biol. 27: 887–899.

Russell, S.D. 1979. Fine structure of megagametophyte development in *Zea mays*. Can. J. Bot. 57: 1093–1110.

Russell, S.D. 1993. The egg cell: developmental role in fertilization and early embryogenesis. Plant Cell 5: 1349–1359.

Russell S.D. 1996. Attraction and transpot of male gametes for fertilization. Sex. Plant Reprod. 9: 337–342.

Sassen, M.M.A. 1974. The stylar transmitting tissue. Acta Bot. Neerl. 23: 99–108.

Schnable, P.S. and Wise, R.P. 1998. The molecular basis of cytoplasmic male sterility and fertility restoration. Trends Plant Sci. 3: 175–180.

Schindler, T., Bergfeld, R. and Schopfer, P. 1995. Arabinogalactan proteins in maize coleoptiles: developmental relationship to cell death during xylem differentiation but not to extension growth. Plant J. 7: 25–36.

Seigel, B.A. and Verbeke, J.A. 1989. Diffusible factors essential for epidermal cell redifferentiation in *Catharanthus roseus*. Science 244: 580–582.

Serpe, M.D. and Nothnagel, E.A. 1994. Effects of Yariv phenylglycosides on *Rosa* cell suspensions: evidence for the involvement of arabinogalactan proteins in cell proliferation. Planta 193: 542–550.

Solomon, M., Belenghi, B., Delledonne, M., Menachem, E. and Levine, A. 1999. The involvement of cysteine proteases and protease inhibitor genes in the regulation of programmed cell death in plants. Plant Cell 11: 431–443.

Sumner, M.J. 1992. Embryology of *Brassica campestris*: the entrance and discharge of the pollen tube in the synergid and the formation of the zygote. Can. J. Bot. 70: 1577–1990.

Susin, S.A., Lorenzo, H.K., Zamzami, N., Marzo, I., Snow, B.E., Brothers, G.M., Mangion, J., Jacotot, E., Costantini, P., Loeffler, M., Larochette, N., Goodlett, D.R., Aebersold, R., Siderovski, D.P., Penninger, J.M. and Kroemer, G. 1999. Molecular characterization of mitochondrial apoptosis-inducing factor. Nature 397: 441–446.

Susin, S.A., Zamazmi, H.K. and Kroemer, G. 1998. Mitochondria as regulator of apoptosis: doubt no more. Biochim. Biophys. Acta. 1366: 151–161.

Thornberry, N.A. and Lazebnik, Y. 1998. Caspases: enemies within. Science 281: 1312–1316.

Tian, H.Q. and Russell, S.D. 1997. Developmental changes in calcium distribution and accumulation in fertilized and unfertilized ovules and embryo sacs of *Nicotiana tabacum*. Planta 202: 93–105.

Tran Thanh Van, K., Toubart, P., Cousson, A., Darvill, A.G., Gollin, D.J., Chelf, P. and Albersheim, P. 1985. Manipulation of the morphogenetic pathway of tobacco explants by oligosaccharins. Nature 314: 615–617.

Vaux, D.L. and Korsmeyer, S.J. 1999. Cell death in development. Cell 96: 245–254.

van Hengel, A.J. 1998. Ph.D. thesis, Department of Molecular Biology, Wageningen Agricultural University.

Vieira, M., Pissarra, J., Verissimo, P., Castanheira, P, Costa, Y., Pereira, S., Pires, E. and Faro, C. 1999. Cardosin B is an aspartic proteinase expressed at the extracellular matrix of the transmitting tissue of *Cynara cardunculus* L. Submitted.

Vierstra, R.D. 1996. Proteolysis in plants: mechanisms and functions. Plant Mol. Biol. 32: 275–302.

Walker, D.B. 1975a. Postgenital carpel fusion in *Catharanthus roseus* (Apocynaceae). I. Light and scanning electron microscopic study of gynoecial onogeny. Am. J. Bot. 62: 457–467.

Walker, D.B. 1975b. Postgenital carpel fusion in *Catharanthus roseus*. II. Fine structure of the epidermis before fusion. Protoplasma 86: 29–41.

Walker, D.B. 1975c. Postgenital carpel fusioin in *Catharanthus roseus*. III. Fine structure of the epidermis during and after fusion. Protoplasma 86: 43–63.

Wang, H., Wu, H.-M. and Cheung, A.Y. 1996. Pollination induces mRNA poly(A) tail-shortening and cell deterioration in flower transmitting tissue. Plant J. 9: 715–727.

Webb, M.C. and Gunning, B.E.S. 1990. Embryo sac development in *Arabidopsis thaliana* I. Megasporogenesis, including the microtubular cytoskeleton. Sex. Plant Reprod. 3: 244–256.

Wemmer, T., Kaufmann, H., Kirch, H-H., Schneider, K., Lottspeich, F. and Thompson, R.D. 1994. The most abundant soluble basic protein of the stylar transmitting tissue in potato (*Solanum tuberosum* L.) is an endochitinase. Planta 194: 264–273.

Winkler, R.G. and Helentjaris, T. 1995. The maize dwarf3 gene encodes a cytochrome p450-mediated early step in gibberellin biosynthesis. Plant Cell 7: 1307–1317.

Wolter-Arts, M., Derksen, J., Kooijman, J.W. and Mariani, C. 1996. Stigma development in *Nicotiana tabacum*. Cell death in transgenic plants as a marker to follow cell fate at high resolution. Sex. Plant Reprod. 9: 243–254.

Wolter-Arts, M., Lush, W.M. and Mariani, C. 1998. Lipids are required for directional pollen tube growth. Nature 392: 818–820.

Worrall, D., Hird, D.L., Hodge, R., Paul, W., Draper, J. and Scott, R. 1992. Premature dissolution of the microsporocyte callose wall causes male sterility in transgenic tobacco. Plant Cell 4: 759–771.

Wu, H.-M., and Cheung, A.Y. 1998. Sexual reproduction: from sexual differentiation to fertilization. Annu. Plant Rev. 1: 181–222.

Wu, H.-M., Wang, H. and Cheung, A.Y. 1995. A pollentube growth stimulatory glycoprotein is deglycosylated by pollen tubes and displays a glycosylation gradient in the flower. Cell 82: 393–403.

Wu, S.S.H., Platt, K.A., Ratnayake, C., Wang, T-W., Ting, J.T.L. and Huang, A.H.C. 1997. Isolation and characterization of neutral-lipid-containing organelles and globuli-filled plastids from *Brassica napus* tapetum. Proc. Natl. Acad. Sci. USA. 94: 12711–12716.

Wu, H.-M., Wong, E., Ogdahl, J. and Cheung, A.Y. 2000. A pollen tube growth-promoting arabinogalactan protein from *Nicotiana alata* is similar to the tobacco TTS protein. Plant J. 22: 165–176.

Wyllie, A.H. 1980. Glucocorticoid-induced thymocyte apoptosis is associated with endogenous endonuclease activity. Nature 284: 555–556.

Xu, F.-X. and Chye, M.-L. 1999. Expression of cysteine proteinase during developmental events associated with programmed cell death in brinjal. Plant J. 17: 321–327.

Xu, W., Purugganan, M.M., Polisensky, D.H., Antosiewicz, D.M., Fry, S.C. and Braam, J. 1995. *Arabidopsis* TCH4, regulated by hormones and the environment, encodes a xyloglucan endotransglycosylase. Plant Cell 7: 1555–1567.

Yamamoto, R., Demura, T. and Fukuda, H. 1997. Brassinosteroids induce entry into the final stage of tracheary element differentiation in cultured *Zinnia* cells. Plant Cell Physiol. 38: 980–983.

Yan, H., Yang, H.Y. and Jensen, W.A. 1991. Ultrastructure of the developing embryo sac of sunflower (*Helianthus annuus*) before and after fertilization. Can. J. Bot. 69: 191–202.

Yao, M.-C. 1996. Programmed DNA deletions in *Tetrahymena*: mechanisms and implications. Trends Genet. 12: 26–30.

Youl, J.J., Bacic, A. and Oxley, D. 1998. Arabinogalactan-proteins from *Nicotiana alata* and *Pyrus communis* contain glycosylphosphatidylinositol membrane anchors. Proc. Natl. Acad. Sci. USA 95: 7921–7926.

Zhang, X.S. and O'Neill, S.D. 1993. Ovary and gametophyte development are coordinately regulated by auxin and ethylene following pollination. Plant Cell 5: 403–418.

Zurek, D.M. and Clouse, S.D. 1994. Molecular cloning and characterization of a brassinosteroid-regulated gene from elongating soybean (*Glycine max* L.) epicotyls. Plant Physiol. 104: 161–170.

Plant Molecular Biology **44:** 283–301, 2000.
E. Lam, H. Fukuda and J. Greenberg (Eds.), Programmed Cell Death in Higher Plants.

Programmed cell death during endosperm development

Todd E. Young* and Daniel R. Gallie
*Department of Biochemistry, University of California, Riverside, CA 92521-0129, USA (*author for correspondence; e-mail: teyoung@citrus.ucr.edu)*

Key words: abscisic acid, endosperm, ethylene, internucleosomal DNA fragmentation, nuclease, programmed cell death

Abstract

The endosperm of cereals functions as a storage tissue in which the majority of starch and seed storage proteins are synthesized. During its development, cereal endosperm initiates a cell death program that eventually affects the entire tissue with the exception of the outermost cells, which differentiate into the aleurone layer and remain living in the mature seed. To date, the cell death program has been described for maize and wheat endosperm, which exhibits common and unique elements for each species. The progression of endosperm programmed cell death (PCD) in both species is accompanied by an increase in nuclease activity and the internucleosomal degradation of nuclear DNA, hallmarks of apoptosis in animals. Moreover, ethylene and abscisic acid are key to mediating PCD in cereal endosperm. The progression of the cell death program in developing maize endosperm follows a highly organized pattern whereas in wheat endosperm, PCD initiates stochastically. Although the essential characteristics of cereal endosperm PCD are now known, the molecular mechanisms responsible for its execution remain to be identified.

Introduction

In cereal seeds, the endosperm is a prominent tissue that persists in the mature seed. The starchy endosperm functions as a storage tissue that exhibits a high level of metabolic activity during development but is fated to die prior to maturation of the seed. The initiation and spread of maize endosperm cell death is highly organized and follows the development of the endosperm (Young *et al.*, 1997). In contrast to maize, the pattern of programmed cell death (PCD) in developing wheat endosperm is not elaborated in an orchestrated manner but rather occurs stochastically (Young and Gallie, 1999). Nevertheless, cell death in the endosperm of both cereals is accompanied by an increase in nuclease activity and the internucleosomal fragmentation of nuclear DNA, both of which appear to be controlled through the action of ethylene. The characteristics of endosperm PCD suggest that the execution of the program may be similar among cereal species, whereas the differences in the onset of the program between maize and wheat suggest differences in aspects of the signaling required for initiating the program.

Overview of seed structure and development

The angiosperm seed consists of four components: (1) the testa (derived from one or both of the integuments of the ovule), (2) the perisperm (derived from the nucellus), (3) the 2N embryo, which develops from the zygote (the product of the fusion of the egg cell with a sperm nucleus), and (4) the 3N endosperm (formed by the fusion of the two polar nuclei with the second sperm nucleus). The extent to which these tissues develop determines to a large extent the structural differences observed among various types of seeds (Bewley and Black, 1978; Boesewinkel and Bouman, 1995). This review will focus primarily on those events associated with cereal endosperm development, with particular attention given to maize and wheat.

The relative contribution that the endosperm makes to the mass of mature seed varies greatly depending upon the species and can range from nonexistent (e.g., many legumes) to consisting of 1 or 2 peripheral layers (e.g., cucurbits and lettuce) to constituting the majority of the seed volume (e.g., cereals, lilies, palms, and most primitive dicots) (Martin, 1946). Moreover, seeds can be grouped as endospermic or non-endospermic depending upon the extent to which the endosperm is retained at maturity although no clear distinction between these types can be made solely on structural considerations (Boesewinkel and Bouman, 1995). In non-endospermic seeds (e.g., beans and peas), the endosperm either terminates early in development or degenerates at a later stage and is reabsorbed during subsequent seed development, either of which results in a mature seed containing little, if any, endosperm. In this case, stored reserves are deposited in the embryo, usually within the cotyledons. In endospermic seeds (e.g., maize, wheat, and rice), the endosperm is retained as a storage tissue in the mature seed which is subsequently subjected to mobilization during germination. In this sense, the reabsorption of the endosperm that occurs during the development of non-endospermic seeds can be, in some endospermic seeds, thought of as delayed until germination. In one type of endospermic seed (e.g., cereals and some endospermic legumes), the endosperm consists of two distinct cell types, non-living reserve cells and living aleurone cells (Bhatnagar and Sawhney, 1981; Bewley and Black, 1994; DeMason, 1994) whereas in others (e.g., lilies, onion, most palms, and castor bean), the endosperm is composed solely of uniform, living reserve cells.

Phases of cereal endosperm development

Three general patterns of endosperm development have been described: nuclear, cellular, and helobial (reviewed in Brink and Cooper, 1947; Vijayaraghavan and Prabhakar, 1984). In the nuclear type, mitosis of the primary endosperm nucleus occurs in the absence of cytokinesis, resulting in a multinucleate cell. Cellularization then follows until every cell in the endosperm becomes uninucleate. With the helobial type, mitosis of the primary endosperm nucleus generates two cells of unequal size. While the larger cell undergoes normal division, division of the smaller cell ceases and the cell either remains uninucleate or undergoes mitosis to generate a multinucleate cell. In the cellular type, the primary endosperm nucleus undergoes normal cell division and remains uninucleate and cellular. Of the three types, nuclear development is the most common and occurs in many dicotyledonous and monocotyledonous species, including cereals (Brink and Cooper, 1947; Vijayaraghavan and Prabhakar, 1984). The development of cereal endosperm passes through four phases: syncytial (i.e., mitosis without cytokinesis), cellularization, differentiation and maturation (Bosnes *et al.*, 1992; Olsen *et al.*, 1992). In maize, the triploid endosperm nucleus undergoes rapid synchronous divisions for several days after fertilization, resulting in a syncytial (i.e., multinucleate) primary endosperm cell (Kowles and Phillips, 1988). Cellularization begins 3–5 days after pollination (DAP; which occurs the same day as fertilization) and progresses in a centripetal pattern until the entire endosperm is cellular and each cell is uninucleate (Kowles and Phillips, 1988). From 5 to 12 DAP, the endosperm grows very rapidly involving both cell division and cell expansion. During this time, the maternally derived nucellar tissue surrounding the embryo sac degenerates and is compressed against the outer edge of the kernel cavity by the expanding endosperm. Cell division ceases within the central endosperm by about 12 DAP but continues in the peripheral regions. Subsequent to this, cells in the interior endosperm increase in size while the outermost cells of the endosperm differentiate into the aleurone layer (Kyle and Styles, 1977). After the cessation of cell division in the central endosperm, nuclear DNA replication continues in a process referred to as endoreduplication resulting in a dramatic increase in nuclear volume and DNA content that peaks at 16–18 DAP (Phillips *et al.*, 1983, 1985; Kowles and Phillips, 1985). Cell division in the remaining endosperm ceases in a wave-like manner starting at the silk-scar region of the kernel near the apex (or crown), and progresses downward toward the base and outward toward the periphery until all cell division (with the exception of the aleurone layer) has terminated by 20–25 DAP (Duvik, 1961).

The period between 12 and 15 DAP marks the transition from endosperm differentiation to maturation, when dry matter begins to accumulate. During this period, the expression of those enzymes required for starch and storage protein synthesis increases rapidly (Tsai *et al.*, 1970; Ozbun *et al.*, 1973; Ou-Lee and Setter, 1985; Doehlert *et al.*, 1988; Giroux *et al.*, 1994) and this grain-filling period continues until the death of the endosperm is completed. The pattern of starch deposition follows the pattern observed for cellular development in that its synthesis occurs in a wave-like fashion that initiates in the central endosperm and spreads toward the periphery, where deposition occurs near the upper periphery before proceeding towards the base. With the exception of the center-most region, all cells within the endosperm fill with starch by maturity and, consequently, this part of the tissue is referred to as the starchy endosperm (Larkins and Hurkman, 1978; Kowles and Phillips, 1988; Doehlert *et al.*, 1994). Storage protein synthesis initiates at ca. 15 DAP in the endosperm and follows the pattern of cellular development and starch deposition in that the synthesis spreads toward the periphery of the endosperm. The extent of protein deposition is greatest in the outer periphery (Larkins and Hurkman, 1978; Kowles and Phillips, 1988; Doehlert *et al.*, 1994) particularly in the subaleurone cells and the aleurone layer which also contains a high level of lipids but little starch (Morrison and Milligan, 1982).

Late maturation involves the shutdown of biosynthetic processes, induction of desiccation, and finally quiescence. The reduction in metabolic activity appears to be due, in part, to the loss of water (Kermode and Bewley, 1985; Comai and Harada, 1990). Coordinate with mid to late development, cereal endosperm undergoes a progressive cell death that engulfs the entire tissue, leaving only the aleurone layer unaffected and viable at maturity (Bartels *et al.*, 1988; Kowles and Phillips, 1988; Lopes and Larkins, 1993; DeMason 1994; Young *et al.*, 1997; Young and Gallie, 1999). The programmed death of the endosperm represents the end point of its development. The regulation of the program is critical as death of the starchy endosperm may be required to facilitate access of those hydrolases expressed and secreted from the aleurone and scutellum into the starchy endosperm during germination in order to provide the rapid mobilization of the storage reserves necessary to support seedling growth. However, premature induction of the cell death program would limit reserve deposition and thus jeopardize germination. Consequently, tight regulation of the cereal endosperm cell death program is essential to ensure sufficient storage reserve synthesis during development and its efficient mobilization during germination.

Developmental progression of PCD in cereal endosperm

The pattern and progression of cell death during maize endosperm development follows that for cellular development and deposition of the storage reserves (Young *et al.*, 1997). Viability staining of developing kernels revealed that cells within the central endosperm die at ca. 16 DAP (Figure 1; also Young *et al.*, 1997). After or concomitant with this, a second wave of cell death initiates at the crown (i.e., the upper region of the endosperm adjacent to the silk scar or attachment site) prior to 20 DAP and proceeds towards the base of the kernel between 24 and 40 DAP (Figure 1; also Young *et al.*, 1997). The basis for two overlapping waves of cell death may be indicative of zones established within the endosperm during its development. For example, cell division activity ceases within the central endosperm by ca. 12 DAP but continues in the peripheral regions followed by similar patterns for endoreduplication and synthesis of storage reserves. The increase and subsequent decline in DNA content within the central endosperm corresponds well with the initiation of cell death in these cells while the subsequent spread of cell death throughout the remaining endosperm closely follows the pattern of starch deposition (Figures 1 and 2). The progressive nature of cell death is consistent with the persistence of at least some degree of gene expression into the late stage of kernel development with the decline in total RNA during development (Figure 2) paralleling the decline in gene expression (Doehlert *et al.*, 1994).

Several tissues in developing wheat seed undergo PCD. Cell death occurs in the nucellus/pericarp early in development (Figure 3) to make room for the rapidly growing endosperm. At ca. 4–6 days after flowering (DAF; which also represents the day of fertilization), a specific set of cells adjacent to the nucellar projection undergoes PCD (Figure 3). This permits the expanding endosperm to separate from the nucellar projection thereby allowing the formation of the endosperm cavity across which photosynthates and nutrients are transported. A third instance of PCD occurs in the starchy endosperm which exhibits significant differences when compared to the organized onset and progression of maize endosperm cell death. As in maize, the PCD program in wheat endosperm begins at ca. 16 DAF (Young and Gallie, 1999). In contrast to maize, however, the cell death program in wheat endosperm does not initiate in a specific location (Figure 3). Although the central cells of wheat

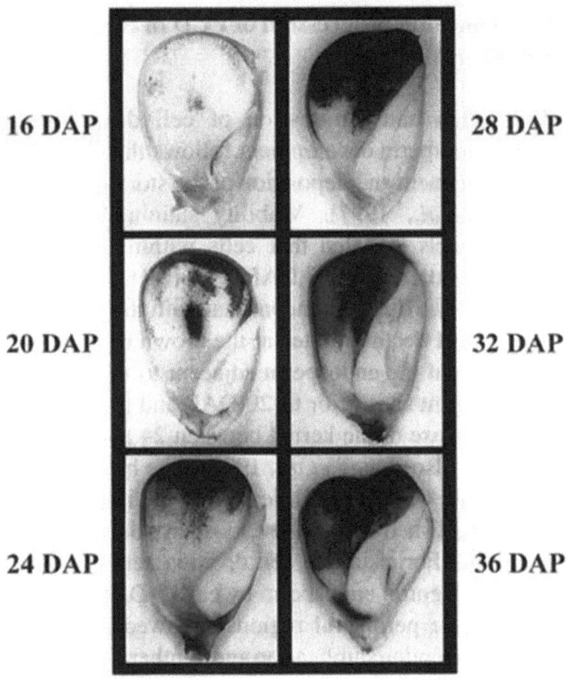

Figure 1. Progression of endosperm cell death in wild-type maize kernels as indicated by Evans Blue stain which identifies those cells no longer able to maintain membrane integrity and exclude the stain. Cell death is indicated by the dark-staining areas. Developmental stages (days after pollination; DAP) are indicated.

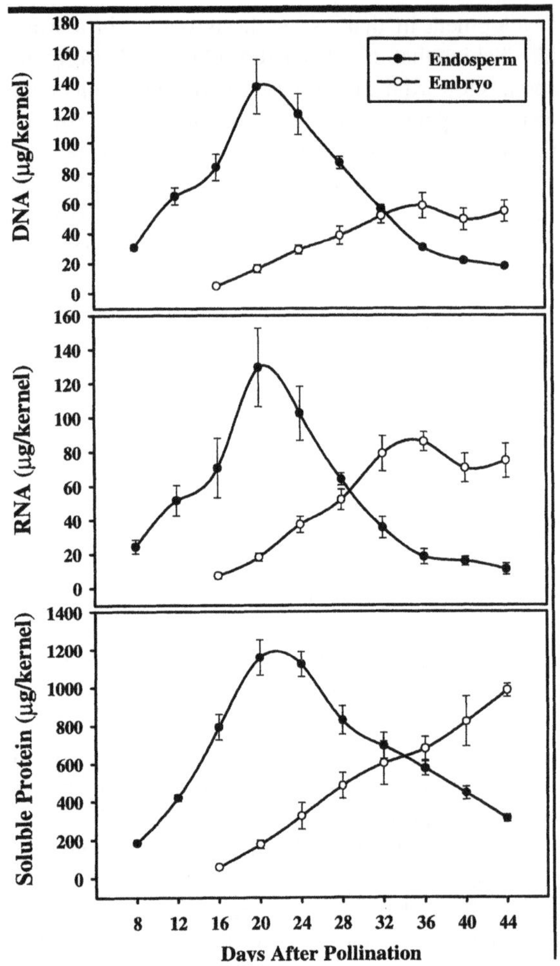

Figure 2. Total content of DNA (top panel), RNA (middle panel), and soluble protein (bottom panel) of maize endosperm and embryos during kernel development. DNA, RNA, and soluble protein was extracted from developing wild-type maize endosperm and embryos, quantitated, and normalized on a per kernel basis. The initial increase in DNA, RNA, and soluble protein in endosperm up to 24 DAP is due to growth of the kernels.

endosperm are also the oldest (Bradbury *et al.*, 1956) and cell death is reproducibly observed at a specific developmental stage, the onset of cell death is not confined to a specific region or cell type but, instead, appears to be a stochastic process. Likewise, the progression of cell death during subsequent development appears to be a random process that continues until the entire endosperm (with the exception of the aleurone layer) is affected by 30 DAF (Young and Gallie, 1999). Differences in the timing and progression of endosperm PCD in wheat compared to maize may be a consequence of the different organization of the wheat endosperm, which is composed of radial cells at the periphery of the starchy endosperm just beneath the aleurone layer, and central endosperm cells which fill the core of the endosperm (Bradbury *et al.*, 1956). The smaller size of the wheat endosperm may also obviate the need for a more orchestrated progression of the cell death process.

Genetic interactions between endosperm, embryo, and maternal tissue influencing the induction of the PCD program

Several mutations that disrupt normal seed development have proven useful in dissecting various aspects of endosperm development, including the interactions between the endosperm, embryo, and maternal tissues. Examples include the *defective seed (de)* and *defective kernel (dek)* mutants in maize (Mangelsdorf, 1926; Lowe and Nelson, 1947; Manzocchi *et al.*, 1980; Neuffer and Sheridan, 1980; Sheridan and Neuffer, 1980; Clark and Sheridan, 1986) and

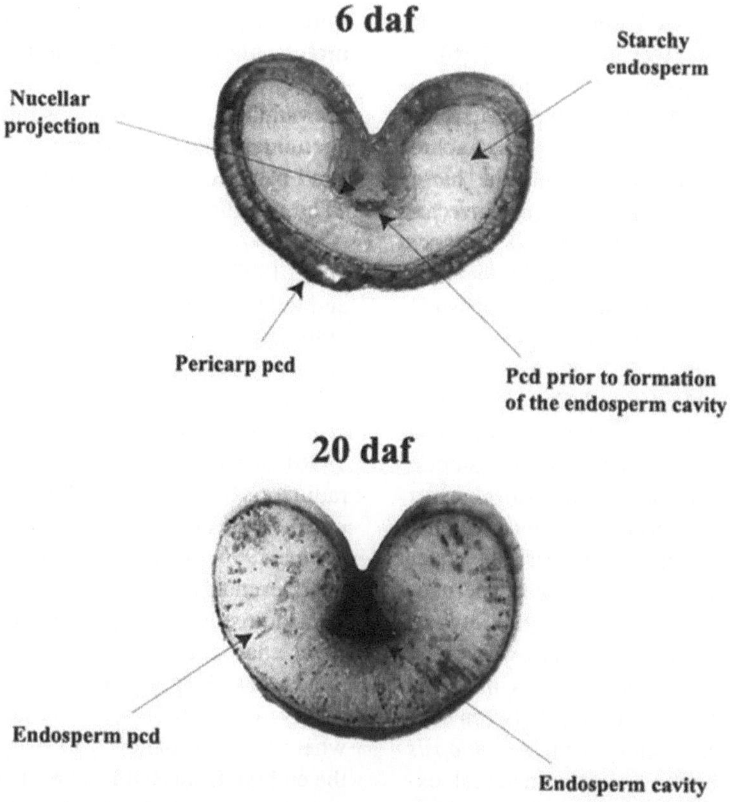

Figure 3. Identification of wheat seed tissues undergoing programmed cell death during development. Top panel: staining of a cross section of a wheat seed with Evans Blue during its early development (6 DAF) indicates that the pericarp and cells adjacent to the nucellar projection have initiated cell death. Bottom panel: the stochastic induction of the PCD program in the endosperm is illustrated after staining of seed at 20 DAF. Cell death is indicated by the dark-staining areas.

the *defective endosperm xenia (dex)* and *shrunken endosperm mutants expressing xenia (sex)* mutants in barley (Ramage and Crandall, 1981a, 1981b; Bosnes *et al.*, 1987, 1992) which lead to the premature death of the endosperm or the ectopic death of the embryo. Mutant analysis has led to at least three conclusions regarding cereal endosperm development. First, the majority of the maize *dek* mutants have been shown to cause abnormal development of both the endosperm and the embryo (Neuffer and Sheridan, 1980), observations consistent with those made with the *sex* mutants of barley (Bosnes *et al.*, 1987). In several cases, the *dek* mutation leads to premature induction of PCD in the endosperm and ectopic induction of the program in the embryo (T. Young and D. Gallie, unpublished observations). These studies, as well as those in barley (Jakobsen *et al.*, 1989), suggest that several seed-specific genes are expressed in both the endosperm and embryo, consistent with the proposal (Friedman, 1990, 1992) that the endosperm evolved from a supernumerary embryo (Olsen *et al.*, 1992).

Second, mutant analysis suggests that interactions between the endosperm and embryo occur during development. For example, Chang and Neuffer (1994) used B-A translocation stocks to test the extent to which the *dek*-conditioned mutant phenotype of the endosperm and embryo could be corrected by interaction with its normal (wild-type) counterpart. Their results indicated that in several mutants, a wild-type embryo could correct an endosperm defect or that a wild-type endosperm could correct an embryo defect, observations suggesting developmental signaling between the two tissues. The basis for these interactions is not currently known. However, in some mutants, the level of auxin and, to a lesser extent, cytokinin was affected, which may account for the abnormal development of both the endosperm and embryo (Lur and Setter, 1993; Torti *et al.*, 1984, 1986). That exogenous auxin treatment of some mutant kernels was sufficient to restore near wild-type seed weight supports this possibility (Lur and Setter, 1993; Torti *et al.*, 1986).

Third, mutant analysis suggests a mutual co-regulation between maternal and endosperm tissues during their development. Examples of this include the maize *minature1 (mn1)* mutant and the barley *shrunken endosperm (seg)* mutant. Loss of invertase activity in the basal endosperm cells and adjacent maternal tissue in *mn1* kernels results in the premature cell death in the maternal cells responsible for transporting sucrose into the developing kernel, thereby limiting the ability of the endosperm to synthesize storage reserves (Miller and Chourey, 1992; Chourey *et al.*, 1995). The premature cell death is a pleiotropic effect that may result from an osmotic imbalance associated with the inability to transport photosynthates to the endosperm (Miller and Chourey, 1992). In *seg* mutants, maternal transcripts appear responsible for mediating the abnormal endosperm development, although the basis for this phenotype is still unresolved (Felker *et al.*, 1985). However, recent mutant analysis in dicot species has provided additional insight into the relationship between maternal tissues and the developing seed. *Arabidopsis* embryonic pattern formation requires maternal expression of the *Short Integument1 (Sin1)* gene (Ray *et al.*, 1996) while in petunia, two MADS box genes, *Floral Binding Protein7 (FBP7)* and *FBP11*, influence endosperm development (Angenent *et al.*, 1995; Colombo *et al.*, 1997). In the latter example, the genes have been shown to be expressed specifically within the sporophytic tissues of the ovule but act in a cell non-autonomous manner on the endosperm. These studies provide direct molecular evidence that sporophytic tissues of the ovule are capable of controlling endosperm and/or embryo development. Conversely, several mutations in *Arabidopsis* have been identified which allow endosperm to develop in the absence of fertilization (Ohad *et al.*, 1996, 1999; Luo *et al.*, 1999). Two of these genes, *Fertilization-Independent Endosperm (FIE)* and *Fertilization-Independent Seed 1 (FIS1)*, encode proteins related to the *Polycomb* group of proteins previously identified in *Drosophila* and mammals involved in long-term repression of homeotic genes (Pirrotta, 1998), an observation suggesting that, if *FIE* and *FIS1* regulate homeotic genes in plants, they may play a similar role to their animal orthologues during seed development.

Potential involvement of reactive oxygen intermediates in endosperm PCD

Several studies have demonstrated that reactive oxygen intermediates (ROI) can activate PCD in animals and plants (reviewed in Jacobson, 1996; Jabs, 1999). The most direct evidence is that low doses of H_2O_2 (Levine *et al.*, 1994; Jacobson and Raff, 1995; Zettl *et al.*, 1997), $^{\bullet}O_2^-$ (Jabs *et al.*, 1996), $^{\bullet}OH$ (Jacobson and Raff, 1995), and lipid peroxides (Sandstrom *et al.*, 1994) are sufficient to induce PCD. In plant tissues, a variety of conditions such as senescence (Pastori and del Rio, 1997), cold (Prasad *et al.*, 1994; Koukalova *et al.*, 1997), osmotic stress (Streb and Feierabend, 1996), ozone fumigation (Overmyer *et al.*, 1998), UV radiation (Green and Fluhr, 1995), salicylic acid (Jabs *et al.*, 1996), and the hypersensitive response (Chandra *et al.*, 1996; Levine *et al.*, 1994) lead to ROI formation and, eventually, PCD.

Indirect evidence also implicates ROI during cereal endosperm PCD. Superoxide dismutase (SOD) and catalase are highly expressed and are localized primarily within the embryo during late embryogenesis where they are believed to play a role in protecting the embryo from ROI produced during the desiccation process (Wadsworth and Scandalios, 1989; Cannon and Scandalios, 1989; Guan and Scandalios, 1998). Interestingly, a product of the reaction catalyzed by SOD is H_2O_2, a potential inducer of PCD, although whether H_2O_2 generated by SOD is produced in sufficient quantities or persists long enough to trigger PCD is unknown. SOD is positively regulated by abscisic acid (ABA) and the steady-state level of SOD transcripts is significantly reduced in *viviparous1* kernels (Guan and Scandalios, 1998) which, as discussed below, exhibit a premature induction of PCD. Finally, expression of the antioxidant peroxiredoxin *(Per1)* gene in barley is restricted to the aleurone and embryo, the only two tissues that remain viable in the mature seed (Stacy *et al.*, 1996).

Programmed cell death in developing endosperm is accompanied by the internucleosomal degradation of nuclear DNA

A hallmark of animal apoptosis is the organized degradation of the genome through internucleosomal fragmentation resulting in the appearance of a ladder of DNA fragments in multimers of 180–200 nt when resolved by agarose gel electrophoresis (Wyllie, 1980;

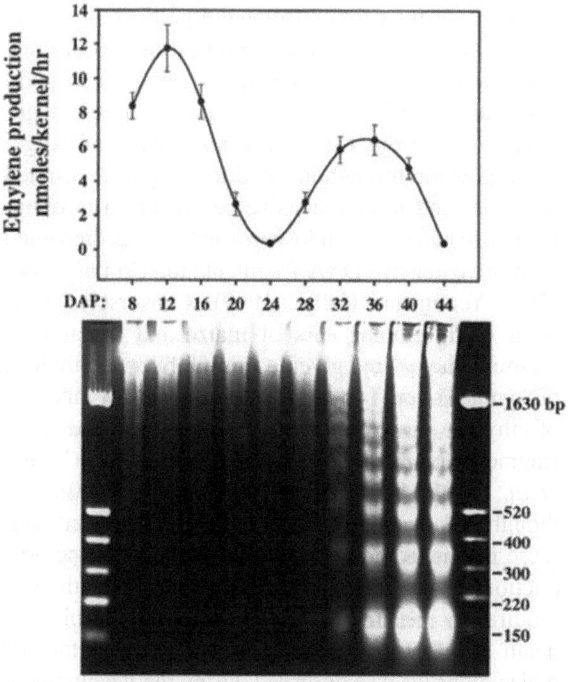

Figure 4. Internucleosomal degradation of nuclear DNA integrity and production of ethylene during maize kernel development. Top panel: ethylene production (nmol/kernel per hour) during kernel development was measured. Each point is the average ± SD of three replicates. Bottom panel: internucleosomal fragmentation of DNA was analyzed during wild-type kernel development. DNA extracted at the developmental stages (days after pollination, DAP) indicated above each lane was resolved on a 1.8% agarose gel and stained with ethidium bromide in order to visualize the DNA. Equal amounts of DNA (20 μg) were loaded in each lane. The sizes (bp) of molecular weight markers (lane S) are indicated to the right of each panel.

Arrends *et al.*, 1990; Barry and Eastman, 1992; Collins *et al.*, 1992). The orderly degradation of the genome during apoptosis is in contrast to the random decay of DNA that follows other types of cell death such as injury-induced necrosis (Kerr *et al.*, 1995). The fragmentation of the genome is considered to be a means by which the cell death program is made irreversible and facilitates the disassembly of the nucleus required for its subsequent reabsorption.

As in animal apoptosis, the internucleosomal fragmentation of nuclear DNA is an intrinsic aspect of PCD in maize and wheat endosperm (Young *et al.*, 1997; Young and Gallie, 1999) as it is in other types of pathogen-induced or development-associated PCD in plants (Wang *et al.*, 1996a, 1996c, 1999; Levine *et al.*, 1996; Mittler *et al.*, 1997). The appearance of internucleosomal DNA fragments is first detected in maize endosperm at 28 DAP and continues to increase throughout development (Figure 4; Young *et al.*, 1997). As internucleosomal fragments represent the completion of genome degradation, the initiation of the degradation process must occur prior to the appearance of these end products. Analysis of genome disassembly during animal apoptosis has demonstrated that larger fragments of DNA are generated prior to the appearance of the internucleosomal multimers (Walker *et al.*, 1991; Dusenberry *et al.*, 1991; Brown *et al.*, 1993) which are specifically related to the structural organization of the chromatin in the nucleus (Filipski *et al.*, 1990). Nuclear DNA is ordered into chromatin loops of ca. 50 kb, six of which are folded into 300 kb rosette structures (Walker and Sikorska, 1994). During animal apoptosis, DNA fragments of 50 to 300 kb are produced well in advance of the characteristic internucleosomal fragments, indicating that endonuclease attack first occurs at the points of folding within the rosette structure and is responsible for the initial stage of chromatin destruction and subsequent change in nuclear morphology (Walker and Sikorska, 1994). Cleavage of the chromatin into 50 kb fragments has also been observed during pathogen-induced PCD in tobacco leaves and bacteria-induced PCD of cultured soybean cells (Levine *et al.*, 1996; Mittler *et al.*, 1997). Analysis of endosperm DNA by pulse-field electrophoresis confirmed that processing of the maize endosperm genome from 300 to 50 kb fragments occurs between 4 and 16 DAP (Figure 5; Young and Gallie, 2000), a period that precedes visible cell death (as determined by viability staining, Figure 1; Young *et al.*, 1997) and the appearance of internucleosomal DNA fragments (Figure 4; Young *et al.*, 1997). Consequently, the initial degradation of endosperm DNA occurs prior to detectable cell death whereas the appearance of internucleosomal DNA fragments (the end point of genome degradation) follows cell death.

The organized degradation of endosperm DNA is specific to this tissue as DNA in the embryo and the aleurone layer remains intact (Young *et al.*, 1997). As a consequence, the DNA content in maize endosperm decreases after 24 DAP (Figure 2) correlating with the internucleosomal degradation of the genome (Figure 4; Young *et al.*, 1997). No similar decrease in DNA content was observed during embryo development (Figure 2). However, in *shrunken 2 (sh2)* kernels, a starch-defective mutation, accumulation of sugars in the endosperm and an elevated production of ethylene in the developing kernel result in the premature cell death of the endosperm and ectopic cell death

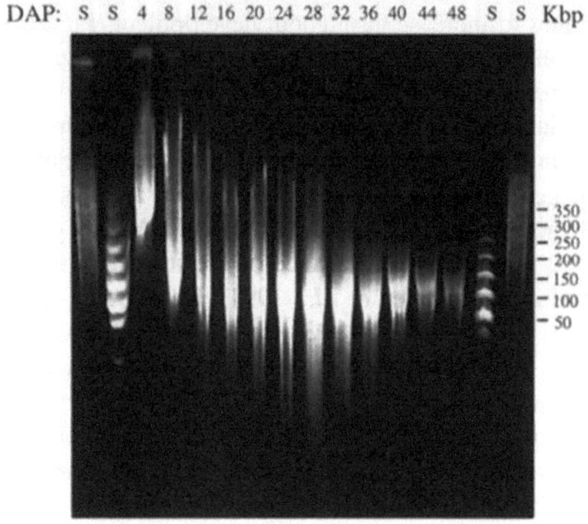

Figure 5. Pulsed-field electrophoretic analysis of DNA integrity during maize kernel development. DNA extracted at the developmental stages (days after pollination, DAP) indicated above each lane was resolved on a 1% agarose gel using pulsed-field electrophoresis (170 V with a pulse time of 45 s for both north/south and east/west electrodes). The gel was then stained with ethidium bromide in order to visualize the DNA. DNA from an equal fresh weight of kernels was loaded in each lane. The sizes (bp) of lambda molecular weight markers (lane S) are indicated to the right.

in the embryo (Young *et al.*, 1997). This premature and ectopic induction of the cell death program is accompanied by the premature appearance and ectopic occurrence of internucleosomal DNA fragmentation in the endosperm and embryo, respectively (Young *et al.*, 1997). In developing wheat endosperm, internucleosomal degradation is detected as early as 20 DAF and, as in maize, continues to increase throughout its remaining development (Young and Gallie, 1999). Although a low level of DNA fragmentation is detectable during early seed development in both maize and wheat, this corresponds to the period in which the pericarp/nucellus undergoes cell death suggesting that these maternal tissues also undergo a similar organized cell death program to make room for the expanding endosperm (Young *et al.*, 1997; Young and Gallie, 1999).

Ethylene as a mediator of endosperm cell death

Ethylene has emerged as a key regulator of PCD during plant development and examples of the critical role that it plays during PCD include formation of lysogenic aerenchyma in maize roots upon exposure to hypoxic conditions (Drew *et al.*, 1979; Campbell and Drew, 1983; He *et al.*, 1996) and in transmitting tract tissue following pollination (Wang *et al.*, 1996b). Ethylene is produced in two discrete peaks during the development of cereal endosperm (for maize, see Figure 4) and it participates in regulating PCD in this tissue. The application of exogenous ethylene throughout seed development resulted in earlier and more extensive cell death in developing maize and wheat endosperm and was accompanied by more extensive DNA fragmentation (Young *et al.*, 1997; Young and Gallie, 1999). Conversely, treatment of developing ears of maize and wheat with 2-aminoethoxyvinyl glycine (an inhibitor of ethylene biosynthesis) or 1-methylcyclopropene (an inhibitor of ethylene perception) reduced cell death and DNA fragmentation during endosperm development (Young *et al.*, 1997; Young and Gallie, 1999, 2000). Although cell death was not completely prevented with these inhibitors, complete inhibition of ethylene production was not achieved and the threshold of ethylene required to initiate PCD is not known. Ectopic cell death and DNA fragmentation in the scutellum of maize embryos was observed during the development of *sh2* kernels in which ethylene levels are elevated or in wild-type embryos of kernels exposed to ethylene (Young *et al.*, 1997). The ectopic induction of the PCD program by elevated ethylene in a tissue such as the scutellum that normally remains viable in the mature kernel establishes a causal relationship between ethylene and PCD in developing cereal seed. However, the observation that ectopic cell death in the embryo does require a higher level of ethylene (e.g., that which is present during *sh2* kernel development) suggests that components involved in the perception of ethylene may be differentially regulated in a tissue-specific or developmental manner. Consistent with this possibility is the fact that ethylene receptors are encoded by a multi-gene family (Wilkinson *et al.*, 1995; Hua and Meyerowitz, 1998) and are subject to differential regulation during *Arabidopsis* seedling development (Hua *et al.*, 1998) and during fruit ripening, flower senescence and abscission in tomato (Wilkinson *et al.*, 1995; Payton *et al.*, 1996). Changes in ethylene sensitivity rather than absolute ethylene levels have also been shown to correlate with senescence of orchid flowers after pollination (Porat *et al.*, 1995) and during lysogenous aerenchyma formation of maize roots exposed to limited nitrogen and phosphate (He *et al.*, 1992) and such changes in ethylene sensitivity may be related to changes in ethylene receptor synthesis and/or activity. As two peaks of ethylene production are observed during maize and wheat seed development whereas endosperm cell death follows an organized (i.e., in maize) or stochastic (i.e., in wheat) progression, it is likely that initiation of cell death is controlled by the timing and location of ethylene receptor synthesis or action and would provide a basis for the selective cell death that occurs within a developing seed.

Interaction of ethylene with other hormones

As mentioned above, ethylene is produced in the maize endosperm during two discrete developmental phases, the first of which coincides with the visible onset of the PCD program in the central endosperm and the second correlating with the increase in nuclease activity and the appearance of the end products of internucleosomal fragmentation of the nuclear DNA (Young *et al.*, 1997). How ethylene production is controlled during endosperm development is not known, although auxin, cytokinin, and ABA are known to influence its production in a wide variety of plant tissues (Morgan and Hall, 1962; Fuchs and Liberman, 1968; Lau and Yang, 1973; Yeong-Biau and Yang, 1979). Moreover, these hormones have been implicated in directing various aspects of cereal grain development (Rademacher and Graebe, 1984; Mengel *et al.*, 1985; Jones and Brenner, 1987; Lee *et al.*, 1989; Lur and Setter, 1993). The observation that auxin and ABA are also produced in two peaks during maize kernel development (Lur and Setter, 1993; Jones and Brenner, 1987) provides the basis for their possible involvement in regulating ethylene production (and/or action) during endosperm development.

Whether hormones other than ethylene also influence PCD in developing cereal endosperm can be addressed genetically using mutants compromised in hormone synthesis or perception. Two *viviparous (vp)* mutants have implicated ABA as a negative regulator of ethylene biosynthesis and/or action during maize endosperm development (Young and Gallie, 2000): *vp9* is deficient in ABA as it lacks a functional *Viviparous 9 (Vp9)* gene that encodes the subunit A of γ-carotene C-7,-8 desaturase, an enzyme required for xanthophyll and ABA synthesis, whereas *vp1* is an ABA-insensitive mutant that lacks a functional *Viviparous 1 (Vp1)* gene that encodes a seed-specific transcriptional activator required for the ABA responses associated with late embryogenesis (McCarty *et al.*, 1989, 1991; Hattori *et al.*, 1992). Both mutants result in precocious germination (i.e., vivipary).

A 2–4-fold increase in ethylene production in *vp1* kernels relative to wild-type kernels correlates with an observed premature onset and acceleration of the cell death program in the mutant endosperm as determined by viability staining of the endosperm and internucleosomal fragmentation of the genomic DNA (Young and Gallie, 2000). Moreover, the observed increases in DNA fragmentation correlate with an elevation in nuclease and RNase activities in the mutant kernels. The appearance of internucleosomal fragments precedes the onset of precocious germination in the *viviparous* kernels by at least 4–12 days of development, suggesting that induction of PCD in the mutants occurs at an even earlier developmental stage. Similarly, treatment of developing wild-type kernels with fluridone, an inhibitor of ABA biosynthesis, recapitulated the elevated synthesis of ethylene and accelerated endosperm cell death that was observed in *vp1* and *vp9* endosperm. These data suggest that ABA participates not only in the acquisition of desiccation tolerance and prevention of vivipary, but also in establishing the appropriate onset and progression of PCD during maize endosperm development.

A similar role for ABA has been recently reported during androgenesis of microspores in barley in which the application of the hormone to developing anthers reduced the occurrence of apoptosis-linked characteristics such as condensation of the chromatin and internucleosomal DNA fragmentation (Wang *et al.*, 1999). ABA has also been shown to prevent the PCD of aleurone cells of germinating barley seed (Wang *et al.*, 1996c; Bethke *et al.*, 1999). Consequently, ABA may play a general protective role that functions to establish the timing and extent of PCD in diverse tissues during cereal development. However, ABA may not play a protective role for all species: the application of ABA to daylily flowers prematurely induced those events that occur during their natural senescence (Panavas *et al.*, 1998) although ethylene does not appear to play any role in daylily flower senescence (Lukaszewski and Reid, 1989; van Doorn and Stead, 1994).

Whether ABA modulates ethylene production during maize endosperm development is not known, although it has been shown to inhibit ethylene production in sunflower hypocotyl segments (Bailly *et al.*, 1992) or during senescence of rice leaves (Chen and Kao, 1992) and inhibits synthesis of ACC in mung bean hypocotyls, the first step in the ethylene biosynthesis pathway (Yoshii and Imaseki, 1981), supporting the possibility that ABA may influence the timing of endosperm PCD through the regulation of ethylene synthesis. However, ABA may function to delay the onset of PCD independently of any possible direct regulation of ethylene biosynthesis. For instance, ABA may alter the sensitivity of the endosperm to ethylene or may regulate one or more steps downstream of ethylene perception. Nevertheless, the control of the onset of cell death in maize endosperm appears to be achieved through a balance between ABA and ethylene.

Factors affecting the timing and progression of endosperm cell death

Endoreduplication is estimated to occur in over 90% of all angiosperms and often accompanies or precedes cell growth and differentiation (reviewed in Traas *et al.*, 1998). Although the purpose of endoreduplication during maize endosperm development is still unresolved, it has been suggested that it provides a means to enhance protein synthesis through the amplification of important genes or, alternatively, serves to store nucleotides for subsequent use during germination (Tsai *et al.*, 1970; Kowles *et al.*, 1986). During maize kernel development, endoreduplication occurs most extensively in the central endosperm and decreases proportionately towards the periphery. The completion of endoreduplication in the central endosperm is followed rapidly by the induction of the cell death program in these same cells, suggesting these two processes may be linked. Further evidence to support this view is that during seed development in lily, the endosperm tissue differentiates without undergoing endoreduplication and the endosperm cells remain alive (D'Amato, 1984). While the precise role that endoreduplication plays remains to be determined, it may constitute a developmental requirement that prepares the entry of the endosperm into the PCD process during endosperm development.

In addition to the developmental mutants (e.g., *dek* mutants) are those which exhibit reduced starch synthesis, storage protein synthesis, and hormone biosynthesis and perception. Endosperm mutations affecting carbohydrate metabolism during maize kernel development result in elevated sugar levels as a result of decreased starch production (Boyer and Shannon, 1983; Douglass *et al.*, 1993; Azanza *et al.*, 1996). In addition to the effect on carbohydrate metabolism, an increase in ethylene biosynthesis and an early onset of PCD is observed (Young *et al.*, 1997). Premature endosperm degeneration in *shrunken2 (sh2)* mutants appears to be causally related to elevated ethylene production as treatment of normally developing maize kernels with exogenous ethylene, as previously mentioned, is capable of inducing a similar effect, while treatment of ethylene inhibitors has the opposite effect (Young *et al.*, 1997). Consequently, the accumulation of sugar in such mutants leads to an elevated production of ethylene which in turn triggers a premature induction and rapid execution of PCD as determined by an accelerated spread of cell death and earlier and more extensive internucleosomal fragmentation of nuclear DNA in the mutant endosperm (Young *et al.*, 1997). Whether the sugar-mediated effect on endosperm PCD is a result of increased ethylene produced in response to the increased osmotic stress caused by the elevated level of sugar in these mutants, or whether there is sugar-mediated regulation of the ethylene biosynthetic machinery or ethylene sensitivity remains to be determined. However, studies on maize kernel abortion induced by shading demonstrated that glucose levels rose in shaded kernels and that these kernels exhibited increased ethylene sensitivity upon ethylene precursor treatment as determined by the rate of abortion (Cheng and Lur, 1996). A recent report demonstrates that the *Arabidopsis glucose insensitive* locus *(GIN1)*, responsible for glucose repression of cotyledon greening and expansion, shoot development, floral transition, and gene expression, acts downstream of the ethylene receptor *ETR1* (Zhou *et al.*, 1998). This study suggests a convergence between the glucose and ethylene signal transduction pathways that, if present in developing cereal endosperm, could provide an additional mechanism to account for the effect that increased sugar levels have on endosperm PCD.

Combination of the storage protein-deficient mutant *opaque2 (o2)* with starch-deficient mutations results in a further elevation of ethylene synthesis and more pronounced cell death during endosperm development (Young, 1997). The accelerated cell death in the *o2* and starch-deficient endosperm mutants is accompanied by increased RNase and nuclease activity (Dalby and Davies, 1967; Wilson and Alexander, 1967; Dalby and Tsai, 1975; Glover *et al.*, 1975; Tsai *et al.*, 1978; T. Young, unpublished results), suggesting that the increase in these activities may be partly responsible for the accelerated execution of PCD in these mutants. With respect to the elevation in ethylene production observed during endosperm development in both starch-deficient and *o2* kernels, it is possible that a common mechanism is operating. One characteristic shared between these two classes of endosperm mutants may be a more negative osmotic potential in the storage tissue due to the elevation in sugar and amino acids, respectively, which in turn may signal an increase in ethylene production (Creech, 1965; Boyer and Shannon, 1983; Lee and Tsai, 1984; Marshall, 1987; Doehlert and Kuo, 1994; Azanza *et al.*, 1996).

Nuclease expression during endosperm PCD

The internucleosomal DNA degradation that occurs during animal apoptosis is accompanied by the activation of endogenous nuclear Ca^{2+}-dependent endonucleases (Wyllie, 1980; Arrends *et al.*, 1990; Collins *et al.*, 1992; Barry and Eastman, 1992). In plants, nucleases are regulated both developmentally and in response to endogenous and exogenous stimuli (Wilson, 1975, 1982; Farkas, 1982; Green, 1994; Sugiyama *et al.*, 2000, this issue). Increases in nuclease activities have been observed in senescent tissues (reviewed in Green, 1994) and during the PCD associated with xylem differentiation (Thelen and Northcote, 1989; Sugiyama *et al.*, 2000, this issue), observations suggesting that these activities may mediate aspects of the cell death process.

Increases in nuclease activities have also been observed during the development of wheat, rice, sorghum, and maize seed (Ingle *et al.*, 1965; Wilson, 1967, 1968; Cruz *et al.*, 1970; Dalby and Cagampang, 1970; Cagampang and Dalby, 1972; Donovan *et al.*, 1977; Johari *et al.*, 1977; Wilson, 1980, 1982; Young *et al.*, 1997; Young, 1997; Young and Gallie, 1999). Although nuclease activity is present throughout cereal seed tissues, the level present in maize and wheat endosperm that is fully engaged in the cell death program (as determined by viability staining) is at least 5- to 10-fold higher than in embryos from the same seed (Young *et al.*, 1997; Young and Gallie, 1999). Correlated with the increase in these activities during late endosperm development is a reduction in both DNA and RNA content specifically in endosperm but not the embryo (Figure 2). Additional evidence to support the relationship between nuclease levels and extent of DNA degradation is that nuclease activity is ca. 2-fold higher in the maize *sh2* endosperm (which is defective in starch biosynthesis, contains a high level of sugars, and exhibits elevated ethylene production) than in wild-type endosperm. Premature and accelerated internucleosomal fragmentation of *sh2* endosperm DNA is observed relative to wild-type endosperm (Young *et al.*, 1997). The observed increases in nuclease activity that occur during endosperm development could be due to an increased synthesis or activation of pre-existing inactive precursors, similar to that during animal apoptosis (Wyllie, 1980; Arrends *et al.*, 1990; Collins *et al.*, 1992; Barry and Eastman, 1992). Whether protein synthesis inhibitors prevent nuclease expression and internucleosomal degradation remains unknown.

As observed for nucleases, RNase activities also increase during maize endosperm development and are substantially higher in the endosperm than in embryos (Young, 1997). The expression pattern of at least one of these activities spatially correlates with the pattern of cell death in developing maize endosperm: total acidic RNase activity was observed to be high in the crown by 18 DAP, more uniformly distributed by 35 DAP, and high in the base during late development (Wilson, 1980), similar to the progression of endosperm PCD (Figure 1). The extent to which observed changes in nuclease activity are causally related to endosperm cell death remains to be determined. However, RNases, nucleases and proteases have all been implicated in the PCD process in both animals and plants (Thelen and Northcote, 1989; Tomei and Cope, 1991, 1994; Mittler *et al.*, 1995; Beers *et al.*, 2000, this issue; Sugiyama *et al.*, 2000, this issue).

Control of the protein synthetic machinery in cereals during PCD

Regulation of the level or activity of the translational machinery has been shown to influence entry into apoptosis in mammalian cells. Overexpression of the eukaryotic initiation factor (eIF) 4E, which is the cap-binding subunit of eIF4F (see Figure 6) that is required to promote translation initiation, was sufficient to signal survival in the absence of serum or growth factors in NIH3T3 fibroblasts (Polunovsky *et al.*, 1996). eIF4G, the large subunit of eIF4F, has been shown to be a target of caspase activity that releases a 76 kDa peptide during apoptosis in BJAB and Jurkat cells (Clemens *et al.*, 1998). A novel homologue of eIF4G, which is unable to interact with eIF4E, has been suggested to function as a positive regulator of γ-interferon-induced PCD in HeLa cells (Levy-Strumpf *et al.*, 1997). The phosphorylation state of eIF2α, the initiation factor responsible for binding the initiator Met-tRNA to the 40S ribosomal subunit (Figure 6), controls the activity of this initiation factor.

That phosphorylation of eIF2α may play a role in animal PCD was suggested by the hyperphosphorylation of this initiation factor, which promoted tumor necrosis factor α (TNF-α)-induced apoptosis in NIH3T3 cells (Srivastava *et al.*, 1998). These diverse observations suggest two possible roles for initiation factors with respect to animal PCD: the level or phosphorylation state of key initiation factor activities can influence entry into PCD and once PCD has started, components of the translational machinery are subject to specific caspase-directed cleavage which may serve to up-regulate expression from selected transcripts whose products are required to prosecute the later stages of the cell death process.

Analysis of the translational machinery during seed development demonstrated that, as observed in animals, several initiation factors undergo changes in their phosphorylation state and/or steady-state level that correlate with the progression of PCD in the endosperm. eIF4B is an initiation factor that interacts with eIF3 (Methot *et al.*, 1996) and, in plants, both eIF4B and eIF4G interact with the poly(A)-binding protein (PABP) (Figure 6; Le *et al.*, 1997). eIF4B also assists the function of eIF4F in binding the 5′ cap structure and the subsequent removal of secondary structure present in the 5′ leader of an mRNA that would otherwise impede the scanning of the 40S ribosomal subunit (Ray *et al.*, 1985). eIF4B is a multiply phosphorylated protein under conditions of active translation but is dephosphorylated under conditions of translational repression in animal and plant cells (Duncan and Hershey, 1984, 1985; Gallie *et al.*, 1997), suggesting that the phosphorylation state regulates eIF4B activity. eIF4B is present in a highly phosphorylated state in developing wheat seed corresponding to the period of greatest protein synthetic activity (up to 25 DAF) but undergoes dephosphorylation during subsequent development (Le *et al.*, 1998). As the bulk of cells have initiated PCD by 30 DAF as determined by viability staining as well as by the appearance of internucleosomal fragmentation of nuclear DNA (Young and Gallie, 1999), the changes in eIF4B phosphorylation correlate with the late stages of wheat endosperm PCD. The dephosphorylation of eIF4B would be expected to reduce its functional interaction with PABP during translation as phosphorylated eIF4B interacts with PABP at least an order more strongly than does dephosphorylated eIF4B (Le and Gallie, manuscript submitted). The loss in the interaction between eIF4B and PABP would be expected to result in a reduction in protein synthetic activity because translational efficiency is dependent on the functional interaction between the cap and poly(A) tail of an mRNA (Gallie, 1991) which is facilitated by protein-protein interactions between the cap-associated initiation factors and PABP (Tarun and Sachs, 1996; Le *et al.*, 1997). This could serve as a global regulatory mechanism to inhibit protein synthesis as part of the execution of a cell death program in plants. However, the translation of some transcripts may escape this means of repression if their translation is not dependent on an interaction between the cap and poly(A) tail: for example, heat-shock mRNAs continue to be translated upon thermal stress in spite of the loss of the functional interaction between the cap and poly(A) tail that occurs as part of the heat shock response (Pitto *et al.*, 1992; Gallie *et al.*, 1995). Instead, their continued translational competence is conferred by the presence of the heat-shock 5′ leader (Pitto *et al.*, 1992). Whether transcripts required for plant PCD are regulated by unique translational mechanisms remains to be determined.

eIF4A, which also associates with eIF4F (Figure 6) and assists in the unwinding of secondary structure in a 5′ leader, is subject to phosphorylation upon heat shock in wheat (Gallie *et al.*, 1997) and hypoxia in maize (Webster *et al.*, 1991), suggesting that its phosphorylation is part of the stress response. A portion of total eIF4A present in wheat seed undergoes phosphorylation early in seed development and may represent that portion present in the nucellar/pericarp tissues which are undergoing PCD at this developmental stage (Le *et al.*, 1998). Interestingly, the factor is not phosphorylated during endosperm PCD suggesting that eIF4A may participate in cell death only in selected tissues or that the observed changes in eIF4A phosphorylation are only a by-product of a tissue-specific PCD program.

As mentioned above, eIF2α activity is controlled by its phosphorylation state such that loss of its activity and repression of protein synthesis occurs upon eIF2α phosphorylation. The phosphorylation state of eIF2α is dynamic during wheat seed development: it is partially phosphorylated during nucellar/pericarp PCD and is less phosphorylated up to mid-development of the endosperm, but between 20 to 25 DAF, it undergoes hyperphosphorylation, correlating with the onset of endosperm PCD (Le *et al.*, 1998). How these changes in eIF2α phosphorylation regulate this factor in plants and the role it might play in plant PCD is unknown.

A substantial decrease in the level of eIFiso4E, several subunits of eIF3, and eIF4A in developing

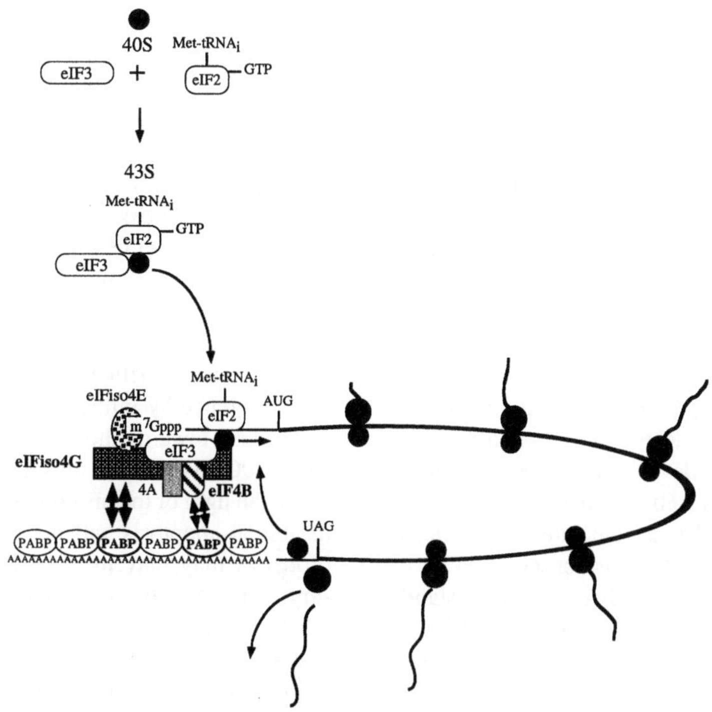

Figure 6. Summary of selected steps during translation initiation. eIF4F is composed of two subunits: eIF4E, the small cap-binding subunit, and eIF4G, the large subunit that interacts with several other initiation factors. In addition to eIF4F, plants also contain eIFiso4F, an isoform of eIF4F, that is composed of eIFiso4E and eIFiso4G. eIF4F (or eIFiso4F) binds to the 5′ cap and recruits eIF4B and eIF4A. Together, these factors unwind any secondary structure present in the 5′ leader. eIF3 and the ternary complex (i.e., eIF2, GTP, and Met-tRNAi) bind to the 40S subunit to form the 43S complex. The interaction between eIF3 and eIF4G (or eIFiso4G) results in 40S subunit binding downstream of the cap followed by scanning of the leader to locate the initiation codon and 60S subunit joining to form the 80S ribosome. The interaction between eIF4B and PABP or eIFiso4G and PABP, represented by a pair of double-headed arrows, mediates the functional interaction between the termini of an mRNA actively engaged in translation in plants. The 60S subunit is shown dissociating from the mRNA upon translation termination, whereas the 40S subunit is depicted as available for re-recruitment to the 5′ terminus.

wheat seed occurs between 25 and 30 DAF (Gallie *et al.*, 1998). Reductions in several other initiation factors, including eIF4B, eIF2α, eIF2β, PABP, eIF4E and eIFiso4G occur during late seed development (40 DAF) which may be part of the general loss of soluble protein that occurs in the endosperm of cereals (see Figure 2). Interestingly, the level of eIF4G and at least one subunit of eIF3 remains constant or actually increases during late seed development (Gallie *et al.*, 1998). These observations demonstrate that components of the translational machinery are differentially regulated during endosperm development. The changes in the phosphorylation state of eIF4B and eIF2α and the changes in the level of eIFiso4E and several eIF3 subunits are consistent with a loss in their activity and correlate with the progression of PCD in the endosperm. However, the reduction in protein synthetic activity, in itself, cannot serve as a signal for PCD in plants, as translational activity in the embryo and aleurone layer of cereal seed is also substantially reduced during late seed development and yet these tissues remain living. If alterations in protein synthetic activity are involved in cereal seed PCD, tissue-specific expression of inhibitors of cell death may protect those tissues, such as the embryo and aleurone layer, that remain living at seed maturity whereas their absence in tissues such as the endosperm may permit PCD to proceed when the appropriate signals are perceived. However, the observation that PCD can be induced ectopically in developing maize embryos, either in *sh2* kernels or in wild-type kernels exposed to ethylene, suggests that the PCD program can be initiated in tissues normally destined to survive when an inducer capable of signaling PCD is present at a sufficiently high level.

Conclusions and future directions

Maize should continue to be an ideal system to study developmental PCD due to the large size of its endosperm and its economic importance. Identification of genes involved in controlling endosperm PCD will be a necessary step in elucidating the program in cereals. Current technology that allows the simultaneous generation of mutants and gene tagging will help to facilitate this process. Moreover, altering the timing and progression of cereal endosperm PCD through the identification of mutants or using transgenic approaches may provide the means by which endosperm PCD can be delayed in order to increase yield by extending the developmental period during which storage reserves can be synthesized.

Sufficient evidence has accumulated to implicate ethylene as key to regulating the onset of PCD in developing cereal endosperm. Given our understanding of the role of ethylene in endosperm PCD, it is now possible to understand how those endosperm mutations that are not directly related to ethylene biosynthesis or perception can affect the onset and progression of cell death in developing endosperm: mutations such as *sh2* and *o2* that accumulate high levels of sugars and amino acids, respectively, and exhibit a premature induction of the cell death program appear to do so through the production of stress-related ethylene. Similarly, other mutants, such as *vp1* or *vp9*, in which ABA synthesis or perception is affected, also produce elevated levels of ethylene and exhibit an early onset of PCD. These examples demonstrate that ethylene is sufficient to induce PCD in cereal endosperm. That maize and wheat endosperm PCD can be delayed by the application of inhibitors of ethylene biosynthesis or perception supports the conclusion that ethylene regulates the onset of PCD in cereals and suggests, but does not prove, that ethylene is necessary for endosperm PCD. A genetic approach, using mutants affected in ethylene biosynthesis or signaling, will be needed to prove that ethylene is necessary for cereal endosperm PCD, albeit no such mutants are known to exist at present.

Although factors such as elevated sugar may be indirectly responsible for increasing ethylene production, those developmental signals responsible for inducing ethylene production and the precise role that this hormone plays remain to be determined. Ethylene may function as the primary signal that, for those endosperm cells competent to perceive the hormone, induces expression or activity of those gene products responsible for initiating and executing the cell death program. That elevated ethylene is required for the ectopic induction of PCD in developing wild-type maize embryos or in *sh2* embryos that are exposed to elevated ethylene produced during mutant kernel development demonstrates that the PCD program can be initiated in a tissue normally destined to live and suggests that developmental differences exist between endosperm and the embryo which affect the sensitivity of these tissues to ethylene. The nature of this developmental difference remains to be determined and the question of whether ethylene may indirectly affect this difference by influencing the sensitization of the endosperm towards those developmental signals that trigger PCD is one that must be considered.

In the light of the prominent role that caspases play in animal apoptosis and the role that other types of proteases may play in plant PCD, it will be necessary to identify those proteases (as well as protease inhibitors) expressed in developing cereal endosperm that are involved in the initiation or execution phases of the endosperm PCD program. Localization of these transcripts and/or proteins during cereal development would help establish a possible link to the pattern of endosperm PCD. Determination of target substrates of any caspase or caspase-like proteases during endosperm PCD would aid in identifying those aspects of the cellular machinery that are involved in the execution phase of the cell death program.

Acknowledgements

Grants supporting the work on programmed cell death (97-35304-4657) and the fate of the translational machinery (97-35301-4404 and 99-35301-7866) during cereal endosperm development have been provided by the NRICGP program of the United States Department of Agriculture.

References

Angenent, G.C., Franken, J., Busscher, M., van Dijken, A., van Went, J.L., Dons, H.J.M. and Tunen, A.J. 1995. A novel class of MADS box genes is involved in ovule development in petunia. Plant Cell 7: 1569–1582.

Arrends, M.J., Morris, R.G. and Wyllie, A.H. 1990. Apoptosis: the role of the endonuclease. Am. J. Path. 136: 593–608.

Azanza, F., Bar-Zur, A. and Juvik, J.A. 1996. Variation in sweet corn kernel characteristics associated with stand establishment and eating quality. Euphytica 87: 7–18.

Bailly, C., Corbineau, F. and Come, D. 1992. The effect of abscisic acid and methyl jasmonate on 1-aminocyclopropane 1-carboxylic acid conversion to ethylene in hypocotyl segments of sunflower seedlings, and their control by calcium and calmodulin. Plant Growth Reg. 11: 349–355.

Barry, M.A. and Eastman, A. 1992. Endonuclease activation during apoptosis: the role of cytosolic Ca^{2+} and pH. Biochem. Biophys. Res. Com. 186: 782–789.

Bartels, D., Singh, M. and Salamini, F. 1988. Onset of desiccation tolerance during development of the barley embryo. Planta 175: 485–492.

Beers, E.P., Woffenden, B.J. and Zhao, C. 2000. Plant proteolytic enzymes: possible roles during programmed cell death. Plant Mol. Biol., this issue.

Bethke, P.C., Lonsdale, J.E., Fath, A. and Jones, R.L. 1999. Hormonally regulated programmed cell death in barley aleurone cells. Plant Cell 11: 1033–1045.

Bewley, J.D. and Black, M. 1978. Physiology and Biochemistry of Seeds, Springer-Verlag, Berlin.

Bewley, J.D. and Black, M. 1994. Seeds: Physiology of Development and Germination, Plenum Press, New York.

Bhatnagar, S.P. and Sawhney, V. 1981. Endosperm: its morphology, ultrastructure, and histochemistry. Int. Rev. Cytol. 73: 55–102.

Boesewinkel, F.D. and Bouman, F. 1995. Seed morphology and development. In: J. Kigel and G. Galili (Eds.), Seed Development and Germination, Marcel Dekker, New York, pp. 1–24.

Bosnes, M., Harris, E., Aigeltinger, L. and Olsen, O.-A. 1987. Morphology and ultrastructure of 11 barley shrunken endosperm mutants. Theor. Appl. Genet. 74: 177–187.

Bosnes, M., Weideman, F. and Olsen, O.-A. 1992. Endosperm differentiation in barley wild-type and sex-mutants. Plant J. 2: 661–674.

Boyer, C.D. and Shannon, J.C. 1983. The use of endosperm genes for sweet corn quality improvement. Plant Breed. Rev. 1: 139–161.

Bradbury, D., MacMasters, M.M. and Cull, I.M. 1956. Structure of the mature wheat kernel. III. Microscopic structure of the endosperm of hard red winter wheat. Cereal Chem. 33: 361–373.

Brink, R.A. and Cooper, D.C. 1947. The endosperm in seed development. Bot. Rev. 13: 423–541.

Brown, D.G., Sun, X.M. and Cohen, G.M. 1993. Dexamethasone-induced apoptosis involves cleavage of DNA to large fragments prior to internucleosomal fragmentation. J. Biol. Chem. 268: 3037–3039.

Cagampang, G.B. and Dalby, A. 1972. Development of ribonuclease activity in nine inbred lines of normal and *opaque2* maize. Can. J. Plant Sci. 52: 901–905.

Campbell, R. and Drew, M.C. 1983. Electron microscopy of gas space (aerenchyma) formation in adventitious roots of *Zea mays* L. subjected to oxygen shortage. Planta 157: 350–357.

Cannon, R.E. and Scandalios, J.G. 1989. Two cDNAs encode two nearly identical Cu/Zn superoxide dismutase proteins in maize. Mol. Gen. Genet. 219: 1–8.

Chandra, S., Martin, G.B. and Low, P.S. 1996. The *Pto* kinase mediates a signaling pathway leading to the oxidative burst in tomato. Proc. Natl. Acad. Sci. USA 93: 13393–13397.

Chang, M.T. and Neuffer, M.G. 1994. Endosperm-embryo interactions in maize. Maydica 39: 9–18.

Chasan, R. 1994. Tracing tracheary element development. Plant Cell 6: 917–919.

Chen, T.C. and Kao, C.H. 1992. Senescence of rice leaves XXXII. Effects of abscisic acid and benzyladenine on polyamines and ethylene production during senescence. J. Plant Physiol. 139: 617–620.

Cheng, C.-Y. and Lur, H.-S. 1996. Ethylene may be involved in abortion of the maize caryopsis. Physiol. Plant. 98: 245–252.

Chourey, P.S., Cheng, W.-H., Taliercio, E.W. and Im, K.H. 1995. Genetic aspects of sucrose-metabolizing enzymes in developing maize seed. In: M.A. Madore and W.J. Lucas (Eds.), Carbon Partitioning and Source-Sink Interactions in Plants, American Society of Plant Physiologists, Rockville, MD, pp. 239–245.

Clark, J.K. and Sheridan, W.F. 1986. Developmental profiles of the maize embryo-lethal mutants *dek-22* and *dek-23*. J. Hered. 77: 83–92.

Clemens, M.J., Bushell, M. and Morley, S.J. 1998. Degradation of eukaryotic polypeptide chain initiation factor (eIF) 4G in response to induction of apoptosis in human lymphoma cell lines. Oncogene 17: 2921–2931.

Collins, R.J., Harmon, B.V., Gobe, G.C. and Kerr, J.F.R. 1992. Internucleosomal DNA cleavage should not be the sole criterion for identifying apoptosis. Int. J. Radiat. Biol. 61: 451–453.

Colombo, L. Franken, J. van der Krol, A., Wittich, P., Dons, H.J.M. and Angenent, G.C. 1997. Downregulation of ovule-specific MADS box genes from petunia results in maternally controlled defects in seed development. Plant Cell 9: 703–715.

Comai, L. and Harada, J.J. 1990. Transcriptional activities in dry seed nuclei indicate the timing of the transition from embryogeny to germination. Proc. Natl. Acad. Sci. USA 87: 2671–2674.

Creech, R.G. 1965. Genetic control of carbohydrate synthesis in maize endosperm. Genetics 52: 1175–1186.

Cruz, L.J., Cagampang, G.B. and Juliano, B.O. 1970. Biochemical factors affecting protein accumulation in the rice grain. Plant Physiol. 46: 743–747.

Dalby, A. and Cagampang, G.B. 1970. Ribonuclease activity in normal, *opaque2* and *floury2* maize endosperm during development. Plant Physiol. 46: 142–144.

Dalby, A. and Davies, I. 1967. Ribonuclease activity in the developing seed of normal and *opaque-2* maize. Science 155: 1573.

Dalby, A. and Tsai, C.Y. 1975. Comparison of lysine and zein and non-zein protein contents in immature and mature maize endosperm mutants. Crop Sci. 15: 513–520.

D'Amato, F. 1984. Role of polyploidy in reproductive organs and tissues. In: B.M. Johri (Ed.), Embryology of Angiosperms, Springer-Verlag, Berlin, pp. 519–566.

DeMason, D.A. 1994. Controls of germination in noncereal monocotyledons. Adv. Struct. Biol. 3: 285–310.

DeMason, D.A. 1997. Endosperm structure and development. In: B.A. Larkins and I.K. Vasil (Eds.), Cellular and Molecular Biology of Plant Seed Development, Kluwer Academic Publishers, Dordrecht, Netherlands, pp. 73–115.

Doehlert, D.C. and Kuo, T.M. 1994. Gene expression in developing kernels of some endosperm mutants of maize. Plant Cell Physiol. 35: 411–418.

Doehlert, D.C., Kuo, T.M. and Felker, F.C. 1988. Enzymes of sucrose and hexose metabolism in developing kernels of two inbreds of maize. Plant Physiol. 86: 1013–1019.

Doehlert, D.C., Smith, L.J. and Duke, E.R. 1994. Gene expression during maize kernel development. Seed Sci. Res. 4: 299–305.

Donovan, G.R., Lee, J.W. and Hill, R.D. 1977. Compositional changes in the developing grain of high- and low-protein wheats. Cereal Chem. 52: 638–645.

Douglass, S.K., Juvik, J.A. and Splittstoesser, W.E. 1993. Sweet corn seedling emergence and variation in kernel carbohydrate reserves. Seed Sci. Technol. 21: 433–445.

Drew, M.C., Jackson, M.B. and Giffard, S. 1979. Ethylene-promoted adventitious rooting and development of cortical air spaces (aerenchyma) in roots may be adaptive response to flooding in *Zea mays* L. Planta 147: 83–88.

Duncan, R. and Hershey, J.W.B. 1984. Heat shock-induced translational alterations in HeLa cells. Initiation factor modifications and the inhibition of translation. J. Biol. Chem. 259: 11882–11889.

Duncan, R. and Hershey, J.W.B. 1985. Regulation of initiation factors during translational repression caused by serum depletion: covalent modification. J. Biol. Chem. 260: 5493–5497.

Dusenbury, C.E., Davis, M.A., Lawrence, T.S. and Maybaum, J. 1991. Induction of megabase DNA fragments by 5-fluorodeoxyuridine in human colorectal tumor (HT29) cells. Mol. Pharmacol. 39: 285–289.

Duvik, D.N. 1961. Protein granules of maize endosperm cells. Cereal Chem. 38: 374–385.

Farkas, G.L. 1982. Ribonucleases and ribonucleic acid breakdown. In: B. Parthier and D. Boulter (Eds.), Encyclopedia of Plant Physiology, Springer-Verlag, Berlin, pp. 224–262.

Felker, F.C., Peterson, D.M. and Nelson, O.N. 1985. Anatomy of immature grains of eight maternal effect shrunken endosperm barley mutants. Am. J. Bot. 72: 248–256.

Filipski, J., Leblanc, J., Youdale, T., Sikorska, M. and Walker, P.R. 1990. Periodicity of DNA folding in higher order chromatin. EMBO J. 9: 1319–1327.

Friedman, W.E. 1990. Double fertilization in *Ephedra*, a nonflowering plant: its bearing on the origin of angiosperms. Science 247: 951–954.

Friedman, W.E. 1992. Double fertilization in nonflowering seed plants and its relevance to the origin of flowering plants. Int. Rev. Cytol. 140: 319–355.

Fuchs, Y. and Lieberman, M. 1968. Effects of kinetin, IAA, and gibberellin on ethylene production, and their interactions in growth of seedlings. Plant Physiol. 42: 2029–2036.

Fukuda, H. and Komamin, A. 1983. Changes in the synthesis of RNA and protein during tracheary element differentiation in single cells isolated from the mesophyll of *Zinnia elegans*. Plant Cell Physiol. 24: 603–614.

Gallie, D.R. 1991. The cap and poly(A) tail function synergistically to regulate mRNA translational efficiency. Genes Dev. 5: 2108–2116.

Gallie, D.R., Caldwell, C. and Pitto, L. 1995. Heat shock disrupts cap and poly(A) tail function during translation and increases mRNA stability of introduced reporter mRNA. Plant Physiol.108: 1703–1713.

Gallie, D.R., Le, H., Caldwell, C., Tanguay, R.L., Hoang, N.X. and Browning, K.S. 1997. The phosphorylation state of translation initiation factors is regulated developmentally and following heat shock in wheat. J. Biol. Chem. 272: 1046–1053.

Gallie, D.R., Le, H., Tanguay, R.L. and Browning, K.S. 1998. Translation initiation factors are differentially regulated in cereals during development and following heat shock. Plant J. 14: 715–722.

Giroux, M.J., Boyer, C., Feix, G. and Hannah, C.L. 1994. Coordinated transcriptional regulation of storage product genes in the maize endosperm. Plant Physiol. 106: 713–722.

Glover, D.V., Crane, P.L., Misra, P.S. and Mertz, E.T. 1975. Genetics of endosperm mutants in maize as related to protein quality and quantity. In: High-Quality Protein Maize, Dowden, Hutchison and Ross, Stroudsburg, PA, p. 228.

Green, P.J. 1994. The ribonucleases of higher plants. Annu. Rev. Plant Physiol. Plant Mol. Biol. 45: 421–445.

Green, R. and Fluhr, R. 1995. UV-B-induced *PR-1* accumulation is mediated by active oxygen species. Plant Cell 7: 203–212.

Guan, L. and Scandalios, J.G. 1998. Two structurally similar maize cytosolic superoxide dismutase genes, *Sod4* and *Sod4A*, respond differentially to abscisic acid and high osmoticum. Plant Physiol. 117: 217–224.

Hattori, T., Vasil, V., Rosenkrans, L., Hannah, L.C., McCarty, D.R. and Vasil, I.K. 1992. The *viviparous-1* gene and abscisic acid activate the *C1* regulatory gene for anthocyanin biosynthesis during seed maturation in maize. Genes Dev. 6: 609–618.

He, C.-J., Morgan, P.W. and Drew, M.C. 1992. Enhanced sensitivity to ethylene in nitrogen- or phosphate-starved roots of *Zea mays* L. during aerenchyma formation. Plant Physiol. 98: 137–142.

He, C.-J., Morgan, P.W. and Drew, M.C. 1996. Transduction of an ethylene signal is required for cell death and lysis in the root cortex of maize during aerenchyma formation induced by hypoxia. Plant Physiol. 112: 463–472.

Hua, J. and Meyerowitz, E.M. 1998. Ethylene responses are negatively regulated by a receptor gene family in *Arabidopsis thaliana*. Cell 94: 261–271.

Hua, J., Sakai, S., Nourizadeh, S., Chen, Q.C., Bleeker, A.B., Ecker, J.R. and Meyerowitz, E.M. 1998. *EIN4* and *ERS2* are members of the putative ethylene receptor gene family in *Arabidopsis thaliana*. Plant Cell 10: 1321–1332.

Ingham, P.W. 1998. Trithorax and the regulation of homeotic gene expression in *Drosophila*: a historical perspective. Int. J. Dev. Biol. 42: 423–429.

Ingle, J., Beitz, D. and Hageman, R.H. 1965. Changes in composition during development and maturation of maize seeds. Plant Physiol. 40: 835–839.

Jabs, T. 1999. Reactive oxygen intermediates as mediators of programmed cell death in plants and animals. Biochem. Pharmacol. 57: 231–245.

Jabs, T., Dietrich, R.A. and Dangl, J.L. 1996. Initiation of runaway cell death in an *Arabidopsis* mutant by extracellular superoxide. Science 273: 1853–1856.

Jacobson, M.D. 1996. Reactive oxygen species and programmed cell death. Trends Biochem 21: 83–86.

Jacobson, M.D. and Raff, M.C. 1995. Programmed cell death and *Bcl-2* protection in very low oxygen. Nature 374: 814–816.

Jakobsen, K., Klemsdal, S., Aalen, R., Bosnes, M., Alexander, D. and Olsen, O.-A. 1989. Barley aleurone cell development: molecular cloning of aleurone-specific cDNAs from immature grains. Plant Mol. Biol. 12: 285–293.

Johari, R.P., Mehta, S.L. and Naik, M.S. 1977. Protein synthesis and changes in nucleic acids during grain development of sorghum. Phytochemistry 16: 19–24.

Jones, R.J. and Brenner, M.L. 1987. Distribution of abscisic acid in maize kernels during grain filling. Plant Physiol. 83: 905–909.

Kermode, A.R. and Bewley, J.D. 1985. The role of maturation drying in the transition from seed development to germination. I. Acquisition of desiccation-tolerance and germinability during development of *Ricinus communis* L. seeds. J. Exp. Bot. 36: 1906–1915.

Kerr, J.F.R., Gobe, G.C., Winterford, C.M. and Harmon, B.V. 1995. Anatomical methods in cell death. In: L.M. Schwartz and B.A. Osborne (Eds.), Cell Death Methods in Cell Biology, Academic Press, San Diego, pp. 1–27.

Koukalova, B., Kovarik, A., Fajkus, J. and Siroky, J. 1997. Chromatin fragmentation associated with apoptotic changes in tobacco cells exposed to cold stress. FEBS Lett. 414: 289–292.

Kowles, R.V. and Phillips, R.L. 1985. DNA amplification patterns in maize endosperm nuclei during kernel development. Proc. Natl. Acad. Sci. USA 82: 7010–7014.

Kowles, R.V. and Phillips, R.L. 1988. Endosperm development in maize. Int. Rev. Cytol. 112: 97–136.

Kowles, R.V., McMullen, M.D. and Phillips, R.L. 1986. Gene expression in developing maize kernels. In: J.C. Shannon, D.P. Knievel and C.D. Boyer (Eds), Regulation of Carbon and Nitrogen Reduction and Utilization in Maize, American Society of Plant Physiologists, Rockville, MD, pp. 189–206.

Kyle, D.J. and Styles, E.D. 1977. Development of aleurone and subaleurone layers in maize. Planta 137: 185–193.

Larkins, B.A. and Hurkman, W.J. 1978. Synthesis and deposition of zein in protein bodies of maize endosperm. Plant Physiol. 62: 256–263.

Lau, O.L. and Yang, S.F. 1973. Mechanism of a synergistic effect of kinetin on auxin-induced ethylene production: suppression of auxin conjugation. Plant Physiol. 51: 1011–1014.

Le, H., Browning, K.S. and Gallie, D.R. 1998. The phosphorylation state of the wheat translation initiation factors eIF4B, eIF4A, and eIF2 is differentially regulated during seed development and germination. J. Biol. Chem. 273: 20084–20089.

Le, H., Tanguay, R.L., Balasta, M.L., Wei, C.-C., Browning, K.S., Metz, A.M., Goss, D.J. and Gallie, D.R. 1997. The translation initiation factors eIFiso4G and eIF-4B interact with the poly(A)-binding protein to increase its RNA binding affinity. J. Biol. Chem. 272: 16247–16255.

Lee, L. and Tsai, C.Y. 1984. Zein synthesis in the embryo and endosperm of maize mutants. Biochem. Genet. 22: 729–737.

Lee, B., Martin, P. and Bangerth, F. 1989. The effect of sucrose on the levels of abscisic acid, indoleacetic acid and zeatin / zeatin-riboside in wheat ears growing in liquid culture. Physiol. Plant. 77: 73–80.

Levine, A., Tenhaken, R., Dixon, R. and Lamb, C. 1994. H_2O_2 from the oxidative burst orchestrates the plant hypersensitive disease resistance response. Cell 79: 583–593.

Levine, A., Pennell, R.I., Alvarez, M.E., Palmer, R. and Lamb, C. 1996. Calcium-mediated apoptosis in a plant hypersensitive disease resistance response. Curr. Biol. 6: 427–437.

Levy-Strumpf, N., Deiss, L.P., Berissi, H. and Kimchi, A. 1997. DAP-5, a novel homolog of eukaryotic translation initiation factor 4G isolated as a putative modulator of gamma interferon-induced programmed cell death. Mol. Cell. Biol. 17: 1615–1625.

Lopes, M.A. and Larkins, B.A. 1993. Endosperm origin, development, and function. Plant Cell 5: 1383–1399.

Lowe, J. and Nelson, O.E. 1947. Minature seed: a study in the development of a defective caryopsis in maize. Genetics 31: 525–533.

Lukaszewski, T.A. and Reid, M.S. 1989. Bulb type flower senescence. Acta. Hort. 261: 59–62.

Luo, M., Bilodeau, P., Koltunow, A., Dennis, E.S. and Peacock, W.J. 1999. Genes controlling fertilization-independent seed development in *Arabidopsis thaliana.* Proc. Natl. Acad. Sci. USA 96: 296–301.

Lur, H.-S. and Setter, T.L. 1993. Role of auxin in maize endosperm development. Plant Physiol. 103: 273–280.

Mangelsdorf, P.C. 1926. The genetics and morphology of some endosperm characters in maize. Conn. Exp. Stn. Bull. 279: 513–614.

Manzocchi, L.A., Daminati, M.G. and Gentinetta, E. 1980. Viable endosperm mutants in maize. II. Kernel weight, nitrogen and zein accumulation during endosperm development. Maydica 25: 199–210.

Marshall, S.W. 1987. Sweet Corn. In: S.A. Watson and P.E. Ramstad (Eds.), Corn: Chemistry and Technology, American Association of Cereal Chemists, St. Paul, MN, pp. 431–445.

Martin, A.C. 1946. The comparative internal morphology of seeds. Am. Midl. Naturalist 36: 513–660.

McCarty, D.R., Carson, C.B., Stinard, P.S. and Robertson, D.S. 1989. Molecular analysis of viviparous-1: an abscisic acid-insensitive mutant of maize. Plant Cell 1: 523–532.

McCarty, D.R., Hattori, T., Carson, C.B., Vasil, V., Lazar, M. and Vasil, I.K. 1991. The *viviparous-1* developmental gene of maize encodes a novel transcriptional activator. Cell 66: 895–905.

Mengel, K., Friedrich, B. and Judel, G.K. 1985. Effect of light intensity on the concentrations of phytohormones in developing wheat grains. J. Plant. Physiol. 120: 255–266.

Methot, N., Song, M.S. and Sonenberg, N. 1996. A region rich in aspartic acid, arginine, tyrosine, and glycine (DRYG) mediates eukaryotic initiation factor 4B (eIF4B) self- association and interaction with eIF3. Mol. Cell. Biol. 16: 5328–5334.

Miller, M.E. and Chourey, P.S. 1992. The maize invertase-deficient *minature-1* seed mutation is associated with aberrant pedicel and endosperm development. Plant Cell 4: 297–305.

Mittler, R., Shulaev, V. and Lam, E. 1995. Coordinate activation of programmed cell death and defense mechanisms in transgenic tobacco plants expressing a bacterial proton pump. Plant Cell 7: 29–42.

Mittler, R., Simon, L. and Lam, E. 1997. Pathogen-induced programmed cell death in tobacco. J. Cell. Sci. 110: 1333–1344.

Morgan, P.W. and Hall, W.C. 1962. Effect of 2,4-dichlorophenoxyacetic acid on the production of ethylene by cotton and grain sorghum. Plant Physiol. 15: 420–427.

Morrison, W.R. and Milligan, T.P. 1982. Lipids in maize starches. In: G. Inglett (Ed.), Maize: Recent Progress in Chemistry and Technology, Academic Press, New York, pp. 1–18.

Neuffer, M.G. and Sheridan, W.G. 1980. Defective kernel mutants of maize. I. Genetic and lethality studies. Genetics 95: 929–944.

Ohad, N. Margossian, L. Hsu, Y.-C., Williams, C. Repetti, P. And Fischer R.L. 1996. A mutation that allows endosperm development without fertilization. Proc. Natl. Acad. Sci. USA 93: 5319–5324.

Ohad, N. Yadegari, R., Margossian, L., Hannon, M., Michaeli, D., Harada, J.J., Goldberg, R.B. and Fischer, R.L. 1999. Mutations in *FIE*, a WD polycomb group gene, allow endosperm development without fertilization. Plant Cell 11: 407–415.

Olsen, O.-A., Potter, R.H. and Kalla, R. 1992. Histo-differentiation and molecular biology of developing cereal endosperm. Seed Sci. Res. 2: 117–131.

Ou-Lee, T.M. and Setter, T.L. 1985. Enzyme activities of starch and sucrose pathways and growth of apical and basal kernels. Plant Physiol. 79: 848–851.

Overmyer, K., Kangasjarvi, J.S., Kuittinen, T. and Saarma, M. 1998. Gene expression and cell death in ozone-exposed plants: is programmed cell death involved in ozone-sensitive *Arabidopsis* mutants? In: L.J. de Kok and I. Stulen (Eds.), Responses of Plant Metabolism to Air Pollution, Backhuys Publishers, Leiden, Netherlands, pp. 403–406.

Ozbun, J.L., Hawker, J.S., Greenberg, E., Lammel, C., Preiss, J. and Lee, E.Y.C. 1973. Starch synthase, phosphorylase, and UDP-Glc pyrophosphorylase in developing maize kernels. Plant Physiol. 51: 1–5.

Panavas, T., Walker, E.L. and Rubinstein, B. 1998. Possible involvement of abscisic acid in senescence of daylily petals. J. Exp. Bot. 49: 1987–1997.

Pastori, G.M. and del Rio, L.A. 1997. Natural senescence of pea leaves. An active oxygen-mediated function for peroxisomes. Plant Physiol. 113: 411–418.

Payton, S., Fray, R.G., Brown, S. and Grierson, D. 1996. Ethylene receptor expression is regulated during fruit ripening, flower senescence and abscission. Plant Mol. Biol. 31: 1227–1231.

Phillips, R.L., Wang, A.S. and Kowles, R.V. 1983. Molecular and developmental cytogenetics of gene multiplicity in maize. Stadler Symp. 15: 105–118.

Phillips, R.L., Kowles, R.V., McMullen, M., Enomoto, S. and Rubenstein, I. 1985. Developmentally timed changes in maize endosperm DNA. In: M. Freeling (Ed.), Plant Genetics, Alan R. Liss, New York, pp. 739–754.

Pirrotta, V. 1998. Polycombing the genome: PcG, trxG, and chromatin silencing. Cell 93: 333–336.

Pitto, L., Gallie, D.R. and Walbot, V. 1992. Role of the leader sequence during thermal repression of translation in maize, tobacco, and carrot protoplasts. Plant Physiol. 100: 1827–1833.

Polunovsky, V.A., Rosenwald, I.B., Tan, A.T., White, J., Chiang, L., Sonenberg, N. and Bitterman, P.B. 1996. Translational control of programmed cell death: eukaryotic translation initiation factor 4E blocks apoptosis in growth-factor-restricted fibroblasts with physiologically expressed or deregulated Myc. Mol. Cell. Biol. 16: 6573–6581.

Porat, R., Halevy, A.H., Serek, M. and Borochov, A. 1995. An increase in ethylene sensitivity following pollination is the initial event triggering an increase in ethylene production and enhanced senescence of *Phalaenopsis* orchid flowers. Physiol. Plant. 93: 778–784.

Prasad, T.K., Anderson, M.D., Martin, B.A. and Stewart, C.R. 1994. Evidence for chilling-induced oxidative stress in maize seedlings and a regulatory role for hydrogen peroxide. Plant Cell 6: 65–74.

Rademacher, W. and Graebe, J.E. 1984. Hormonal changes in developing kernels of two spring wheat varieties differing in storage capacity. Ber. Dtsch. Bot. Ges. 97: 167–181.

Ramage, R.T. and Crandall, C.L. 1981a. A proposed gene symbol for defective endosperm mutants that express *xenia*. Barley Genet. Newsl. 11: 30–31.

Ramage, R.T. and Crandall, C.L. 1981b. *Defective endosperm xenia (dex)* mutants. Barley Genet. Newsl. 11: 32–33.

Ray, B.K., Lawson, T.G., Kramer, J.C., Cladaras, M.H., Grifco, J.A., Abramson, R.D., Merrick, W.C. and Thach, R.E. 1985. ATP-dependent unwinding of messenger RNA structure by eukaryotic initiation factors. J. Biol. Chem. 260: 7651–7658.

Ray, S., Golden, T. and Ray, A. 1996. Maternal effects of the *short integument* mutation on embryo development in *Arabidopsis*. Dev. Biol. 180: 365–369.

Sandstrom, P.A., Tebbey, P.W., Van Cleave, S. and Buttke, T.M. 1994. Lipid hydroperoxides induce apoptosis in T cells displaying a HIV-associated glutathione peroxidase deficiency. J. Biol. Chem. 269: 798–801.

Sheridan, W.F. and Neuffer, M.G. 1980. Defective kernel mutants of maize. II. Morphology and embryo culture studies. Genetics 95: 945–960.

Srivastava, S.P., Kumar, K.U. and Kaufman, R.J. 1998. Phosphorylation of eukaryotic translation initiation factor 2 mediates apoptosis in response to activation of the double-stranded RNA-dependent protein kinase. J. Biol. Chem. 273: 2416–2423.

Stacy, R.A.P., Munthe, E., Steinum, T., Sharma, B. and Aalen, R.B. 1996. A peroxiredoxin antioxidant is encoded by dormancy-related gene, *Per1*, expressed during late development in the aleurone and embryo of barley grains. Plant Mol. Biol. 31: 1205–1216.

Streb, P. and Feierabend, J. 1996. Oxidative stress responses accompanying photoinactivation of catalase in NaCl-treated rye leaves. Bot. Acta. 109: 125–132.

Sugiyama, M. Ito, J., Aoyagi, S. and Fukuda, H. 2000. Endonuclease. Plant Mol. Biol, this issue.

Tarun, S.Z. and Sachs, A.B. 1996. Association of the yeast poly(A) tail binding protein with translation initiation factor eIF-4G. EMBO J. 15: 7168–7177.

Thelen, M.P. and Northcote, D.H. 1989. Identification and purification of a nuclease from *Zinnia elegans* L.: a potential marker for xylogenesis. Planta 179: 181–195.

Tomei, L.D. and Cope, F.O. 1991. Apoptosis: The Molecular Basis of Cell Death, Cold Spring Harbor Laboratory Press, Plainview, NY.

Tomei, L.D. and Cope, F.O. 1994. Apoptosis II: The Molecular Basis of Apoptosis in Disease, Current Communications in Cell and Molecular Biology, Cold Spring Harbor Laboratory Press, Plainview, NY.

Torti, G., Lombardi, L. and Manzocchi, L.A. 1984. Indole-3-acetic acid content in viable defective endosperm mutants of maize. Maydica 29: 335–343.

Torti, G., Manzocchi, L.A. and Salamini, F. 1986. Free and bound indole-acetic acid is low in the endosperm of the maize mutant *defective endosperm-B18*. Theor. Appl. Genet. 72: 602–605.

Traas, J., Hulskamp, M., Gendreau, E. and Hofte, H. 1998. Endoreduplication and development: rule without dividing? Curr. Opin. Plant Biol. 1: 498–503.

Tsai, C.Y., Salamini, F. and Nelson, O.E. 1970. Enzymes of carbohydrate metabolism in the developing endosperm of maize. Plant Physiol. 46: 299–306.

Tsai, C.Y., Larkins, B.A. and Glover, D.V. 1978. Interaction of the *opaque2* gene with starch- forming mutant genes of the synthesis of zein in maize endosperm. Biochem. Genet. 16: 883–896.

van Doorn, W.G. and Stead, A.D. 1994. The physiology of petal senescence which is not initiated by ethylene. In: R.J. Scott and A.D. Stead (Eds.), Molecular and Cellular Aspects of Plant Reproduction, Cambridge University Press, Cambridge, UK, pp. 239–254.

Vijayaraghavan, M.R. and Prabhakar, K. 1984. The endosperm. In: B.M. Johri (Ed.), Embryology of Angiosperms, Springer-Verlag, Berlin, pp. 319–376.

Wadsworth, G.J. and Scandalios, J.G. 1989. Differential expression of the maize catalase genes during kernel development: the role of steady-state mRNA levels. Dev. Genet. 10: 304–310.

Walker, P.R. and Sikorska, M. 1994. Endonuclease activities, chromatin structure, and DNA degradation in apoptosis. Biochem. Cell. Biol. 72: 615–623.

Walker, P.R., Smith, C., Youdale, T., Leblanc, J., Whitfield, J.F. and Sikorska, M. 1991. Topoisomerase II-reactive chemotherapeutic drugs induce apoptosis in thymocytes. Cancer Res. 51: 1078–1085.

Wang, H., Li, J., Bostock, R.M. and Gilchrist, D.G. 1996a. Apoptosis: a functional paradigm for programmed cell death induced by a host-selective phytotoxin and invoked during development. Plant Cell 8: 375–391.

Wang, H., Wu, H.M. and Cheung, A.Y. 1996b. Pollination induces mRNA poly(A) tail shortening and cell deterioration in flower transmitting tissue. Plant J. 9: 715–727.

Wang, M., Oppedijk, B.J., Lu, X., van Duijn, B. and Schilperoort, R.A. 1996c. Apoptosis in barley aleurone during germination and its inhibition by abscisic acid. Plant Mol. Biol. 32: 1125–1134.

Wang, M., Hoekstra, S., van Bergen, S., Lamers, G.E.M., Oppedijk, B.J., van der Heijden, M.W., de Priester, W. and Schilperoort, R.A. 1999. Apoptosis in developing anthers and the role of ABA in this process during androgenesis in *Hordeum vulgare* L. Plant Mol. Biol. 39: 489–501.

Webster, C., Gaut, R.L., Browning, K.S., Ravel, J.M. and Roberts, J.K.M. 1991. Hypoxia enhances phosphorylation of eukaryotic

initiation factor 4A in maize root tips. J. Biol. Chem. 266: 23341–23346.
Wilkinson, J.Q., Lanahan, M.B., Yen, H.-C., Giovannoni, J.J. and Klee, H.J. 1995. An ethylene-inducible component of signal transduction encoded by *never-ripe*. Science 270: 1807–1809.
Wilson, C.M. 1967. Purification of corn ribonuclease. J. Biol. Chem. 242: 2260–2263.
Wilson, C.M. 1968. Plant nucleases. I. Separation and purification of two ribonucleases and one nuclease from corn. Plant Physiol. 43: 1332–1338.
Wilson, C.M. 1975. Plant nucleases. Annu. Rev. Plant Physiol. 26: 187–208.
Wilson, C.M. 1980. Plant nucleases. VI. Genetic and developmental variability in ribonuclease activity in inbred and hybrid corn endosperms. Plant Physiol. 66: 119–125.
Wilson, C.M. 1982. Plant nucleases: biochemistry and development of multiple molecular forms. Isozymes 6: 33–54.
Wilson, C.M. and Alexander, D.E. 1967. Ribonuclease activity in normal and *opaque2* mutant endosperm of maize. Science 155: 1575.
Wyllie, A.H. 1980. Cell death: the significance of apoptosis. Int. Rev. Cytol. 68: 251–306.
Yeong-Biau, Y. and Yang, S.F. 1979. Auxin-induced ethylene production and its inhibition by aminoethoxyvinylglycine and cobalt ion. Plant Physiol. 64: 1074–1077.
Yoshii, H. and Imaseki, H. 1981. Biosynthesis of auxin-induced ethylene. Effects of indole-3-acetic acid, benzyladenine and abscisic acid on endogenous levels of 1-aminocyclopropane-1-carboxylic acid (ACC) and ACC synthase. Plant Cell Physiol. 22: 369–379.
Young, T.E. 1997. Pleiotropic effects of starch-deficient endosperm mutations on maize kernel and seedling development, Ph.D. dissertation, University of California, Riverside, CA.
Young, T.E. and Gallie, D.R. 1999. Analysis of programmed cell death in wheat endosperm reveals differences in endosperm development between cereals. Plant Mol. Biol. 39: 915–926.
Young, T.E. and Gallie, D.R. 2000. Regulation of programmed cell death in maize endosperm by abscisic acid. Plant Mol. Biol, this issue.
Young, T.E., Gallie, D.R. and DeMason, D.A. 1997. Ethylene-mediated programmed cell death during maize endosperm development of *Su* and *sh2* genotypes. Plant Physiol. 115: 737–751.
Zettl, U.K., Mix, E., Zielasek, J., Stangel, M., Hartung, H.P. and Gold, R. 1997. Apoptosis of myelin-reactive T cells induced by reactive oxygen and nitrogen intermediates in vitro. Cell Immunol. 178: 1–8.
Zhou, L., Jang, J.-C., Jones, T.L. and Sheen, J. 1998. Glucose and ethylene signal transduction crosstalk revealed by an *Arabidopsis* glucose-insensitive mutant. Proc. Natl. Acad. Sci. USA 95: 10294–10299.

Plant Molecular Biology **44:** 303–318, 2000.
E. Lam, H. Fukuda and J. Greenberg (Eds.), Programmed Cell Death in Higher Plants.

Regulation of cell death in flower petals

Bernard Rubinstein
Biology Department and Plant Biology Graduate Program, University of Massachusetts, Amherst, MA 01003-5810, USA (fax (413) 545-3243; e-mail: bernrub@bio.umass.edu)

Key words: flower petals, hormones, membrane permeability, pollination, programmed cell death, senescence, senescence-associated genes

Abstract

The often rapid and synchronous programmed death of petal cells provides a model system to study molecular aspects of organ senescence. The death of petal cells is preceded by a loss of membrane permeability, due in part to increases in reactive oxygen species that are in turn related to up-regulation of oxidative enzymes and to a decrease in activity of certain protective enzymes. The senescence process also consists of a loss of proteins caused by activation of various proteinases, a loss of nucleic acids as nucleases are activated, and enzyme-mediated alterations of carbohydrate polymers. Many of the genes for these senescence-associated enzymes have been cloned. In some flowers, the degradative changes of petal cells are initiated by ethylene; in others, abscisic acid may play a role. External factors such as pollination, drought and temperature stress also affect senescence, perhaps by interacting with hormones normally produced by the flowers. Signal transduction may involve G-proteins, calcium activity changes and the regulation of protein phosphorylation and dephosphorylation. The efficacy of the floral system as well as the research tools now available make it likely that important information will soon be added to our knowledge of the molecular mechanisms involved in petal cell death.

Introduction

Flowers are the structures responsible for sexual reproduction, and thus play a crucial role in the perpetuation of the earth's most dominant group of plants. Many flowers have evolved elaborate, complex corollas which serve to entice the pollinators that aid in reproduction, but the larger and more attractive the petals may be, the more resources are needed for their preservation (Ashman and Schoen, 1997) and the more likely the chances are for infection (Shykoff *et al.*, 1996). Indeed, flowers with the most showy petals often have the shortest life-times (Ashman and Schoen, 1994). The importance for the plant of strictly regulating the death of its flowers makes it obvious that a sensitive, tightly controlled program for cell death must exist. This review will describe some components of such a program for the petals. Besides being able to explain a vital stage of plant development, an understanding of the events relating to petal cell death may also provide cultural and bioengineering approaches that will have important economic and aesthetic ramifications. Petal senescence has been reviewed previously (e.g. Lesham *et al.*, 1986; Borochov and Woodson, 1989; Stead and van Doorn, 1994), so literature only over the past five years will be emphasized.

For the purposes of this review, the term 'senescence' will consist of those events that are part of a genetically based program leading to programmed cell death (PCD). 'Apoptosis', on the other hand, is a term used for animal cells to describe morphological changes such as condensation, shrinkage and blebbing of the nucleus, as well as decreases in cell size, and changes at the cell surface that allow phagocytosis (Jacobson *et al.*, 1997). While some of the modifications occurring during PCD in plants, especially DNA fragmentation, may resemble those in apoptotic animal cells (e.g. Mittler and Lam, 1995a, b; Wang *et al.*, 1996; Young *et al.*, 1997; O'Brien *et al.*, 1998), there is little evidence that the molecular and biochemical events responsible for the altered morphology and/or cell death are identical in plant and animal systems. In fact, it is likely that plant cells, surrounded by complex cell walls and containing large central vacuoles, have unique mechanisms for bringing about cell death. This may be especially true for the senescence of entire organs like petals, which has no analogous events in most animals.

The termination of a flower involves two sometimes overlapping mechanisms. In one, the petals abscise before the majority of their cells initiate a cell death program (e.g. Clark *et al.*, 1997). Abscission may occur before or during the mobilization of food reserves to other parts of the plant. In another mechanism, the petals are more persistent, so that cell deterioration and food remobilization occur while the petals are still part of the flower (Lesham *et al.*, 1986; Stead and van Doorn, 1994). This review will emphasize the PCD of petal cells that takes place before abscission.

Events associated with petal cell death

Changes in membranes

Structural and biophysical changes

For the cells of biological organisms, survival depends on the cell's ability to regulate its content of various nutrients by selective transport and to preserve internal compartmentalization. In other words, the structure and function of cellular membranes must be maintained. At the light and electron microscope levels, obvious changes occur at the membranes that portend cell death. In morning glory petals, for example, the tonoplast invaginates during senescence in a manner that indicates an endocytotic uptake of cytosolic components. Meanwhile the cytoplasm becomes only a thin layer, but later increases in volume at the expense of the vacuole. This is evidence of a loss of differential permeability (Matile and Winkenbach, 1971; Phillips and Kende, 1980).

In carnation petals, vesiculation in vacuolar and cytosolic compartments was observed before the senescence-associated climacteric evolution of ethylene. A smaller volume of cytoplasm along with vacuolar deposits are characteristic of senescent (climacteric) cells, and at later stages (post-climacteric), the membranes rupture and cytoplasmic debris is evident in the intracellular spaces (Smith *et al.*, 1992).

By the time the daylily flower opens, degeneration of organelles has occurred in the very thin layer of cytoplasm, and the tonoplast of epidermal cells appears to rupture. Autolysis of the cells is obvious 24 h after opening (Stead and Van Doorn, 1994). The cells of living sections of daylily petals exclude Evans Blue, a stain whose penetration indicates loss of membrane permeability, up to 12 h after flower opening, but the dye enters the cells 24 h after opening (Panavas *et al.*, 1998).

Biochemical and molecular changes

Along with structural alterations, measurements of physical and biochemical properties of the membranes indicate that important changes occur before the leakage of nutrients is evident (Lesham *et al.*, 1986; Thompson, 1988; Paliyath and Droillard, 1992; Thompson *et al.*, 1997). For example, the fluidity of membranes from a variety of different flower petals decreases, usually before senescence becomes obvious (Paliyath and Droillard, 1992; Thompson *et al.*, 1997). There is also a lateral phase separation of lipids in the membrane bilayer detected in carnations by X-ray diffraction (Faragher *et al.*, 1987) and in roses by ^{2}H-NMR (Itzhaki *et al.*, 1995), resulting in the formation of gel phase lipid. This, in turn, probably leads to the loss of water from the cells of many types of flower petals during the later stages of senescence. Evidence is accumulating that the phase changes are related, at least in part, to the inability of the membrane to remove metabolites by a blebbing of lipid-protein particles that occurs during normal membrane turnover (Thompson *et al.*, 1997) . The cause for this impairment is under investigation, and its outcome may have important implications for understanding PCD in petals as well as in other systems.

Biochemical changes at the membrane include simultaneous declines in all classes of phospholipids (PL) and increases in neutral lipids (Paliyath and Droillard, 1992). The sterol/PL ratio also increases, which may contribute to the decreased fluidity mentioned above. The reasons for these alterations in PL content are related in part to greater activities of phospholipases and acyl hydrolases, which would not only lower PL, but would also lead to increases in neutral lipids (Paliyath and Droillard, 1992). In rose petals, however, where 50% of the total membrane PL is phosphatidyl choline (PC), a study of CDP-choline phosphotransferase, the enzyme catalyzing the ultimate step in PC biosynthesis, indicates that the amount of enzyme and its V_{max} decrease during senescence (Itzhaki *et al.*, 1998). In daylily, too, the data suggest that phospholipid synthesis is blocked early in senescence (Bieleski and Reid, 1992). An mRNA coding for

Table 1. Some cDNA clones up-regulated during petal senescence. The genes enumerated below were identified by an increase in steady-state levels of their mRNA on gel blots or the RNase protection assay. It is not meant to be a comprehensive list. In some cases, there is no information on whether the gene product is translated.

Presumed function	Gene name	Accession number	Source	Putative identity	Reference
Related to membrane	SR8	M64268	carnation	glutathione *S*-transferase	Meyer *et al.*, 1991
(lipid	PAC01	U66299	*Phalaenopsis*	acyl CoA oxidase	Do and Huang, 1997
metabolism	DSA3	AF082028	daylily	in-chain fatty acid hydrolase	Panavas *et al.*, 1999
and/or oxidation)	DSA12	AF082032	daylily	allene oxide synthase	Panavas *et al.*, 1999
Protein	SEN11	U12637	daylily	cysteine proteinase	Valpuesta *et al.*, 1995
degradation	SEN102	X74406	daylily	cysteine proteinase	Valpuesta *et al.*, 1995
	DCCP1	U17135	carnation	cysteine proteinase	Jones *et al.*, 1995
	DSA4	AF082029	daylily	aspartic proteinase	Panavas *et al.*, 1999
Nucleic acid degradation	DSA6	AF082031	daylily	S1-type nuclease	Panavas *et al.*, 1999
Ethylene-	ACC2	X59145	tomato	ACC synthase	Rottman *et al.*, 1991
related	CAR ACC3	M66619	carnation	ACC synthase	Park *et al.*, 1992
	CPG217	U64804	petunia	ACC synthase	Michael *et al.*, 1993
	CPG214	Z18952	carnation	ACC synthase	Michael *et al.*, 1993
	CPG320	–	carnation	ACC oxidase	Michael *et al.*, 1993
	SR120	–	carnation	ACC oxidase	Woodson, 1994
	ACO1	L21976	petunia	ACC oxidase	Tang *et al.*, 1994
	ERS	AJ005829	pea	ethylene response sensor	Orzaez *et al.*, 1999
Other	SR132	L11598	carnation	carboxy PEP mutase	Wang *et al.*, 1993
	SR5	–	carnation	β-glucosidase	Woodson, 1994
	SR12	–	carnation	β-galactosidase	Woodson, 1994
	TORE-NIA-1	AB010953	*Torenia*	unknown	Aida *et al.*, 1998
	DAS15	AF0082033	daylily	fatty acid elongase	Panavas *et al.*, 1999

a carboxyphospho*enol*pyruvate/phospho*enol*pyruvate mutase is up-regulated in carnation petals during senescence (Table 1); the enzyme could be involved in membrane turnover (Wang *et al.*, 1993). Thus, both a down-regulation of enzymes responsible for PL synthesis and an up-regulation of hydrolytic enzymes may cause the membrane breakdown that is an important component of PCD.

Another senescence-associated event that leads to a loss of membrane permeability is the oxidation of existing membrane components (Figure 1). For example, lipid peroxidation as measured by thiobarbituric acid reactive substances (TBARS) increases during senescence in carnation (Sylvestri *et al.*, 1989; Bartoli *et al.*, 1995) and daylily (Panavas and Rubinstein, 1998) petals. Peroxidation may occur in part by the action of lipoxygenase (LOX), which oxidizes fatty acids liberated from membranes (Siedow, 1991). LOX activity increases before senescence becomes obvious in carnation petals (Sylvestre *et al.*, 1989), and in daylily, the specific activity of LOX increased before the flowers even open, which is about the same time that increases of TBARS occur (Panavas and Rubinstein, 1998).

In two orchid species, however, even though linoleic and linolenic acids, two substrates of LOX, promoted senescence, no increases in LOX specific activity were detected over time, and inhibitors of LOX were without effect (Porat *et al.*, 1995). So, even though the peroxidation of lipids that may result from LOX activity precedes loss of membrane integrity, the activity of lipid oxidases other than LOX may be limiting factors for membrane degradation in petals of these species of orchid.

Two cDNA clones from daylily have been sequenced whose gene products may play a role in oxidizing membrane lipids and whose message levels are up-regulated prior to or during senescence (Table 1) (Panavas *et al.*, 1999). One clone, detected mainly in petals, shows the greatest similarity to an in-chain

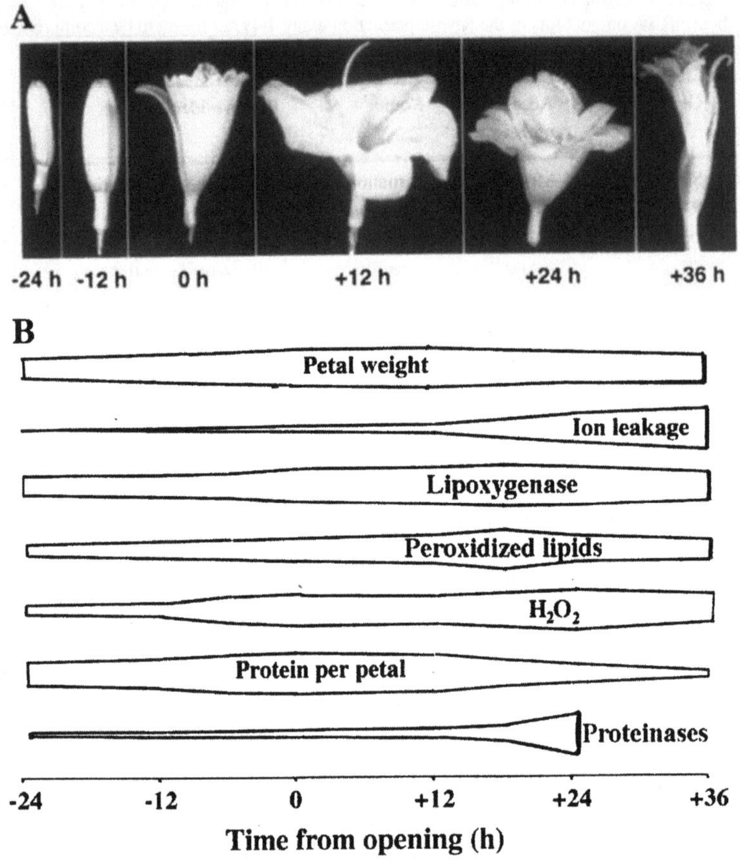

Figure 1. Characteristics of the progression of senescence and programmed cell death in daylily petals. A. Petal growth (−24 h to 0 h), flower opening (0 h), and visible degradation of the petals (+24 h). Maximum size of the petals is 5 cm and opening occurs at 21:00. B. Representation of some features that change during senescence of daylily petals. Data shown are a per cent of maximum, indicated by the widest portion of each bar.

fatty acid hydroxylase bound to cytochrome P450. The translated product of this mRNA may modify fatty acids, leading to their degradation (Cabello-Hurtado *et al.*, 1998). The other clone is highly similar to an allene oxide synthase, which converts fatty acid hydroperoxides to allene epoxides, and eventually results in molecules that may have signaling capabilities (Song and Brash, 1991). The increase in peroxidized lipids already described makes it likely that substrates would be available for this enzyme. No data are available as to whether these messages are translated in daylily petals in proportion to their abundance.

Reactive oxygen species (ROS) have been linked to stress-induced and normal death of animal and plant cells, including petal cells (Borochov and Woodson, 1989). In fact, ion leakage of daylily petals does not occur under non-lethal conditions of hypoxia (Panavas and Rubinstein, 1998), suggesting that the oxygen is required for reactions leading to ROS, as has been proposed for leaf senescence of tobacco (Mittler *et al.* 1996) and oats (Trippi, 1985). ROS may also be a by-product of LOX activity (Siedow, 1991), which, as mentioned above, increases during senescence of carnation and daylily petals (Panavas and Rubinstein, 1998). Furthermore, the ROS-generating herbicide paraquat hastens the appearance of ion leakage in whole carnation flowers (Bartoli *et al.*, 1996).

Highly reactive oxygen radicals can be formed from H_2O_2 during various oxidase reactions (Halliwell, 1989). In daylily petals, exogenous H_2O_2 stimulates ion leakage, and endogenous levels of H_2O_2 increase along with TBARS before flower opening. H_2O_2 also increases in the petals after treatments that accelerate senescence, and it decreases when senescence is retarded by the antioxidant sodium benzoate (Panavas and Rubinstein, 1998).

Since higher levels of endogenous H_2O_2 would help stimulate senescence, it is important to determine the activity of enzymes that serve to regulate levels of H_2O_2. One well-known reaction involves reduction of H_2O_2 by ascorbate peroxidase (APX) and the consequent oxidation of ascorbate to dehydroascorbate (Asada, 1992). In carnation petals, APX actually increases during senescence (Bartoli *et al.*, 1995), but in daylily petals, APX activity decreases sharply, especially as the flower opens, which is about 18 h before cell deterioration becomes obvious (Panavas and Rubinstein, 1998).

Another enzyme that can lower H_2O_2 levels is catalase (CAT), which converts H_2O_2 to H_2O and O_2 (Scandalius, 1993). Overexpression of the gene for this enzyme protects leaves against ROS (Zelitch *et al.*, 1991) and CAT-deficient plants are more sensitive to a variety of stresses (Willekins *et al.*, 1997). In carnation, CAT activity increases during flower aging and then remains unchanged (Bartoli *et al.*, 1995), but in daylily, the specific activity of the only form of CAT detected on activity gels decreases steadily from about 6 h before flower opening. When senescence is induced prematurely, CAT activity decreases earlier (Panavas and Rubinstein, 1998).

Superoxide dismutase (SOD) may also protect against buildup of ROS, but because the product of SOD is H_2O_2, an effective scavenging system must be present if cell damage is to be avoided. SOD decreases in activity in carnation petals (Sylvestre *et al.*, 1989) and increases in daylily (Panavas and Rubinstein, 1998), although in both cases the changes occur rather late in the progression of senescence. Since both CAT and APX decrease when daylily flowers open, the H_2O_2 build-up observed in these petals may in part be a result of increased SOD activity. Thus, in daylily, at least, SOD may be hastening cell death.

Peroxidase activity uses H_2O_2 as a substrate for several reactions and its specific activity increases in both carnation (Bartoli *et al.*, 1995) and daylily (Panavas and Rubinstein, 1998) during senescence. The transcript for an enzyme that detoxifies peroxidized lipids, glutathione *S*-transferase (GST) (Meyer *et al.* 1991), and the corresponding enzyme activity (Sylvestri, 1989) is also up-regulated in carnation petals (Table 1). These data suggest, however, that POX and GST activities are responding to conditions of oxidative stress that result from H_2O_2 accumulations, and are thus not directly responsible for cell death.

Because lipid metabolism and ROS are so important to PCD, one must consider a role for peroxisomes. These organelles are the site of lipid breakdown and can produce O_2^- and H_2O_2, especially if protective enzymes are absent. Furthermore, as these organelles are converted to glyoxysomes, a rapid oxidation of lipids would ensue. There is an increase in numbers of peroxisomes during senescence of carnation petals (del Rio *et al.*, 1996). Recently, a cDNA up-regulated during orchid petal senescence has been cloned that appears to code for a peroxisomal acyl-CoA oxidase (Table 1) (Do and Huang, 1997). This particular enzyme may not be a cause of PCD, but it could serve to oxidize the fatty acids resulting from lipid breakdown at the membranes.

Natural antioxidants, such as ascorbate, glutathione and α-tocopherol, also are present in flower petals. All of these substances decline in carnation petals, although α-tocopherol decreases only after visual symptoms of senescence appear (Bartoli *et al.*, 1997). In daylily, ascorbate drops by 50% even before the flowers open (Panavas and Rubinstein, 1998). Inhibiting senescence of carnation petals with an ethylene synthesis inhibitor maintained higher levels of α-tocopherol and glutathione, and both antioxidants decreased when senescence was accelerated by ethylene or paraquat (Bartoli *et al.*, 1996). Thus, with some exceptions (Bartoli *et al.*, 1997), α-tocopherol levels are correlated with changes occurring later in senescence, but in both carnation and daylily, the decrease in ascorbate precedes many of the parameters associated with senescence. However, it is not possible to determine the involvement of antioxidants or the protective enzymes mentioned above in PCD until more information becomes available about their localization in the cell and the degree of increase or decrease in activity that is needed to affect a particular cellular process.

Membrane proteins may also play a role in petal senescence. Membrane protein and the content of thiol groups decline during aging of carnation petals (Borochov and Woodson, 1989). Furthermore, in older compared to non-senescent petals, there is a large decrease in vanadate-sensitive ATPase activity, presumably the plasma membrane H^+ pump. But smaller decreases also occur in activities of cytochrome *c* oxidase, cytochrome *c* reductase and the nitrate-sensitive ATPase, which is likely the tonoplast H^+ pump (Beja-Tal and Borochov, 1994). Taken as a whole, the data indicate that the membrane proteins mediating transport and redox reactions decrease in activity during aging. The question remains, however, if these decreases in activity are causally related to PCD or are just a result of previously triggered degradative processes.

Loss of cellular proteins

Evidence for protein degradation

A common feature of PCD is the loss of protein during organ senescence (Smart, 1994; Callis, 1995). The breakdown products of these proteins are then transported from the senescing organ to other portions of the plant. In fact, after the daylily flower opens, it changes abruptly from a sink to a source for transported materials, and carbohydrates and amino acids continue to be exported many hours after the flower has wilted (Bieleski, 1995).

Protein synthesis may decline after flower opening (Woodson and Handa, 1987), but separation of petal proteins by PAGE reveals a decrease in higher-molecular-weight proteins during senescence while lower-molecular-weight proteins increase (Woodson and Handa, 1995; Courtney *et al.*, 1994). The decrease in protein levels in daylily petals appears to occur in both membrane-bound proteins and those in the soluble fraction (Stephenson and Rubinstein, 1998) and, in the case of carnations, in the chloroplast (Mayak *et al.*, 1998).

Proteinase activity during PCD

The preceding evidence suggests that a range of different proteinases may be active simultaneously at multiple sites in the cell (Figure 1). Indeed, an increase in the activity of several classes of proteinases is detected using different inhibitors and substrates as criteria, and there are numerous proteinase isoforms within the classes (Stephenson and Rubinstein, 1998). The appearance of most of the proteinase activity is prevented by the protein synthesis inhibitor cycloheximide (CH) (Stephenson and Rubinstein, 1998), suggesting that the proteinases are newly translated during senescence and/or that they are activated by proteins that are newly translated. It is interesting that in morning glory, the drop in protein level in the petals is not correlated with an increase in proteinase activity (Matile and Winkenbach, 1971). In this case, the proteinases may be released from the vacuolar compartment as the differential permeability of the tonoplast is lost.

Several senescence-induced cDNAs have been sequenced whose identities suggests close homology to cloned proteinase genes (Table 1). For example, a cysteine proteinase was cloned from carnation by amplifying a specific cDNA by PCR (Jones *et al.*, 1995). RNA gel blot analyses indicate that the transcript increases after the pollination-induced burst of ethyene production that leads to senescence. Furthermore, appearance of the transcript is stimulated by adding ethylene to pre-senescent petals, and the ethylene action inhibitor NBD prevents the appearance of the transcript (Jones *et al.*, 1995).

Differential screening of a cDNA library yields two cDNAs from daylily petals whose derived amino acid sequences show a strong homology to cysteine proteinases (Valpuesta *et al.*, 1995; Guerrero *et al.*, 1998) (Table 1). They appear to have an ER retention signal and an ERFNIN sequence that is consistent with activation by cleavage (Guerrero *et al.*, 1998). Northern blot analysis indicates that message levels have two peaks, one at 12 h and one at 19 h after flower opening. These transcripts were also found in leaves, but the mRNAs decreased at senescence (Guerrero *et al.*, 1998).

Certain cysteine proteinases, collectively called caspases, are enzymes leading to apoptosis in animal cells, regardless of the initiating factor or the nature of the signal transduction pathway (Cryns and Yuan, 1998). Caspase-like activity is detected in tobacco leaves and caspase-specific inhibitors prevent the hypersensitive response (del Pozo and Lam, 1998). But a study using caspase-specific inhibitors on daylily petals showed no effects on rates of ion leakage (Stephenson and Rubinstein, unpublished results).

A cDNA for a putative aspartic proteinase with several family members has been cloned from daylily petals (Panavas *et al.*, 1999) (Table 1). The steady-state level of the message as determined by the RNase protection assay increases steadily from flower opening and the message is present only at much smaller levels in roots and leaves; in the latter, there is no correlation with senescence. Perhaps aspartic proteinases, whose message levels also increase in senescing leaves (Buchanan-Wollaston, 1997; Griffiths *et al.*, 1997), are involved in enzyme precursor processing as they are in some other plant systems (Mutlu and Gal, 1999). However, it remains to be shown if aspartic proteinase activity increases during petal senescence.

The ubiquitin-proteosome pathway

The ubiquitin-proteosome pathway is a major cytoplasmic proteolytic system for the degradation of short-lived and abnormal proteins (Vierstra, 1993). In plants, changes in the level of ubiquitinated proteins have been associated with cell death during the development of vascular tissue (Bachmeier *et al.*, 1990; Stephenson *et al.*, 1996; Woffenden *et al.*, 1998) as well as during senescence of leaves (Garbarino *et al.*, 1995; Pinedo *et al.*, 1996). In immunoblots of daylily petals, protein-ubiquitin conjugates were detected, some of which increased and some decreased during senescence. RNA gel blots revealed several ubiquitin transcripts that either increased, decreased or stayed unchanged during flower opening and senescence (Courtney *et al.*, 1994). Analysis in daylily of enzymatic components of the ubiquitin-proteosome pathway (ubiquitin-activating enzyme, ubiquitin-conjugating enzymes, a subunit of the multiubiquitin-binding protein) show detectable but steadily decreasing quantities during senescence when expressed on a per petal basis (Stephenson and Rubinstein, 1998). Because proteinase activity increases during senescence, but banding patterns on PAGE gels remain sharp and well defined, the ubiquitin-proteosome system may be efficiently breaking down the cleaved proteins to amino acids rather than initiating senescence. However, inhibitors of proteosome activity prevent cell death leading to tracheid formation (Woffenden *et al.*, 1998) and delay the ion leakage associated with senescence in daylily petals (Stephenson and Rubinstein, 1998). This implies that there may be a causal role for the ubiquitin-proteosome system in petal senescence.

Losses of nucleic acids: evidence for changes in nucleic acids

A decrease in extractable nucleic acids is a common feature of flower senescence (Lesham, 1986) as it is in other senescing organs (Smart, 1994). Furthermore, degradation of nuclear DNA is observed in apoptotic animal cells (Jacobson *et al.*, 1997) as well as in dying plant cells (Mittler and Lam, 1995a, b; Wang *et al.*, 1996; Young *et al.*, 1997; Orzaez and Granell, 1997; O'Brien *et al.*, 1998), including daylily petals (T. Panavas, R. LeVangie, J. Mistler, P.D. Reid and B. Rubinstein, in preparation). These observations make it likely that nuclease activities play important roles in PCD.

Indeed, the up-regulation of enzymes degrading RNA and DNA seems to be a common feature of the later stages of senescence in plants (Lesham, 1986; Green, 1994). For example, endonuclease activity increases in tobacco leaves during the hypersensitive response (HR) (Mittler and Lam, 1995a, b), and the activities of several different RNases have been detected in senescing wheat leaves (Blank and McKeon, 1991) and in petals of morning glory (Lesham *et al.*, 1986) and daylily (Panavas and Rubinstein, 1998). Furthermore, senescence-associated RNases have been cloned from leaves and petals of *Arabidopsis* (Taylor *et al.*, 1993) and tomato (Lers *et al.*, 1998). And a cDNA has been cloned from daylily petals, with similarity to fungal S1- and P-type endonucleases that degrade both single-stranded DNA and RNA (Table 1). The steady-state message level for this putative nuclease increases at flower opening and continues to increase during senescence. There is only a single copy of the gene and the message is petal-specific (Panavas *et al.*, 1999).

Carbohydrate/wall polymer alterations

Observations concerning the structures of petal cells have correlated loss of dry weight during senescence with reduced size of starch bodies as well as declines in content of starch (Lesham *et al.*, 1986). Furthermore, the levels of mRNA for two putative carbohydrate hydrolases, β-glucosidase and β-galactosidase, increase during senescence of carnation petals (Woodson, 1994) (Table 1). In daylily, carbohydrates are the main product transported from the petals (Bieleski, 1995), with 95% of the soluble carbohydrate lost after flower opening (Bieleski, 1993).

Interestingly, petal cells often undergo profound changes in size and shape prior to and during senescence, indicating that constituents of the wall are being altered. Morning glory columnar cells become rounded and then the wall is reduced in thickness and protrudes into the nearby lumen (Phillips and Kende, 1980). The tightly packed circular cells in the cortex of daylily petals assume a reticulate pattern due to the formation of irregular cell shapes and the decrease of cell-to-cell attachments (Stead and van Doorn, 1994; Panavas *et al.*, 1998). There are also decreases in wall constituents in carnation (de Vetten and Huber, 1990), and significant variations in wall-based enzymes prior to flower opening and senescence of daylily (Panavas *et al.*, 1998). A causal role in senescence for these wall changes, if any, remains unknown, but perhaps an intact, more rigid cell wall helps to maintain membrane structure (Paliyath and Droillard, 1992).

Signals for senescence-associated events

Hormones

Ethylene

Senescence of leaves is often associated with ethylene (Smart, 1994), as is flower senescence of many species, including commercially important ones (Stead and van Doorn, 1994). Thus, additions of ethylene or its precursor, ACC, should initiate prematurely the same anatomical, biophysical, biochemical and molecular events that occur during normal senescence (e.g. Borochov and Woodson, 1989). Another characteristic of ethylene-sensitive flowers is an autocatalytic evolution of this hormone before or during senescence (van Altvorst and Bovy, 1995). In that case, inhibitors of ethylene synthesis, such as amino oxyacetic acid (AOA) or norbornidiene (NBD), or compounds that interfere with ethylene binding, such as CO_2 or silver thiosulfate (STS), greatly prolong the life of ethylene-sensitive flowers (van Altvorst and Bovy, 1995). There are also other substances that delay senescence by inhibiting ethylene production (Midoh *et al.*, 1996; Lee *et al.*, 1997; Podd and van Staden, 1998) or ethylene action (Serek *et al.*, 1995), leading to cultural practices that substantially prolong the life of economically valuable flowers.

The increased ethylene production may be due to higher amounts of mRNAs for enzymes responsible for ethylene synthesis. cDNA clones from preclimacteric carnation petals have a deduced amino acid sequence with a high homology to ACC synthase (Michael *et al.*, 1993; Rottmann *et al.*, 1991; van Altvorst and Bovy, 1995; Jones and Woodson, 1999), the enzyme that converts SAM to ACC, the precursor of ethylene (Table 1). Another clone from petal material is homologous to ACC oxidase, which converts ACC to ethylene (Michael *et al.*, 1993; van Altvoorst and Bovy, 1995) (Table 1). RNA gel blot analyses show that the steady-state message levels for ACC synthase transcripts are tissue-specific (Jones and Woodson, 1999) and increase most strikingly before the rise in ethylene evolution (Woodson *et al.*, 1992; Oraez *et al.*, 1999).

When an ACC oxidase cDNA fragment is inserted in the antisense orientation in carnation, one transformant yields considerably less ethylene and petal senescence is significantly delayed (Michael *et al.*, 1993). Flower longevity is extended up to five days in transformants of *Torenia* when a fragment of the ACC oxidase gene is inserted into the genome in either the sense or antisense orientation. Transcription of ACC oxidase, as determined by RNA gel blots, is inhibited by both constructs, and ethylene evolution is halved (Aida *et al.*, 1998); the inhibition of transcription in the sense orientation is probably due to co-suppression. Finally, flower senescence of tomato is stimulated when ACC synthase is over-expressed (Lanahan *et al.*, 1994). All of these data suggest that ethylene is an essential inducer of petal senescence in certain species, and that ethylene evolution is regulated by the level of mRNA for enzymes that result in synthesis of the hormone.

There is also evidence that petals become more sensitive to ethylene as they age (Lawton *et al.*, 1990). A mechanism for this response is being studied using clones for the putative ethylene-response sensor, i.e. an ethylene receptor (Table 1). The steady-state message level for this gene in pea petals increases along with ethylene production, and the amounts of the mRNA are reduced in the presence of STS, the inhibitor of ethylene binding (Orzaez *et al.*, 1999). The mRNA for the ethylene receptor in *Arabidopsis* is not detected in 'pre-senescent' or 'late senescent' flowers, but is found in 'early senescent' flowers (Payton *et al.*, 1996). These data imply that the ethylene response receptor may be limiting for senescence and that the amount of this protein is related to the level of its mRNA, which may be regulated in part by the presence of ethylene.

Abscisic acid

The hormone abscisic acid (ABA) has been implicated in leaf senescence (Smart, 1994). Additions of ABA to certain flowers hasten flower senescence (Borochov and Woodson, 1989; Vardi and Mayak, 1989), but in roses a relationship between this hormone and senescence is not strong. ABA levels determined by radioimmunoassay decreased for about two days after cutting the flowers and then increased about 7 days later, but this was at the end of the vase life. In flowers still attached to the plant, a similar pattern was recorded: ABA levels fell continuously until just before flower abscission and then they increased (Le Page-Degivry *et al.*, 1991). However, a correlation between rose longevity and ABA content was suggested for flowers responding to light of different qualities (Garello *et al.*, 1995).

In carnation flowers, ABA levels increase before the losses of fresh weight that are usually associated with senescence, and the effects of ABA are delayed by pretreatment with an inhibitor of ethylene binding. ABA applications also lead to early increases in ethylene evolution and a greater sensitivity to ethylene, which may account for the PCD that follows (Borochov and Woodson, 1989).

Unlike carnation petals mentioned above, where a relationship between endogenous ABA and senescence may be due to an effect on ethylene production and sensitivity, the senescence of other flowers (e.g. Amaryllidaceae, Liliaceae, Iridaceae and Asteraceae) does not appear to be regulated by ethylene (Woltering and van Doorn, 1988; van Doorn, 1994). In daylily, which is an example of an ethylene-insensitive flower, ABA addition hastens senescence-associated events, such as ion leakage, lipid peroxidation, and both the overall activities and patterns of proteinases and nucleases on activity gels. This similarity also exists for mRNA levels, in which 80% of the cDNA bands on differential display resemble those that increase or decrease during normal flower development and senescence, but after ABA treatment, the bands appear (or disappear) 24 h earlier than normal (Panavas *et al.*, 1998). Furthermore, five of six cDNAs cloned from the differential display are prematurely up-regulated 3- to 45-fold by ABA (Panavas *et al.*, 1999).

Endogenous ABA levels in daylily petals increase slowly from 24 h before flower opening and then more rapidly starting 12 h before opening. When the internal ABA level was increased by slowing the uptake of water into developing petals, all of the senescence-associated parameters mentioned above, including many of the same cDNAs, increased prematurely. An elevated level of endogenous ABA caused by a water stress to petals has been reported previously (Orlandini, 1991; Vardi and Mayak, 1989), and this treatment also leads to a stimulation of senescence in petunia (Vardi and Mayak, 1989). There are no known ABA mutants for ephemeral flowers, regardless of whether they are sensitive to ethylene, so precise information cannot be obtained about the role of ABA in PCD, but for daylily, at least, the evidence suggests that endogenously produced ABA is important for the progression of senescence.

Other hormones

The effects of auxin and cytokinin on petal senescence have been reviewed by Borochov and Woodson (1989). In general, cytokinins appear to inhibit senescence of petals, perhaps by acting to prevent ethylene synthesis, or by decreasing the sensitivity of the cells to ethylene. Applications of auxins, which are known to enhance ethylene production, stimulate senescence of some ethylene-sensitive flowers (Stead, 1992). In daylily, exogenous auxin delays petal senescence, and benzyl adenine, gibberellic acid and brassinolides are without a detectable effect (Panavas and Rubinstein, unpublished data).

Environmental effects

Pollination

An example of a very striking response of flower petals to an environmental stimulus is the senescence and cell death caused by pollination (Stead, 1992; Woltering *et al.*, 1994; O'Neill, 1997; van Doorn, 1997). Senescence is accelerated by pollination in at least 60 genera, most of which are suspected of being sensitive to ethylene (van Doorn, 1997; O'Neill, 1997). The response in some orchids is quite extreme. In *Phalaenopsis*, the flowers may live for as long as 3 months, but they begin to die within one day after pollination (Halevy, 1998). Since ethylene initiates petal senescence, but no ethylene is evolved by the petals themselves just after pollination, one can ask what is the nature of the communication that must exist between the pollinated stigma and the corolla tissues.

The signaling results in ethylene evolution, but this need not be restricted to growth of the pollen since tube growth does not occur until five days after senescence begins in *Phalaenopsis* (Halevy, 1998). In carnation petals, the signal also does not arise from pollen tube growth *per se*, but from some aspect of the compatible reaction between pollen and stigma (Larsen *et al.*, 1995). However, tube growth may be important in petunia. In this flower, a burst of ethylene occurs regardless of whether the pollen is compatible or incompatible, which may be due to ACC in the pollen grain. Senescence, however, occurs after a second burst of ethylene, which is seen only when compatible pollen is applied (Singh *et al.*, 1992).

RNA gel blot analysis of orchid petals after pollination shows high levels of an mRNA for ACC oxidase, the ethylene-forming enzyme, but no ACC synthase mRNA is detected. Because the mRNAs for both enzymes are present in other parts of the flower, a model has been proposed in which pollination results in the transport of ACC, the precursor of ethylene, to the petals with the subsequent production of senescence-inducing quantities of ethylene (O'Neil *et al.*, 1993; O'Neill, 1998). This model is supported by the detection from pollen diffusates of ACC along with four other compounds that induce petal senescence (Porat *et al.*, 1998). An ACC synthase gene was subsequently characterized that was up-regulated by pollination and auxin in the stigma. This further expands the model by implicating ethylene as well as ACC as a mobile factor leading to ethylene-induced cell death of perianth tissues (Bui and O'Neill, 1998).

In contrast to these results, there is no detectable transport of radiolabeled ACC in *Cymbidium* or petunia styles (Woltering *et al.*, 1995, 1997). On the basis of these and other experiments, the authors conclude that ethylene is the mobile signal in the orchid (Woltering *et al.*, 1995), but the signal in petunia is unknown (Woltering *et al.*, 1997).

There is also an ethylene-independent increase in sensitivity to the hormone after pollination, quite apart from an increase in ethylene production (Whitehead, 1994). This sensitivity change may be mediated by short-chain fatty acids, which not only mimic the senescence-inducing effects of pollination in carnation (Whitehead and Vasiljevic, 1993) and *Phalaenopsis* (Halevy and Porat, 1996), but are also found to increase a few hours after pollination (Halevy and Porat, 1996).

It appears, then, that ethylene is an important component of the pollination-induced mechanism leading to petal cell death, but the signal transduction pathway depends on the species being investigated. The effect of pollination may be due both to an increase in ethylene production and to an increase in sensitivity to ethylene. It would be interesting to learn if pollination-induced changes in sensitivity to ethylene also involve increases in the ethylene receptor. This is an ethylene-related phenomenon in pea flowers, but senescence of this flower is not accelerated by pollination (Orzaez *et al.*, 1999).

Abiotic stresses

Petal senescence is often hastened by wounding of the flower parts. This damage results in the evolution of ethylene which may then induce degradative events. For example, the *Portulaca* flower lives for only about 8 h after opening, and wounding the filaments, but not the petals themselves, halves corolla lifetime. The wounding effect is mimicked by ethylene applications to unwounded flowers and is eliminated by adding NBD, the inhibitor of ethylene action, to wounded flowers (Ichimura, 1998). Work with petunia suggests that senescence-inducing signals related to wounding are different from those initiated by pollination. In this flower, ethylene produced by the wounded stigma/style leads directly to ethylene production by the petals. The signal for this response appears to be ethylene made by the wounded tissue, since no transport of ACC is detected (Woltering *et al.*, 1997).

Drought stress shortens the lifetime of certain flowers, perhaps by causing increases in ABA levels. In carnation, withholding water leads to an earlier production of ethylene and to increases in ion leakage. When protein synthesis is inhibited by CH, the water stress-induced senescence is delayed (Beja-Tal *et al.*, 1995). This suggests that ethylene is a signal for the stress and that protein synthesis is necessary for the accelerated response. In daylily, when sorbitol is added to isolated petals to reduce water uptake, the appearance of many of the parameters associated with natural senescence is hastened (Panavas *et al.*, 1998). These are the same parameters that also appear prematurely in petals treated with ABA (Panavas and Rubinstein, 1998), so this hormone may be an important hormonal intermediate signal for responses to drought in ethylene-insensitive flowers such as daylily.

No complete studies relating petal senescence to light quality have been published, but the vase-life of roses is extended using high-pressure sodium lights compared to metal halide lamps (Garello *et al.*, 1995). In carnation, ACC oxidase activity in excised gynoecia is higher in the light than in the dark, an effect related to photosynthetic activity (Woltering *et al.*, 1997). It is intriguing, however, that the flowers of many different species open only at a specific time of day and the petal cells subsequently begin to die at fixed times (e.g., morning glory, daylily). This suggests that a relationship exists beween the onset of senescence and photoperiod. A photoperiodic regulator of PCD would then involve the circadian clock and, equally important, phytochrome as a regulator of that clock. Yet, little work has been done on this aspect of the signalling system for petal senescence.

Heat shock is known to turn off the synthesis of many proteins and to up-regulate the production of heat shock proteins (Vierling, 1991). A 24 h heat shock to carnation flowers slows senescence and delays ethylene evolution and the elevation of steady-state message levels of both ACC synthase and ACC oxidase (Verlinden and Woodson, 1998). A 3 h heat shock administered 9 h before opening of daylily flowers up-regulates two classes of heat shock proteins and delays parameters associated with senescence by 12 h (Panavas *et al.*, 1998). These effects of heat shock may be related to the prevention of suicide gene expression and/or to the protective effects of heat shock proteins.

Signal transduction

G-protein-mediated lipid metabolism

Great strides have been made in understanding the signal transduction components involved in developmental events, but the intermediate steps between the signals that initiate petal senescence and the events most closely related to PCD must still be investigated. In one model pathway, phospholipase *c* hydrolyzes the membrane component phosphatidylinositol diphosphate (PIP_2) to yield inositol triphosphate (IP_3) and diacylglycerol (DAG); IP_3 then acts as a second messenger for elevating calcium levels in the cytosol. Consistent with this pathway in flowers, activities of phospholipase *a* and *c* increase in petals of roses and petunia before the onset of senescence (Borochov *et al.*, 1994). Furthermore, DAG levels are enhanced prior to the burst of ethylene; ethylene evolution is also stimulated by the phorbol ester PMA, which activates a kinase that requires DAG for activity (Borochov *et al.*, 1997).

Phospholipase *c* activity is often linked to activation of a heterotrimeric G-protein. In the orchid *Phalaenopsis*, a 42 kDa peptide was detected on immunoblots using an antiserum that reacts with the nucleotide-binding site of G-proteins in animals. What is more, cholera toxin and GTPγS that maintain G-proteins in their active state accelerate senescence by 20 to 30%, but only in the presence of ethylene. There is also a 20% increase in GTPγS binding to membranes isolated from pollinated *Phalaenopsis* flowers as they become senescent. The greater binding is observed regardless of whether the flowers were pretreated with AOA, an inhibitor of ethylene synthesis (Porat *et al.*, 1994).

Calcium

Another second messenger of great importance to developmental systems is calcium. Endogenous calcium levels increase in leaves concomitant with senescence (Huang *et al.*, 1997), and in the orchid *Phalaenopsis*, 1 mM calcium added together with the calcium ionophore A23187 increases flower sensitivity to applied ethylene, i.e. senescence is accelerated, while applied EGTA, a calcium chelator, decreases this sensitivity (Porat *et al.*, 1994). Furthermore, calcium (albeit at high concentrations of 200 to 250 μM) stimulates *in vitro* phosphorylation of microsomal membranes (Porat *et al.*, 1994). The possibility thus exists that senescence is controlled in part by increases in cytosolic calcium, which may result in the calcium-dependent phosphorylation of proteins that are necessary for the up-regulation of suicide proteins.

Phosphorylation/dephosphorylation

The kinases and phosphatases that control protein phosphorylation are often involved in signal transduction cascades leading to various developmental responses (Soporny and Munshi, 1998). Both total phosphorylation and phosphorylation of a specific peptide measured *in vitro* with ^{32}P-ATP increases during aging of *Phalaenopsis* petals (Porat *et al.*, 1994). In these same flowers, okadaic acid, an inhibitor of certain phosphatases, mimics the effects of pollination by stimulating the wilting associated with senescence. H-7, on the other hand, an inhibitor of protein kinase C, has no effect (Porat *et al.*, 1994).

We have preliminary evidence (Panavas and Rubinstein, unpublished results) that ion leakage from daylily petals is retarded by 5.0 μM staurosporine, an inhibitor of some protein kinases, and stimulated by 0.3 μM okadaic acid. The effects of these inhibitors on an important parameter of PCD parallel those that occur when measuring the PCD of cortical cells in maize roots (He *et al.*, 1996). Receptor-like protein kinases have been cloned from *Arabidopsis* flowers, but no developmental studies were performed (Takahashi *et al.*, 1998). These results, together with the calcium data and the increases of DAG mentioned above, suggest that kinases participate in aspects of senescence.

Polyamines

The senescence of plant organs has been attributed to certain polyamines, in part because they share the same precursor, SAM, as the senescence-inducing hormone, ethylene, and because added polyamines often retard senescence (Evans and Malmberg, 1989). The senescence of carnation petals is inhibited by 1 mM spermine, which may be due to a corresponding inhibition of ethylene synthesis. In fact, addition of an inhibitor of polyamine synthesis leads to elevated levels of transcripts for ACC synthase and ACC oxidase as well as to increased ethylene production. The relationship of endogenous polyamine levels to events occurring during senescence is not as clear, however (Lee *et al.*, 1997).

Jasmonic acid

A signal transduction pathway of importance to pathogen resistance involves the production of jasmonic acid. The products of LOX activity (an enzyme mentioned previously that may be associated with petal senescence) are components of the pathway to jasmonic acid, so this pathway may be important for petal senescence. In fact, jasmonic acid and several other metabolites promote senescence of two orchid species, presumably by elevating ACC, thereby stimulating ethylene production (Porat *et al.*, 1993a). However, neither LOX activity nor jasmonic acid content changes in orchid petals for 50 h after pollination-induced senescence (Porat *et al.*, 1995).

Summary/conclusions/future research

Petal senescence provides a facile system for studying PCD of plant organs. In many cases isolated petals behave similarly to those on the plant, and the process occurs relatively quickly, synchronously and uniformly throughout the tissue. The most extensively investigated flowers, however, are generally not the species that lend themselves readily to genetic manipulation, mutagenesis, or transformation. But it is likely that more tractable systems such as *Arabidopsis*, tomato or tobacco can also be employed to further analyze the role of particular pathways in petal PCD.

The fact that CH delays petal senescence demonstrates that the synthesis of particular suicide proteins orchestrates the cell death program. It is necessary, however, to show that these proteins and their products are not just correlated with PCD, but actually play a causal role. For example, a particular protein may serve only to mobilize materials during transport out of the dying cell, or a substance may be produced as a result of a perceived stress associated with senescence, and there is little direct effect on those processes most closely related to cell death. After a causal role is established, one must then determine if the substance acts alone or in concert with others.

With this in mind, several enzymatic reactions have been identified in petals that may directly affect senescence. These reactions result in the oxidation of membrane components and the breakdown of proteins and nucleic acids. Genes for several enzymes that mediate these events have been cloned and the mRNAs may be up-regulated prior to, or during, senescence. If the higher steady-state message level is a result of transcriptional regulation, techniques exist to examine the regulatory regions of these genes for consensus sequences that may indicate the factor(s) that enhances transcription. Other regulatory mechanisms that may be investigated include senescence-associated transcription factors, and a role for protease inhibitor genes (Solomon *et al.*, 1999). Characterization of genes that are down-regulated during senescence should also yield useful information. For example, expression of the DAD gene, which appears to protect cells against cell death, decreases in pea petals during senescence (Orzaez and Granell, 1997b).

The endogenous and environmental triggers that initiate suicide genes are also being identified. Senescence in many flowers is controlled in part by ethylene, but whereas other factors may also play a role, one must determine for the particular flower if only the production of ethylene is a limiting factor for senescence or if the sensitivity to ethylene is also important. An enhanced evolution of ethylene may be related to higher levels of the mRNAs for the enzymes responsible for synthesis of this hormone. The up-regulation of the ethylene-binding protein also implicates this process in an increased sensitivity to ethylene. The crucial question remains, however, as to how these messages are regulated.

There are also flowers in several families that are insensitive to ethylene. Even though ABA may be the senescence-inducing factor, many species should be tested before a generalization can be made. Furthermore, it is important to show that depletion of ABA delays senescence in these flowers. There may also be ABA-responsive sequences on the regulatory regions of the genes that directly or indirectly cause PCD.

Finally, pollination, injury, dehydration and light are environmental triggers of petal senescence that may act first on other parts of the flower. The transportable signal, which may be ACC or ethylene, leads to elevation of ethylene evolution (or ABA production in the case of water stress) in the petals. Different mechanisms may exist for different species.

An area that is ripe for exploration is the identification of components of the signal transduction chain resulting in PCD. Various factors and processes known to be important for developmental processes, including PCD, have been implicated in petal senescence, such as G-protein activation, inositol phosphate metabolism, alterations in cytosolic free calcium, and phosphorylation/dephosphorylation reactions. However, events such as calcium level changes need to be investigated on the cell level in living material, and genes involved in producing signal transduction components such as kinases and phosphatases must be identified and their regulation studied. It is important to keep in mind that different species may have different signal transduction systems.

The impact of characterizing PCD in petals allows an understanding of a process crucial to the survival of plants and other plant organs as well as providing economic benefits. The ease of working with flowers and the exciting advancements made in techniques at both the molecular and cell levels should lead to the identification of particular mutants, the cloning of more suicide genes, and the production of transformed plants, all of which make it very likely that great strides will be made in the next few years.

References

Aida, R., Yoshida, T., Ichimura, K., Goto, R. and Shibata, M. 1998. Extension of flower longevity in transgenic torenia plants incorporating ACC oxidase transgene. Plant Sci. 138: 91–101.

Asada, K. 1992. Ascorbate peroxidase – a hydrogen peroxide-scavenging enzyme in plants. Physiol. Plant. 85: 235–241.

Ashman, T.-L. and Schoen, D.J. 1994. How long should flowers live? Nature 371: 788–791.

Ashman, T.-L. and Schoen, D.J. 1997. The cost of floral longevity in *Clarkia tembloriensis*: an experimental investigation. Evol. Ecol. 11: 289–300.

Bachmair, A., Becker, F., Masterson, R.V. and Schell J. 1990. Perturbation of the ubiquitin system causes leaf curling, vascular tissue alterations and necrotic lesions in a higher plant. EMBO J. 9: 4543–4549.

Bartoli, C.G., Simontacchi, M., Guiamet, J., Montaldi, E. and Puntarulo, S. 1995. Antioxidant enzymes and lipid peroxidation during aging of *Chrysanthemum morifolium* RAM petals. Plant Sci. 104: 161–168.

Bartoli, C.G., Simontacchi, M., Montaldi, E. and Puntarulo, S. 1996. Oxidative stress, antioxidant capacity and ethylene production during ageing of cut carnation (*Dianthus caryophyllus*) petals. J. Exp. Bot. 47: 595–601.

Bartoli, C.G., Simontacchi, M., Montaldi, E. and Puntarulo, S. 1997. Oxidants and antioxidants during ageing of chrysanthemum petals. Plant Sci. 129: 157–165.

Baumgartner, B., Kende, H. and Matile, P. 1975. Ribonuclease in senescing morning glory. Purification and demonstration of *de novo* synthesis. Plant Physiol. 55: 734–737.

Beja-Tal, S. and Borochov, A. 1994. Age-related changes in biochemical and physical properties of carnation petal plasma membranes. J. Plant Physiol. 143: 195–199.

Beja-Tal, S., Borochov, A., Gindin, E. and Mayak, S. 1995. Transient water stress in cut carnation flowers: effects of cycloheximide. Scient. Hort. 64: 167–175.

Bieleski, R.L. 1993. Fructan hydrolysis drives petal expansion in the ephemeral daylily flower. Plant Physiol. 103: 213–219.

Bieleski, R.L. 1995. Onset of phloem export from senescent petals of daylily. Plant Physiol. 109: 557–565.

Bieleski, R.L. and Reid, M.S. 1992. Physiological changes accompanying senescence in the ephemeral daylily flower. Plant Physiol. 98: 1042–1049.

Blank, A. and McKeon, T.A. 1991. Expression of three RNase activities during natural and dark-induced senescence of wheat leaves. Plant Physiol. 97: 1409–1413.

Borochov, A. and Woodson, W.R. 1989. Physiology and biochemistry of flower petal senescence. Hort. Rev. 11: 15–43.

Borochov, A., Cho, M.H. and Boss, W.F. 1994. Plasma membrane lipid metabolism of petunia petals during senescence. Physiol. Plant 90: 279–284.

Borochov, A., Spiegelstein, H. and Philosoph-Hadas, S. 1997. Ethylene and flower petal senescence: interrelationship with membrane lipid catabolism. Physiol. Plant. 100: 606–612.

Buchanan-Wollaston, V. 1997. The molecular biology of leaf senescence. J. Exp. Bot. 48: 181–199.

Buchanan-Wollaston, V. and Ainsworth, C.A. 1997. Leaf senescence in *Brassica napus*: cloning of senescence related genes by subtractive hybridisation. Plant Mol. Biol. 33: 821–834.

Bui, A.Q. and O'Neill, S.D. 1998. Three 1-aminocyclopropane-1-carboxylate-synthase genes regulated by primary and secondary pollination signals in orchid flowers. Plant Physiol. 116: 419–428.

Cabello-Hurtado, F., Batard, Y., Salaun, J.P., Durst, F., Pinot, F. and Werck-Reichart, D. 1998. Cloning, expression in yeast and functional characterization of CYP81B1, a plant cytochrome P450 that catalyzes in-chain hydroxylation of fatty acids. J. Biol. Chem. 273: 7260–7267.

Callis, J. 1995. Regulation of protein degradation. Plant Cell 7: 845–857.

Celikel, F.G. and van Doorn, W.G. 1995. Solute leakage, lipid peroxidation and protein degradation during senescence of iris tepals. Physiol. Plant. 94: 515–521.

Clark, D.G., Richards, C., Hilioti, Z., Lind-Iversen, S. and Brown, K. 1997. Effect of pollination on accumulation of ACC synthase and ACC oxidase transcripts, ethylene production and flower petal abscission in geranium (*Pelargonium* × *hortorum* LH Bailey). Plant Mol. Biol. 34: 855–865.

Courtney, S.E., Rider, C.C. and Stead, A.D. 1994. Changes in protein ubiquitination and the expression of ubiquitin-encoding transcripts in daylily petals during floral development and senescence. Physiol. Plant. 91: 196–204.

Cryns, V. and Yuan, J. 1998. Proteases to die for. Genes Dev. 12: 1551–1570.

del Pozo, O. and Lam, E. 1998. Caspases and programmed cell death in the hypersensitive response of plants to pathogens. Curr. Biol. 8: 1129–1132.

del Rio, L.A., Palma, J.M., Sandalio, L.M., Corpas, F.J., Pastori, G.M., Bueno, P. and Lopez-Huertas, E. 1996. Peroxisomes as a source of superoxide and hydrogen peroxide in stressed plants. Biochem. Soc. Trans. 24: 434–438.

del Rio, L.A., Pastori, G.M., Palma, J.M., Sandalio, L.M., Sandalio, F., Sevilla, F., Corpas, F.J., Jimenez, A., Lopez-Huertas, E. and Hernandez, J.A. 1998. The activated oxygen role of peroxisomes in senescence. Plant Physiol. 116: 1195–1200.

de Vetten, N. and Huber, D.J. 1990. Cell wall changes during the expansion and senescence of carnation (*Dianthus caryophyllus*) petals. Physiol. Plant. 78: 447–454.

Do, Y.-Y. and Huang, P.L. 1997. Gene structure of PAC01, a petal senescence-related gene from *Phalaenopsis* encoding peroxisomal acyl-CoA oxidase homolog. Biochem. Mol. Biol. Int. 41: 609–617.

Evans, P.T. and Malmberg, R.L. 1989. Do polyamines have roles in plant development? Annu. Rev. Plant Physiol. Plant Mol. Biol. 40: 235–269.

Faragher, J.D., Wachtel, E. andMayak, S. 1987. Changes in the physical state of membrane lipids during senescence of rose petals. Plant Physiol. 83: 1037–1042.

Garbarino, J.E., Oosumi, T. and Belknap, W.R. 1995. Isolation of a polyubiquitin promoter and its expression in transgenic potato plants. Plant Physiol. 109: 1371–1378.

Garello, G., Menard, C., Dansereau, B. and LePage-Degivry, M.T. 1995. The influence of light quality on rose flower senescence: involvement of abscisic acid. Plant Growth Regul. 16: 135–139.

Green, P.J. 1994. The ribonucleases of higher plants. Annu. Rev. Plant Physiol. Plant Mol. Biol. 45: 421–445.

Griffiths, C.M., Hosken, S.E., Oliver, D., Chojecki, J. and Thomas, H. 1997. Sequencing, expression pattern and RFLP mapping of a senescence-enhanced cDNA from *Zea mays* with high homology to oryzain γ and aleurain. Plant Mol. Biol. 34: 815–821.

Guerrero, C., de la Calle, M., Reid, M.S. and Valpuesta, V. 1998. Analysis of the expression of two thiolprotease genes from daylily (*Hemerocallis* spp.) during flower senescence. Plant Mol. Biol. 36: 656–571.

Halevy, A.H. 1998. Recent advances in postharvest physiology of flowers. J. Korean Soc. Hort. Soc. 39: 652–655.

Halevy, A.H., Porat, R., Spiegelstein, H., Borochov, A., Botha, L. and Whitehead, C.S. 1996. Short-chain fatty acids in the regulation of pollination-induced ethylene sensitivity of *Phalaenopsis* flowers. Physiol. Plant. 97: 469–474.

Halliwell, B. and Gutteridge, J.M.C. 1989. Free Radicals in Biology and Medicine. Clarendon Press, Oxford, UK, pp. 450–499.

He, C.-J., Morgan, P.W. and Drew, M.C. 1996. Transduction of an ethylene signal is required for cell death and lysis in the root cortex of maize during aerenchyma formation induced by hypoxia. Plant Physiol. 112: 463–472.

Huang, F.-Y., Philosoph-Hadas, S., Meir, S., Callahan, D.A., Sabato, R., Zelcer, A. and Hepler, P.K. 1997. Increases in cytosolic Ca^{2+} in parsley mesophyll cells correlated with leaf senescence. Plant Physiol. 115: 51–60.

Ichimura, K. and Suto, K. 1998. Role of ethylene in acceleration of flower senescence by filament wounding in *Portulaca* hybrid. Physiol. Plant. 104: 603–607.

Itzhaki, H., Davis, J.H., Borochov, A., Mayak, S. and Pauls, K.P. 1995. Deuterium magnetic resonance studies of senescence-related changes in the physical properties of rose petal membrane lipids. Plant Physiol. 108: 1029–1033.

Itzhaki, H., Mayak, S. and Borochov, A. 1998. Phosphatidylcholine turnover during senescence of rose petals. Plant Physiol. Biochem. 36: 457–462.

Jacobson, M., Weil, M. and Raff, M.C. 1997. Programmed cell death in animal development. Cell 88: 347–354.

Jones, M.L. and Woodson, W.R.1999. Differential expression of three members of the 1-aminocyclopropane-1-carboxylate synthase gene family in carnation. Plant Physiol. 119: 755–764.

Jones, M.L., Larsen, P.B. and Woodson, W.R. 1995. Ethylene-regulated expression of a carnation cysteine proteinase during flower petal senescence. Plant Mol. Biol. 28: 505–512.

Lanahan, M.B., Yen, H.-C., Giovannoni, J.J. and Klee, H.J. 1994. The Never Ripe mutation blocks ethylene perception in tomato. Plant Cell 6: 521–530.

Larsen, P.B., Ashworth, E.N., Jones, M.L. and Woodson, W.R. 1995. Pollination-induced ethylene in carnation. Plant Physiol. 108: 1405–1412.

Laughton, K., Raghothama, K.G., Goldsbrough, P.B. and Woodson, W.R. 1990. Regulation of senescence-related gene expression in carnation flower petals by ethylene. Plant Physiol. 93: 1370–1375.

Lay-Yee, M., Stead, A.D. and Reid, M.S. 1992. Flower senescence in daylily (*Hemerocallis*). Physiol. Plant. 86: 308–314.

Lee, M., Lee, S.H.and Park, K.Y. 1997. Effects of spermine on ethylene biosynthesis in cut carnation (*Dianthus caryophyllus* L.) flowers during senescence. J. Plant Physiol. 151: 68–73.

LePage-Degivry, M.T., Orlandini, M., Carello, G., Barthe, P. and Gudia, 1991. Regulation of ABA levels in senescing petals of rose flowers. Plant Growth Regul. 10: 67–72.

Lers, A., Khalchitski, A., Lomaniec, E., Burd, S. and Green, P.J. 1998. Senescence-induced RNases in tomato. Plant Mol. Biol. 36: 439–449.

Leshem, Y., Halevy, A.H. and Frenkel, C. 1986. Process and control of plant senescence. Dev. Crop Sci. 8: 142–161.

Matile, P. and Winkenbach, F. 1971. Function of lysosomes and lysosomal enzymes in the senescing corolla of the morning glory. J. Exp. Bot. 22: 759–771.

Mayak, S., Tirosh, T., Thompson, J.E. and Ghosh, S. 1998. The fate of ribulose-1,5-bisphosphate carboxylase subunits during development of carnation petals. Plant Physiol. Biochem. 36: 835–841.

Meyer, R.C., Goldsbrough, P.B. and Woodson, W.R. 1991. An ethylene-responsive flower senescence-related gene from carnation encodes a protein homologous to glutathione *S*-transferases. Plant Mol. Biol. 17: 277–281.

Michael, M.Z., Savin, K.W., Baudinette, S.C., Graham, M.W., Chandler, S.F., Lu, C.-Y., Caesar, C., Gautrais, I., Young, R., Nugent, C.D., Stevenson, K.R., O'Connor, E.L.-J., Cobbett, C.S., Cornish, E.C. 1993. Cloning of ethylene biosynthetic genes involved in petal senescence of carnation and petunia, and their antisense expression in transgenic plants. In: J.C. Pech, A. Latche and C. Balague (Eds.) Cellular and Molecular Aspects of the Plant Hormone Ethylene, Kluwer Academic Publishers, Dordrecht, Netherlands, pp. 298–303.

Midoh, N., Saijou, Y., Matsumoto, K. and Iwata, M. 1996. Effects of 1,1-dimethyl-4-(phenylsulfonyl) semicarbazide (DPSS) on carnation flower longevity. Plant Growth Regul. 20: 195–199.

Mittler, R. and Lam, E. 1995a. Identification, characterization, and purification of a tobacco endonuclease activity induced upon hypersensitive response cell death. Plant Cell 7: 1951–1962.

Mittler, R. and Lam, E. 1995b. In situ detection of nDNA fragmentation during the differentation of tracheary elements in higher plants. Plant Physiol. 108: 489–493.

Mittler, R., Shulaev, V., Seskar, M. and Lam, E. 1996. Inhibition of programmed cell death in tobacco plants during pathogen-induced hypersensitive response at low oxygen pressure. Plant Cell 8: 1991–2001.

Mutlu, A. and Gal, S. 1999. Plant aspartic proteinases: enzymes on the way to a function. Physiol. Plant. 105: 569–576.

O'Brien, I.E.W., Baguley, B.C., Murray, B.G., Morris, B.A.M. and Ferguson, I.B. 1998. Early stages of the apoptotic pathway in plant cells are reversible. Plant J. 13: 803–814.

O'Neill, S.D. 1997. Pollination regulation of flower development. Annu. Rev. Plant Physiol. Plant Mol. Biol. 48: 547–574.

O'Neill, S.D., Nadeau, J.A., Zhang, X.S., Bui, A.Q. and Halevy, A.H. 1993. Interorgan regulation of ethylene biosynthetic genes by pollination. Plant Cell 5: 419–432.

Orlandini, M., Arene, L. and LePage-Degivry, M.T. 1991. The relationship between petal water potential and levels of abscisic acid in rose flower. Acta. Hort. 298: 161–163.

Orzaez, D. and Granell, A. 1997a. DNA fragmentation is regulated by ethylene during carpel senescence in *Pisum sativum*. Plant J. 11: 137–144.

Orzaez, D. and Granell, A. 1997b. The plant homologue of the defender against apoptotic death gene is down-regulated during senescence of flower petals. FEBS Lett. 404: 275–278.

Orzaez, D., Blay, R. and Granell, A. 1999. Programme of senescence in petals and carpels of *Pisum sativum* L. flowers and its control by ethylene. Planta 208: 220–226.

Paliyath, G. and Droillard, M.J. 1992. The mechanisms of membrane deterioration and disassembly during senescence. Plant Physiol. Biochem. 30: 789–812.

Panavas, T. and Rubinstein, B. 1998. Oxidative events during programmed cell death of daylily (*Hemerocallis* hybrid) petals. Plant Sci. 133: 125–138.

Panavas, T., Reid, P.D. and Rubinstein, B. 1998. Programmed cell death of daylily petals: activities of wall-based enzymes and effects of heat shock. Plant Physiol. Biochem. 36: 379–388.

Panavas, T., Pikula, A., Reid, P.D., Rubinstein, B. and Walker, E.L. 1999. Identification of senescence-associated genes from daylily petals. Plant Mol. Biol. 40: 237–248.

Park, K.Y., Drory, A. and Woodson, W.R. 1992. Molecular cloning of a 1-aminocyclopropane-1-carboxylase synthase from senescing carnation flower petals. Plant Mol. Biol. 18: 377–386.

Payton, S., Fray, R.G., Brown, S. and Grierson, D. 1996. Ethylene receptor expression is regulated during fruit ripening, flower senescence and abscission. Plant Mol. Biol. 31: 1227–1231.

Phillips, H.L. Jr. and Kende, H. 1980. Structural changes in flowers of *Ipomea tricolor* during flower opening and closing. Protoplasma 102: 199–215.

Pinedo, M.L., Goicoechea, S.M., Lamattina, L. and Conde, R.D. 1996. Estimation of ubiquitin and ubiquitin mRNA content in dark senescing wheat leaves. Biol. Plant. 38: 321–328.

Podd, L.A. and van Staden, J. 1999. Is acetaldehyde the causal agent in the retardation of carnation flower senescence by ethanol? J. Plant Physiol. 154: 351–354.

Porat, R., Borochov, A. and Halevy, A.H. 1993. Enhancement of petunia and *Dendrobium* flower senescence by jasmonic acid methyl ester is via the promotion of ethylene production. Plant Growth Regul. 13: 297–301.

Porat, R., Borochov, A. and Halevy, A.H. 1994. Pollination-induced senescence in *Phalaenopsis* petals. Relationship of ethylene sensitivity to activity of GTP-binding proteins and protein phosphorylation. Physiol. Plant. 90: 679–684.

Porat, R., Reuveny, Y., Borochov, A. and Halevy, A.H. 1993b. Petunia flower longevity: the role of sensitivity to ethylene. Physiol. Plant. 89: 291–294.

Porat, R., Reiss, N., Atzorn, R., Halevy, A.H. and Borochov, A. 1995. Examination of the possible involvement of lipoxygenase and jasmonates in pollination-induced senescence of *Phalaenopsis* and *Dendrobium* orchid flowers. Physiol. Plant. 94: 205–210.

Porat, R., Nadeau, J.A., Kirby, J.A., Sutter, E.G. and O'Neill, S.D. 1998. Characterization of the primary pollen signal in the post pollination syndrome of *Phalaenopsis* flowers. Plant Growth Regul. 24: 109–117.

Reid, M.S. and Wu, M.-J. 1992. Ethylene and flower senescence. Plant Growth Regul. 11: 37–43.

Rottman, W.H., Peter, G.F., Oeller, P.W., Keller, J.A., Shen, N.F., Nagy, B.P., Taylor, L.P., Campbell, A.D. and Theologis, A. 1991. 1-aminocyclopropane-1-carboxylate synthase in tomato is encoded by a multigene family whose transcription is induced during fruit and floral senescence. J. Mol. Biol. 222: 937–961.

Scandalios, J.G. 1993. Oxygen stress and superoxide dismutases. Plant Physiol. 101: 7–12.

Serek, M., Tamari, G., Sisler, E.C. and Borochov, A. 1995. Inhibition of ethylene-induced cellular senescence symptoms by 1-methylcyclopropene, a new inhibitor of ethylene action. Physiol. Plant. 94: 229–232.

Shykoff, J.A., Bucheli, E. and Kaltz, O. 1996. Flower lifespan and disease risk. Nature 379: 779–780.

Siedow, J.N. 1991. Plant lipoxygenase: structure and function. Annu. Rev. Plant Physiol. Plant Mol. Biol. 42: 145–188.

Singh, A., Evenson, K.B. and Kao, T.-H. 1992. Ethylene synthesis and floral senescence following compatible and incompatible pollinations in *Petunia inflata*. Plant Physiol. 99: 38–45.

Smart, C. 1994. Gene expression during leaf senescence. New Phytol. 126: 419–448.

Smith, M.T., Saks, Y. and van Staden, J. 1992. Ultrastructural changes in the petals of senescing flowers of *Dianthus caryophyllus* L. Ann. Bot. 69: 277–285.

Solomon, M., Belenghi, B., Delledonne, M., Menachem, M. and Levine, A. 1999. The involvement of cysteine proteases and protease inhibitor genes in the regulation of programmed cell death in plants. Plant Cell 11: 431–443.

Song, W.C. and Brash, A.R. 1991. Purification of an allene oxide synthase and identification of the enzyme as a cytochrome P-450. Science 253: 781–784.

Sopory, S. and Munshi, M. 1998. Protein kinases and phosphatases and their role in cellular signalling in plants. Crit. Rev. Plant Sci. 17: 245–318.

Stead, A.D. 1992. Pollination-induced flower senescence: a review. Plant Growth Regul 11: 13–20.

Stead, A.D. and van Doorn 1994. Strategies of flower senescence – a review. In: R.J. Scott and A.D. Stead (Eds.) Molecular and Cellular Aspects of Plant Reproduction, Cambridge University Press, Cambridge, UK, pp. 215–238.

Stephenson, P., Collins, B.A., Reid, P.D. and Rubinstein, B. 1996. Localization of ubiquitin to differentiating vascular tissues. Am. J. Bot. 83: 140–147.

Stephenson, P. and Rubinstein, B. 1998. Characterization of proteolytic activity during senescence in daylilies. Physiol. Plant. 104: 463–473.

Sylvestre, I., Droillard, M.-J., Bureau, J.-M. and Paulin, A. 1989. Effects of the ethylene rise on the peroxidation of membrane lipids during the senescence of cut carnations. Plant Physiol. Biochem. 27: 407–413.

Takahashi, T., Mu, J.-H., Gasch, A. and Chua, N.-H. 1998. Identification by PCR of receptor-like kinases from arabidopsis. Plant Mol. Biol. 37: 587–596.

Tang, X., Gomes, A.M.T.R., Bhatia, A. and Woodson, W.R. 1994. Pistil-specific and ethylene-regulated expression of 1-aminocyclopropane-1-carboxylate oxidase genes in petunia flowers. Plant Cell 6: 1227–1239.

Taylor, C.B., Bariola, P.A., DelCardayre, S.B., Raines, R.T. and Green, P.J. 1993. A senescence-associated RNase of *Arabidopsis* that diverged from the S-RNases before speciation. Proc. Natl. Acad. Sci. USA 90: 5118–5122.

Thompson, J.E. 1988. The molecular basis for membrane deterioration during senescence. In: L.D. Nooden and A.C. Leopold (Eds.) Senescence and Aging in Plants, Academic Press, New York, pp. 51–83.

Thompson, J.E., Froese, C.D., Hong, Y., Hudak, K.A. and Smith, M.D. 1997. Membrane deterioration during senescence. Can. J. Bot. 75: 867–879.

Valpuesta, V., Lange, N.E., Cuerrero, C. and Reid, M.S. 1995. Upregulation of a cysteine protease accompanies the ethylene-insensitive senescence of daylily (*Hemerocallis*) flowers. Plant Mol. Biol. 28: 575–582.

van Altvorst, A.C. and Bovy, A.G. 1995. The role of ethylene in the senescence of carnation flowers: a review. Plant Growth Regul. 16: 43–53.

van Doorn, W.G. 1997. Effects of pollination on floral attraction and longevity. J. Exp. Bot. 48: 1615–1622.

van Doorn, W.G. and Stead, A.D. 1994. The physiology of petal senescence which is not initiated by ethylene. In: R.J. Scott and A.D. Stead (Eds.) Molecular and Cellular Aspects of Plant Reproduction, Cambridge University Press, Cambridge, UK, pp. 239–254.

Vardi, Y. and Mayak, S. 1989. Involvement of abscisic acid during water stress and recovery in petunia flowers. Acta. Hort. 261: 107–112.

Verlinden, S. and Woodson, W.R. 1998. The physiological and molecular responses of carnation flowers to high temperature. Postharvest Biol. Techn. 4: 185–192.

Vierling, E. 1991. The roles of heat shock proteins in plants. Annu. Rev. Plant Physiol. Plant Mol. Biol. 42: 579–620.

Vierstra, R.D. 1993. Protein degradation in plants. Annu. Rev. Plant Physiol. Plant Mol. Biol. 44: 385–410.

Wang, H., Brandt, A.S. and Woodson, W.R. 1993. A flower senescence-related mRNA from carnation encodes a novel protein related to enzymes involved in phosphonate biosynthesis. Plant Mol. Biol. 22: 719–724.

Wang, H., Li, J., Bostock, R.M. and Gilchrist, D.G. 1996. Apoptosis: a functional paradigm for programmed plant cell death induced by a host-selective phytotoxin and invoked during development. Plant Cell 8: 375–391.

Whitehead, C.S. 1994. Ethylene sensitivity and flower senescence. In: R.J. Scott and A.D. Stead (Eds.) Molecular and Cellular Aspects of Plant Reproduction, Cambridge University Press, Cambridge, UK, pp. 269–284.

Whitehead, C.S. and Vasiljevic, D. 1993. Role of short-chain saturated fatty acids in the control of ethylene sensitivity in senescing carnation flowers. Physiol. Plant. 88: 243–250.

Wiemken-Gehrig, V., Wiemken, A. and Matile, P. 1974. Cell wall breakdown in wilting flowers of *Ipomea tricolor* Cav. Planta 115: 297–307.

Willekens, H., Chamnongpol, S., Davey, M., Schrauder, M., Langebartels, C., Van Montagu, M., Inzé, D. and Van Camp, W. 1997. Catalase is a sink for H_2O_2 and is indispensable for stress defence in C_3 plants. EMBO J. 16: 4806–4816.

Woffenden, B.J., Freeman, T.B. and Beers, E.P. 1998. Proteasome inhibitors prevent tracheary element differentiation in *Zinnia* mesophyll cell cultures. Plant Physiol. 118: 419–430.

Woltering, E.J. and van Doorn, W.G. 1988. Role of ethylene and senescence of petals: morphological and taxonomical relationships. J. Exp. Bot. 39: 1605–1616.

Woltering, E.J., de Vrije, T., Harren, F. and Hoekstra, F.A. 1997. Pollination and stigma wounding: same response, different signal? J. Exp. Bot. 48: 1027–1033.

Woltering, E.J., Somhorst, D. and van der Veer, P. 1995. The role of ethylene in interorgan signalling during flower senescence. Plant Physiol. 109: 1219–1225.

Woltering, E.J., ten Have, A., Larsen, P.B. and Woodson, W.R. 1994. Ethylene biosynthetic genes and interorgan signalling during flower senescence. In: R.J. Scott and A.D. Stead (Eds.) Molecular and Cellular Aspects of Plant Reproduction, Cambridge University Press, Cambridge, UK, pp. 285–307.

Woodson, W.R.. 1994. Molecular biology of flower senescence in carnation. In: R.J. Scott and A.D. Stead (Eds.) Molecular and Cellular Aspects of Plant Reproduction, Cambridge University Press, Cambridge, UK, pp. 255–267.

Woodson, W.R. and Handa, A.K. 1987. Changes in protein patterns and *in vivo* protein synthesis during presenescence and senescence of *Hibiscus* petals. J. Plant Physiol. 128: 67–75.

Woodson, W.R., Park, K.Y., Drory, A., Larsen, P.B. and Wang, H. 1992. Expression of ethylene biosynthetic pathway transcripts in senescing carnation flowers. Plant Physiol. 99: 526–532.

Young, T.E., Gallie, D.R., DeMason, D.A. 1997. Ethylene-mediated programmed cell death during maize endosperm development of wild-type and shrunken 2 genotypes. Plant Physiol. 115: 737–751.

Zelitch, I., Havir, E.A., McGonigle, B., McHale, N.A. and Nelson, T. 1991. Leaf catalase mRNA and catalase-protein levels in a high-catalase tobacco mutant with O_2-resistant photosynthesis. Plant Physiol. 97: 1592–1595.

SECTION 2

INDUCED CELL DEATH MODELS

Plant Molecular Biology **44:** 321–334, 2000.
E. Lam, H. Fukuda and J. Greenberg (Eds.), Programmed Cell Death in Higher Plants.

Hypersensitive response-related death

Michèle C. Heath
Botany Department, University of Toronto, Toronto, Ontario, Canada M5S 1A1 (fax: 1-416-978 5878; e-mail: heath@botany.utoronto.ca)

Key words: hypersensitive response, pathogens, plants, programmed cell death, proteases, reactive oxygen species, resistance genes

Abstract

The hypersensitive response (HR) of plants resistant to microbial pathogens involves a complex form of programmed cell death (PCD) that differs from developmental PCD in its consistent association with the induction of local and systemic defence responses. Hypersensitive cell death is commonly controlled by direct or indirect interactions between pathogen avirulence gene products and those of plant resistance genes and it can be the result of multiple signalling pathways. Ion fluxes and the generation of reactive oxygen species commonly precede cell death, but a direct involvement of the latter seems to vary with the plant-pathogen combination. Protein synthesis, an intact actin cytoskeleton and salicylic acid also seem necessary for cell death induction. Cytological studies suggest that the actual mode and sequence of dismantling the cell contents varies among plant-parasite systems although there may be a universal involvement of cysteine proteases. It seems likely that cell death within the HR acts more as a signal to the rest of the plant rather than as a direct defence mechanism.

Abbreviations: HR, hypersensitive response; PARP, poly(ADP-ribose) polymerase; PCD, programmed cell death; ROS, reactive oxygen species

Introduction

The hypersensitive response (HR) has been a source of interest and controversy since its recognition over 80 years ago. However, in the past few years, the combination of molecular genetics, computer-enhanced microscopy, new cytochemical techniques and a better understanding of plant signal transduction and cell biology have shed new light on this seemingly universal plant response. Each review of the HR tends to reflect the biases of its authors; for differently orientated reviews on this topic, readers are directed to Mittler *et al.* (1997b), Morel and Dangl (1997) and Gilchrist (1998).

Can the hypersensitive response and hypersensitive cell death be defined?

Classical definition of the hypersensitive response

'Hypersensitive' was a term first applied by Stakman (1915) to describe the rapid and localized plant cell death induced by rust fungi in rust-resistant cereals. The subsequent realization that such death was a common expression of disease resistance in plants, regardless of the type of inducing pathogen, led to its designation as the hypersensitive response, usually defined as 'the rapid death of plant cells in association with the restriction of pathogen growth' (Goodman and Novacky, 1994). The HR is generally recognized by the presence of brown, dead cells at the infection site and, depending on the pathogen, their number may vary from one to many. The HR may or may not be restricted to cells physically invaded by, or having direct contact with, the pathogen. A visible brown lesion may develop if sufficient cells die.

Definitive features of the HR

This seemingly straightforward definition of the HR masks a high level of underlying complexity. For non-biotrophic pathogens that do not require their host cells to stay alive, cell death alone cannot restrict pathogen growth. This role is ascribed to the many induced defence responses that typically occur within the dying cells and in their adjacent living neighbours. Therefore, the HR encompasses both cell death and 'defence gene' expression. However, it is important to note that disease resistance and all of the inducible defence responses currently associated with the HR can occur in plants in the absence of cell death. Moreover, pathogens may also cause cell death and trigger defence responses while successfully growing in susceptible tissue. A further complication is the fact that genetic defects, or treatments unlikely to resemble those causing cell death during a natural HR, can cause a cell death that mimics the HR morphologically and in the induction of defence responses (Rahe and Arnold, 1975; Hu *et al.*, 1998; Molina *et al.* 1999). As a result, there are no features that currently can unequivocally identify the HR in the absence of a plant-pathogen interaction.

Recent attempts to rectify this problem have resulted in the discovery of two genes in tobacco (with related genes in tomato), *HIN1* and *HRS203J*, that have the potential for being early marker genes for the HR (Gopalan *et al.*, 1996; Pontier *et al.* 1998), and which have been used to distinguish hypersensitive cell death from natural leaf senescence (Pontier *et al.*, 1999). However, *HSR203J* also is expressed during cell death caused by successful pathogenesis (Pontier *et al.*, 1998) and the expression of both genes is induced by heavy-metal salts (Pontier *et al.* 1998; 1999) which do not trigger features typical of hypersensitive cell death in other systems (Meyer and Heath, 1988a, b; Ryerson and Heath, 1996). Therefore, it is still too early to tell whether or not unique marker genes for the HR exist.

Is hypersensitive cell death a form of programmed cell death?

The ubiquity of the HR, its requirement for plant cell metabolism (Aist and Bushnell, 1991; Zeyen *et al.*, 1995; Heath *et al.*, 1997; Mansfield *et al.*, 1997), and a few shared features (discussed later) with a form of mammalian programmed cell death (PCD) known as apoptosis support the assumption that hypersensitive cell death is endogenously programmed. Although its strong and consistent association with defence responses distinguishes it from forms of plant PCD that occur during plant development (Heath, 1998a), there seems to be significant 'cross-talk' between developmental PCD and the HR. For example, some defence genes may be activated during leaf senescence (Buchanan-Wollaston, 1997) and the *SAG12* marker gene for senescence in *Arabidopsis* is expressed in low levels in cells surrounding TMV- and bacteria-induced HR lesions in transgenic tobacco (Pontier *et al.*, 1999). Similarly, the putative HR marker gene, *HIN1*, is expressed in a late stage of leaf senescence (Pontier *et al.*, 1999). Likewise, the promoter for the senescence-inducible *Brassica napus LSC54* gene that codes for a metallothionein-like protein is active in transgenic *Arabidopsis* prior to the HR caused by the oomycete *Peronospora parasitica* (Buchanan-Wollaston, 1997).

Genetic control of the HR

Genes for resistance and avirulence

Plant genotypes within an otherwise susceptible plant species may exhibit resistance, and the HR, against specific genotypes of a pathogen. This resistance is generally, but not always (Heath, 1996), controlled by single, parasite-specific resistance (*R*) genes. For biotrophic fungal pathogens in particular (Heath, 1997), the HR requires the pathogen to have an avirulence (*avr*) gene that 'matches' the *R* gene in a 'gene-for-gene' relationship. *R* and *avr* genes appear to have a more complex relationship for bacterial pathogens, with single *R* genes 'matching' more than one *avr* gene (Hammond-Kosack and Jones, 1997).

Whether the HR expressed in non-host plants has the same type of genetic control is controversial (Heath, 1991) and conceptually most likely when plants are both hosts and non-hosts of related pathogens (Heath, 1991, 1997). The evolution of plant-bacteria interactions is further complicated by the ease of gene transfer between different bacteria (Collmer, 1998; Gabriel, 1999).

avr genes from different fungi or bacteria have little homology (Hammond-Kosack and Jones, 1997;

Laugé and de Wit, 1998). These genes have long been assumed to have had other functions prior to their products being co-opted as resistance-inducing recognitional factors by the plant (Person and Mayo, 1973; Heath, 1997). Some appear to be involved in pathogenicity (Collmer, 1998; Laugé and de Wit, 1998) and some from the bacterium *Pseudomonas syringae* cause cell death when heterologously expressed in plants lacking a cognate *R* gene (Collmer, 1998). It has been suggested that many are maladapted pathogenicity genes originally evolved as determinants of host species range and horizontally transferred into bacteria in which their function is gratuitous or detrimental (Gabriel, 1999).

R genes, in contrast, are often allelic, exist in clusters in the plant genome, and have been suggested to resemble the vertebrate major histocompatibility complex in organization and evolution (Michelmore and Meyers, 1998). Cloned *R* genes share common themes (Hammond-Kosack and Jones, 1997) and plant genomes appear to contain large numbers of genes with similar sequences (Michelmore and Meyers, 1998). Most of their predicted proteins have leucine-rich repeats (LRRs) or a serine-threonine kinase domain and are assumed to be components of signalling systems as one might expect of molecules that detect a specific pathogen's presence. The predominant class of predicted *R* gene products have a nucleotide-binding site (NB) as well as LRRs and may have a leucine zipper (LZ) motif at the amino terminus or a TIR sequence similar to the cytoplasmic portions of the *Drosophila Toll* gene and the mammalian interleukin 1 transmembrane receptor (Hammond-Kosack and Jones, 1997). Some *R* genes have a predicted transmembrane sequence, but others appear to be totally intracellular, although the predicted sequence may be misleading as the protein product of *RPM1* that was assumed to be soluble has been found to be membrane-associated (Boyes *et al.*, 1998). Significantly, there is no relationship between the category of *R* gene and the type of organism against which it confers resistance (Hammond-Kosack and Jones, 1997). There may be some cross-talk between *R* genes, however, as a mutation in *RPS5*, which controls *avrPphB*-mediated resistance to *Pseudomonas syringae* pv. *tomato*, affects the function of several other *R* genes that confer resistance to different bacterial isolates and to the oomycete *Peronospora parasitica* (Warren *et al.*, 1998). Although it is generally assumed that *R* genes are constitutively expressed, there is one report that an *R* gene may be induced during the HR (Seehaus and Tenhaken, 1998).

R genes that are not involved in *avr-R* gene complementarity do not resemble those that are. Thus, the *R* gene *HM1* of maize codes for an enzyme that detoxifies the host-selective toxin produced by the fungus *Cochliobolus carbonum* (Johal and Briggs, 1992), thereby allowing the metabolism that leads to the HR (which is not controlled by *HM1*) to occur (Heath, 1997). The *mlo* resistance gene of barley, on the other hand, controls race-unspecific resistance to powdery mildew due to mutations in a novel membrane-bound protein, Mlo (see Peterhänsel *et al.*, 1997). Like *HM1*, *mlo* does not control the HR, and resistance to powdery mildew is expressed during fungal penetration without involving any plant cell death; however, *mlo* plants show spontaneous necrotic cells in the absence of a pathogen, a condition known as lesion mimicry.

Mutations that cause such lesion mimicry can, in theory, be informative about death-controlling genes in plants. Significantly, mutations in *Rp1* (Hu *et al.*, 1996) and *Pto* (Rathjen *et al.*, 1999) resistance genes result in cell death in maize and tomato respectively in the absence of any pathogen. As well, the *Pto*-related gene *Fen* mediates a hypersensitive-like cell death in response to the insecticide fenthion even though it does not confer disease resistance (Zhou *et al.*, 1998). Many lesion mimic mutants, however, are not defective in genes related to *R* genes. An interesting example is the *Arabidopsis* lesion mimic mutant *lsd5*; some mutants of this mimic in which cell death is suppressed also show complex differential patterns of disease resistance modifications, including delaying the HR towards the oomycete *Peronospora parasitica* (Morel and Dangl, 1999). Such data suggest the involvement of common genes in *lsd5*-controlled cell death and in responses to some pathogens. However, the cause of cell death in some other mutants where the causal mutation is in a gene not related to *R* genes is unlikely to mimic that of the HR (Hu *et al.*, 1998; Molina *et al.*, 1999). (A review of lesion mimic mutants is given by Shirasu and Schulze-Lefert in this issue.)

Interestingly, several *R* gene products have some similarity to the nematode protein CED-4 and the mammalian homologue, APAF-1, involved in animal PCD. This raises the possibility that *R* gene regulation of the HR may share some similarities with PCD regulation in animals (van der Biezen and Jones, 1998).

Other genes involved in the HR

Although *R* genes were the first genes to be identified by Mendelian genetics to control the HR in resistant plants, mutation studies have revealed that the HR also depends on additional genes that presumably are present in both resistant and susceptible members of host species and which confer the ability of all plants to undergo an HR even in non-gene-for-gene situations. These RDR (required for disease resistance) genes (see Morel and Dangl, 1999) may be different for different *R* genes, irrespective of the type of pathogen against which they act. In *Arabidopsis*, for example, mutations in *NDR1*, a gene that codes for a putatively membrane-associated protein, suppresses resistance mediated by the LZ-NB-LRR but not the TIR-NB-LRR class of resistance genes while the reverse is true for mutations in *EDS1*, a putative L-family lipase (Falk *et al.*, 1999). These data suggest that there may be several signalling pathways leading to hypersensitive cell death, and that the one activated depends more on the class of *R* gene than the type of inducing pathogen. Similarly, RDR genes in barley involved in *mlo*-controlled spontaneous cell death differ from those involved in the race-specific, *R* gene-controlled, HR triggered by a powdery mildew fungus (Peterhänsel *et al.*, 1997).

Despite the similarity between predicted *R* gene products and animal CED-4 and APAF-1, attempts to find homologues of other critical genes involved in animal PCD have generally failed (Dangl *et al.*, 1996). Only putative homologues of animal apoptosis supressor genes *Bcl-2* (Dion *et al.*, 1997) and *DAD* (Dong *et al.*, 1998) have been detected in plants by immunocytochemistry and gene cloning, respectively. However, the fact the *DAD* product is a key enzyme in the general synthesis of glycoproteins means that its presence in all eukaryotes is to be expected and there is no clear evidence in plants of its involvement in PCD (Heath, 1998a).

Elicitors and receptors involved in the HR

Specific elicitors

avr gene products might be expected to trigger the HR *only* in plants that contain a matching *R* gene, but few such 'specific elicitors' have been isolated. For viruses, specific elicitors have been identified as coat proteins, the helicase domain of a replicase gene, or a movement protein (Dawson, 1999). For fungi, specific elicitors are primarily peptides of unknown function (Ebel and Scheel, 1997; D'Silva and Heath, 1997) that are known (Laugé and de Wit, 1998) or assumed to be products of *avr* genes and are secreted only under specific conditions (Laugé and de Wit, 1998) or stages of development (Chen and Heath, 1992). Many more *avr* genes have been cloned from bacteria than from fungi, but the identification of their products has been hampered by the fact that they appear to be secreted directly into the plant cell via a type III secretion system, components of which are encoded by *hrp* (hypersensitive reaction and pathogenicity) genes (He, 1998). *avr* gene products alone seem sufficient to cause cell death since the latter is induced when *avr* genes are expressed in transgenic plants containing the corresponding *R* genes (He, 1998). *avrD* from *Pseudomonas syringae* pathovars seems to direct the production of syringolides, glycolipid elicitors of *Rpg4*-mediated cell death in soybean (Ji *et al.*, 1998).

Non-specific elicitors

In addition to *avr* gene products, fungal and oomycete pathogens have a variety of components or secretory products, such as arachidonic acid, cell wall carbohydrates, glycoproteins and proteins, that can elicit plant defence responses and, in some cases, cell death (Ebel and Scheel, 1997; Chen and Heath, 1994). These 'non-specific elicitors' of cell death kill cells in a wide range of plants, often including those susceptible to the pathogen, and their binding-sites generally seem to be associated with the plant plasma membrane (Ebel and Scheel, 1997). Although proof of a role for these elicitors in the HR is generally lacking, an involvement in the HR of non-host plants seems likely. Direct evidence comes from the case of transformants of the potato pathogen, *Phytophthora infestans,* in which the lack of INF1, a 10 kDa protein of the death-eliciting elicitin family, is associated with a loss of ability to trigger the HR in one of three non-host *Nicotiana* species (Kamoun *et al.*, 1998). For this oomycete, and for *P. capsici*, it also has been suggested that pathogen wall components act as non-specific elicitors of cell death in host species, and that cell death is suppressed in susceptible host genotypes by cultivar-specific suppressors (Doke *et al.*, 1998). If this is the case, then *R* genes against these oomycetes may be involved in interfering with these suppressors, rather than directly triggering the HR (Heath, 1982).

In comparison to fungi and oomycetes, few non-specific elicitors of cell death have been isolated from bacteria. The exception is a family of glycine-rich, cysteine-lacking proteins known as harpins which are secreted by bacteria when grown in minimal medium in which *hrp* genes are derepressed. Harpins elicit an apparent HR when introduced in high concentrations into the intercellular spaces of plant leaves (Alfano and Collmer, 1996; He, 1998), and induce the putative HR marker gene *HIN1* in tobacco (Gopalan *et al.*, 1996). However, the membrane potential responses elicited by harpin differ from those elicited by bacteria (Pike *et al.*, 1998). Recent evidence suggests that the site of action of harpins may be the cell wall and, overall, there is still some doubt as to their natural role in the HR (Collmer, 1998).

Do *R* genes code for *avr* gene receptors?

Although *R* genes have been assumed to code for receptors of *avr* gene products, recent data question this assumption. Syringolide elicitors have soluble proteinaceous 34 kDa receptors in plants with or without the cognate *Rpg4* gene (Ji *et al.*, 1998), and the specific elicitor, AVR9, produced by the fungal pathogen *Cladosporium fulvum* has plasma membrane-bound binding sites in tomato plants with or without the *Cf-9* resistance gene (Laugé and de Wit, 1998). Also, yeast two-hybrid analysis has failed to demonstrate an interaction between RPM1 protein and the corresponding bacterial *avr* gene product (Boyes *et al.*, 1998). In fact, in only the tomato/bacterial pathogen *Pseudomonas syringae* pv. *tomato* system, in which the *R* gene *Pto* product has been identified as a serine-threonine kinase, is there evidence of direct interaction between *avr* and *R* gene products (Zhou *et al.*, 1998).*

Induction of hypersensitive cell death

Cell death and defence gene expression during the HR may involve separate pathways

In tomato, the yeast two-hybrid assay has been used to identify 10 classes of plant cDNAs that interact with *R* gene-coded Pto kinase (Zhou *et al.*, 1998). Their products include another protein kinase, Pti1, which accelerates HR development when expressed in tobacco and challenged with a tobacco bacterial pathogen carrying *avrPto*. Three other classes appear to be transcription factors that regulate pathogenesis-related (PR) genes, a sub-group of defence genes. These data suggest that hypersensitive cell death and defence gene activation may involve separate signalling pathways, both initially activated by *avr-R* gene product interaction (Zhou *et al.*, 1998). For fungal pathogens, the widespread sensitivity of plants to their non-specific elicitors makes it quite likely that the latter induce defence responses in addition to any such induction by *avr-R* gene product interactions. Further evidence that hypersensitive cell death and defence gene activation are not mandatorily linked comes from their separation by mutation (Jakobek and Lindgren, 1993; Yu *et al.*, 1998) and inhibitor studies (del Pozo and Lam, 1998).

Ion fluxes appear to be an early step in the HR

Plants commonly respond to external stimuli, including microbial elicitors of cell death and/or defence responses, by calcium influx into the cell (Ebel and Scheel, 1997; Pike *et al.*, 1998; Higgins *et al.*, 1998). An increase in cytosolic calcium precedes, and seems necessary for, hypersensitive cell death triggered by rust fungi (Xu and Heath, 1998) and the calcium channel blocker La^{3+} prevents bacterial-induced HR in soybean leaves (Levine *et al.*, 1996). An influx of calcium also seems necessary for the PCD that occurs during the differentiation of tracheary elements in *Zinnia* (Groover and Jones, 1999). However, the use of kinetin to raise cytosolic calcium to levels equivalent to those seen prior to the HR in cowpea did not result in cell death (Xu and Heath, 1998) suggesting that additional signals are involved or that the calcium signal needs a specific feature, such as a periodicity 'signature' (Malhó *et al.*, 1998).

Calcium signals are assumed to be translated through protein phosphorylation and such activity has been implicated in cell culture responses to the bacterial non-specific death elicitor harpin (Pike *et al.*, 1998), the oligopeptide non-specific elicitor of *Phytophthora sojae* (Ligterink *et al.*, 1997), the xylanase non-specific fungal elicitor from *Trichoderma viride* (Suzuki *et al.*, 1999) and the specific elicitor produced by *C. fulvum* (Blumwald *et al.*, 1998). The elicitor from *P. sojae* and another from *C. fulvum* activate mitogen-activated protein (MAP) kinases in signalling pathways that seem independent of the oxidative burst (described below) (Ligterink *et al.*, 1997; Romeis *et al.*, 1999).

For bacterial (Pike *et al.*, 1998) and oomycete (Doke *et al.*, 1998) pathogens, one of the earliest signs of the HR is membrane dysfunction. In cell suspensions, incompatible bacteria cause membrane depolarization, potassium efflux, and alkalinization of the external medium (Pike *et al.*, 1998) and the resulting K^+/H^+ exchange seems dependent on H^+-ATPase activity (Atkinson and Baker, 1989). In tomato suspension cells, the activation of a H^+-ATPase by a *C. fulvum* specific elicitor has been suggested to be responsible for inducing the opening of plasma membrane calcium channels (Blumwald *et al.*, 1998). In contrast, non-specific elicitors of defence responses often inhibit H^+-ATPase activity (Blumwald *et al.*, 1998). Interestingly, inhibitors or stimulators of H^+-ATPase activate different defence pathways in tomato suspension cells (Schaller and Oecking, 1999), but the link between proton fluxes and cell death is still unclear. For example, mutant bacteria that cannot induce a HR in tobacco leaves or death in cell suspensions still elicit H^+/K^+ exchange (Glazener *et al.*, 1996).

In the *C. fulvum*-tomato (Blumwald *et al.*, 1998) and *Phytophthora infestans*-potato (Kawakita and Doke, 1994) systems, ion flux responses to elicitor-receptor interaction seem to be mediated via heterotrimeric G proteins; correspondingly, mastoparan, a G-protein activator, elicits cell death, extracellular alkalinization, and an oxidative burst in isolated asparagus mesophyll cells (Allen *et al.*, 1999).

Role of reactive oxygen species

In aged potato slices, a burst of extracellular superoxide generation precedes membrane depolarization and electrolyte leakage prior to the HR caused by *Phytophthora infestans* (Doke, 1983; Doke *et al.*, 1998). Since the discovery of this oxidative burst, transient inter- or intra-cellular generation of reactive oxygen species (ROS) has been demonstrated in cell suspension cultures in response to incompatible bacteria (Baker and Orlandi, 1995; Chandra *et al.*, 1996b) or oomycete (Naton *et al.*, 1996) pathogens, as well as to non-specific (Lamb and Dixon, 1997) and specific (Higgins *et al.*, 1998) elicitors and mechanical cell perturbation (Gus-Mayer *et al.*, 1998). While they are not involved in eliciting some forms of plant developmental PCD (Groover *et al.*, 1997), ROS also have been implicated in animal apoptosis (Green and Reed, 1998) and, in the light of the close relationship between apoptosis and the cell cycle (Gilchrist, 1998), it may be significant that oxidative stress in plants can cause cell cycle arrest (Reichheld *et al.*, 1999). The cumulative effect of these data is that it is widely accepted that ROS are involved in triggering and/or executing the HR.

However, it cannot be ignored that some studies do not support a role for ROS in the induction of hypersensitive cell death. The oxidative burst cause by bacteria occurs in two distinct phases of which phase II appears to be mandatory for a HR (Baker and Orlandi, 1995; Chandra *et al.*, 1996b); nevertheless, this phase is triggered by mutant bacteria that do not induce an HR in tobacco leaves or death in cell suspensions (Glazener *et al.*, 1996). Also, because intact plants may not respond to elicitors or pathogens as do suspension cells (Pike *et al.* 1998), it is important to demonstrate an association of ROS with the HR *in planta*. Such studies generally show that an initial extracellular ROS generation is followed by intracellular generation (Doke, 1983; Thordal-Christensen *et al.*, 1997; Bestwick *et al.*, 1998; Heath, 1998b; Lu and Higgins, 1998) which, in at least one case (Heath, 1998b), is probably the result of cellular decompartmentalization. Plants have a number of ways of generating extracellular ROS (Bolwell and Wojtaszek, 1997) and either wall-bound peroxidase (Bestwick *et al.*, 1998) or plasma membrane-bound NADPH oxidase (Higgins *et al.*, 1998) have been implicated as important in different systems. Also, different elicitors of the oxidative burst seem to operate via different signalling pathways (Chandra *et al.*, 1996a). Hydrogen peroxide, in particular, has been suggested to be the more physiologically important ROS, but although exogenous H_2O_2 can cause plant cell death in some situations (Lamb and Dixon, 1997; Kazan *et al.*, 1998) and can induce transcripts of the putative HR marker gene, *HSR203J* (Baudouin *et al.*, 1999), it does not induce death in tomato in concentrations resembling those elicited *in planta* by a cell-death elicitor of the fungus *C. fulvum* (Lu and Higgins, 1999). Possibly, its effect is potentiated by the accompanying generation of nitric oxide (Delledonne *et al.*, 1998; Durner *et al.*, 1998) as seen in the mammalian immune system (see Delledonne *et al.*, 1998) or by the presence of Fe^{2+} (Lu and Higgins, 1999). Nevertheless, cytological studies of barley-powdery mildew interactions suggest that ROS, although present, are not a requirement for HR elicitation (Hückelhoven and Kogel, 1998), and equivalent studies of the cowpea-rust system could not detect any ROS generation prior to the onset of hypersensitive cell death (Heath, 1998b). The variable role of ROS in triggering hypersensitive cell death is also demonstrated by the fact that ROS scavengers can inhibit elicitor-induced cell death in some situations (Lamb and Dixon, 1997), but not in others (Yano *et al.*, 1999). The few studies that have examined ROS scavenger effects on pathogen-induced HR reveal that they may inhibit the HR induced by viruses (e.g. Kato and Misawa, 1976) or bacteria (Keppler and Novacky, 1987), but not by rust fungi in resistant host (Heath, 1998b) or non-host (Mellersh and Heath unpublished) plants.

Taken together, these data suggest that the role of ROS in inducing hypersensitive cell death may differ in different plant-pathogen combinations.

The role of salicylic acid

Transgenic *Arabidopsis* NahG plants expressing the bacterial enzyme salicylate hydroxylase cannot accumulate SA and do not express *R* gene-mediated HR against bacteria or biotrophic oomycete pathogens (Delaney *et al.*, 1994) or non-host HR against rust fungi (Mellersh and Heath, unpublished). NahG *Arabidopsis* plants also do not exhibit a HR-like response to ozone, unlike normal ecotypes (Rao and Davis, 1999), and virus-induced HR is replaced by spreading lesions in NahG tobacco (Chivasa and Carr, 1998). Exogenous SA will accelerate *R* gene-controlled cell death in soybean cell cultures caused by *Pseudomonas syringae* pv. *glycinea* (Shirasu *et al.*, 1997) and that caused in transgenic tobacco and canola plants by H_2O_2 generated by the expression of fungal glucose oxidase (Kazan *et al.*, 1998). Therefore, SA appears to play a central role in many types of plant cell death, including the HR, possibly related to its inhibition of mitochondrial function (Xie and Chen, 1999). (The role of SA is discussed more fully by Alvarez in this issue.)

The need for protein synthesis and the involvement of the cytoskeleton in the induction of hypersensitive cell death

Transcription and translation is necessary for the induction of the HR caused by rust fungi in cowpea (Heath *et al.*, 1997; Mould and Heath, 1999) whereas only translation is needed for that caused by downy mildew in lettuce (Mansfield *et al.*, 1997). Although the relevant protein products have not been identified, studies of genes that are expressed during the HR reveal putative genes involved in defence and basic metabolism (Seehaus and Tenhaken, 1998), as well as some that might be involved in protein and DNA degradation during hypersensitive cell death (Birch *et al.*, 1999).

Interestingly, an intact actin cytoskeleton seems to be necessary for HR induction by rust (Škalamera and

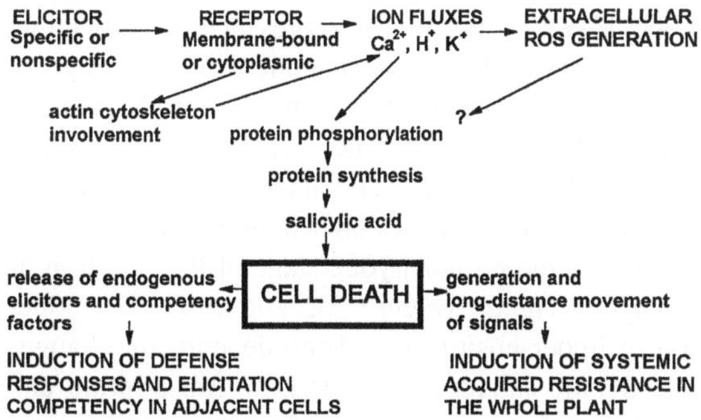

Figure 1. Factors for which there is, in at least some plant-pathogen interactions, current evidence of involvement in the induction of hypersensitive cell death, or in the consequences of such death. Note that single arrows do not preclude the presence of multiple, parallel, pathways. ROS, reactive oxygen species.

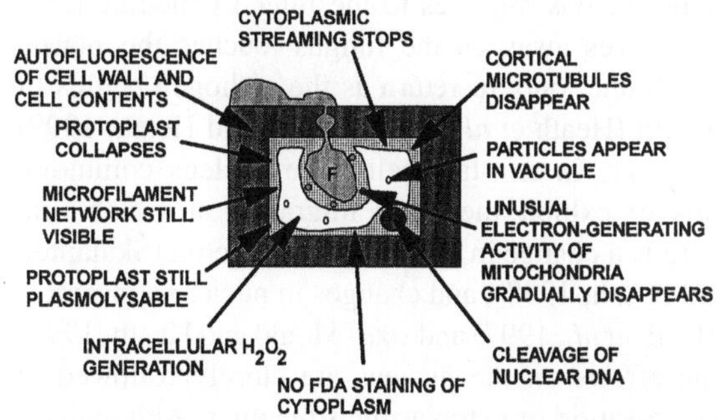

Figure 2. Diagrammatic representation of the changes that occur in a resistant cowpea epidermal cell during the execution phase of hypersensitive cell death caused by the cowpea rust fungus (F). This phase lasts for several hours and the clock-wise listing of events roughly follows the order in which they occur; cell death becomes irreversible just after cytoplasmic streaming stops.

Heath, 1998) and powdery mildew (Aist and Bushnell, 1991) fungi and by the oomycete *Phytophthora infestans* (Tomiyama *et al.*, 1982). Such results suggest either that actin-mediated organelle re-arrangements are needed for hypersensitive cell death, or that the actin cytoskeleton is involved in the signalling system that leads to this response.

A summary of the factors that may be involved in the induction of hypersensitive cell death is shown in Figure 1.

Execution of hypersensitive cell death

Changes in cellular components in hypersensitively dying cells

Relatively few plant-pathogen systems allow cells undergoing the HR to be watched under the light microscope. However, those that do present some intriguing observations and suggest that there may be significant differences in the execution of the death process. One of the most comprehensive correlative studies of the HR using living and fixed cells involves the *R* gene-mediated HR of cowpea epidermal vein cells elicited by the monokaryotic stage of the rust fungus *Uromyces vignae* (Figure 2). Although the induction of hypersensitive cell death depends on changes in cytosolic calcium levels and on transcription and translation that occur while the fungus is still growing through the plant wall (Heath *et al.*, 1997; Xu and Heath, 1998; Mould and Heath, 1999), the first observable sign of the HR occurs a few hours after the fungus enters the cell. In susceptible cells, the plant nucleus migrates to the fungal penetration site but moves away as the fungus touches the plasma membrane, only to return as the pathogen begins tip growth (Heath *et al.*, 1997; Mould and Heath, 1999). However, in resistant cells, the nucleus commonly fails to exhibit the latter migration at which time there is a change in the actin cytoskeleton (Škalamera and Heath, 1998), and changes in nuclear appearance (Heath *et al.*, 1997) and size (Mould and Heath. 1999). The effects on the nucleus are closely followed by the cessation of cytoplasmic streaming. Although cytoplasmic streaming is generally dependent on actin microfilaments (Aist and Bushnell, 1991; Gross *et al* 1993; Škalamera and Heath, 1998) rather than microtubules, actin networks and cables persist in the cytoplasm (Škalamera and Heath, 1998) but cortical or perinuclear microtubules disappear (Škalamera and Heath, 1998; Mould and Heath, 1999) depending on the host cultivar. At this point, cell death becomes irreversible and the cell then undergoes a protracted dismantling of the protoplast which may take several hours and involves sequentially the appearance of particles showing Brownian motion in the plant vacuole, the loss of fungus-induced electron-generating activity in the plant mitochondria, cleavage of nuclear DNA, the loss in ability of the protoplast to stain with the vital stain fluorescein diacetate and the intracellular generation of hydrogen peroxide (Heath *et al.*, 1997; Heath, 1998b). Finally, the plasma membrane loses its semipermeability and the protoplast shrinks in a manner suggestive of contraction rather than loss of turgor (Heath *et al.*, 1997). The cell then becomes autofluorescent and brown, presumably due to the accumulation and oxidation of phenolic compounds secreted by surrounding cells (Heath, 1998b; Škalamera and Heath, 1998) and granules of nucleic acid appear in the plant nucleus (Mould and Heath 1999). The fungus ceases its growth just after the cessation of cytoplasmic streaming but before plant protoplast collapse (Heath *et al.*, 1997).

This process resembles that caused in epidermal cells by powdery mildew fungi (Aist and Bushnell, 1991) during which there is disruption in long-distance streaming, a halt in streaming, a quiescent period of 1–2 h, and nucleus collapse just preceding protoplast collapse. Although there is a reduction in plasma membrane permeability after streaming stops, it retains its semipermeability until protoplast collapse. However, unlike the rust system, the fungus collapses a few minutes before or after plant cell collapse.

Visually different from either the rust- or powdery mildew-induced HR is that caused by the oomycete *Phytophthora infestans* in resistant potato vein epidermal cells in which the plant cell cytoplasm conglomerates around the pathogen and then suddenly expands and collapses within 26 s (Freytag *et al.*, 1994). Such rapid cellular disruption and changes in the volume of cellular compartments suggest rapid changes in ion fluxes across membranes. In this respect, this oomycete-induced cell death may more closely resemble bacteria-induced HR, in which changes in membrane potential and rapid irreversible plasma membrane damage involving lipid peroxidation (Pike *et al.*, 1998) and the degradation of the membrane-bound resistance gene product RPM1 (Boyes *et al.*, 1998) have been demonstrated. Irreversible membrane damage also is the first sign of hypersensitive cell death detected in lettuce cells caused by the oomycete *Bremia lactucae* (Mansfield *et al.* 1997). Interestingly, the rapidity of cell disruption in the *P. infestans*-induced HR also has some resemblance to that seen during the PCD in differentiating tracheary elements in which the cell dies through a rapid disruption of the plant vacuole (Groover *et al.*, 1997; Groover and Jones, 1999).

Cells killed via the HR usually become autofluorescent and then dark brown, attributable to the accumulation and oxidation of phenolic compounds (Nicholson and Hammerschmidt, 1992) and it is these features that are commonly used to identify the HR. However, autofluorescence and browning are the final steps in the death process, and they may occur hours after the death has become irreversible. They also may be inhibited independent of cell death (Mansfield *et al.*, 1997; Heath, 1998b) raising potential problems of data interpretation if hypersensitive cell death is identified only by cell autofluorescence and browning (Heath, 1998b).

Although apoptosis was first recognized in mammalian cells by its strikingly distinctive morphology, particularly as seen by electron microscopy (see Heath, 1998a), no equivalently consistent morphology has been recognized for the HR (e.g. Bestwick *et al.*, 1995; Mould and Heath, 1999) although some recent studies and reviews have tended to generate such consistency by emphasizing any apoptosis-like feature. Another hallmark of apoptosis is the cleavage of nuclear DNA into oligonucleosomal fragments, producing a 'ladder' of DNA when run on agarose gels. Plant DNA cleavage has been reported for fungus-, bacteria- and virus-induced HR (Levine *et al.*, 1996; Ryerson and Heath, 1996; Heath *et al.*, 1997; Mittler *et al.*, 1997a) but its timing during the death process varies and it results in a DNA ladder only during the fungus-induced response (Ryerson and Heath, 1996).

Involvement of proteases

During the death process in some forms of mammalian apoptosis, the release of cytochrome *c* from mitochondria causes the cascading activation of a family of cysteine proteases known as caspases that results in the degradation of a number of cell components including poly(ADP-ribose) polymerase (PARP) (Green and Reed, 1998). That plant cells contain similar proteases is indicated by the fact that mouse liver nuclei in cytosolic extracts of suspension-cultured carrot cells can be induced by cytochrome *c* to undergo features of apoptosis inhibitable by cysteine protease inhibitors (Zhao *et al.*, 1999). Hypersensitive cell death in tobacco caused by a bacterial pathogen of beans can be inhibited by specific inhibitors of caspases (del Pozo and Lam, 1998), and that caused in cowpea by a rust fungus can be delayed by a cysteine protease inhibitor that also prevents the degradation of animal PARP in extracts from cells undergoing the HR; inhibitor studies of this PARP cleavage suggested the involvement of both caspase-like cysteine proteases and those with no defined specificity (D'Silva *et al.*, 1998). Caspase-like activity can also be detected in tobacco tissue undergoing a virus-induced HR (del Pozo and Lam, 1998). All of these data suggest that the dismantling of the cell during the HR involves the activation of cysteine proteases as it does during mammalian apoptosis. However, plants can produce a variety of proteolytic enzymes and in the virus (del Pozo and Lam, 1998) system at least, it seems likely that other proteases are involved as they are in developmental forms of PCD in plants (Buchanan-Wollaston, 1997; Groover and Jones, 1999) and some forms of elicitor-induced cell death (Yano *et al.*, 1999). The role of caspases in plant PCD is discussed in more detail by Lam and del Pozo in this issue.

Is hypersensitive cell death a single phenomenon?

The combined evidence available so far suggests that, like mammalian apoptosis, there are a variety of elicitors and signalling pathways that can lead to hypersensitive cell death. There also is considerable phenotypic diversity among responses that are defined as the HR in different systems and compelling evidence that the actual mode and sequence of dismantling of the cell may vary. The HR also may differ between different plant-pathogen combinations in the timing of irreversible plant membrane damage, responses to heat shock, and the effects of metabolic inhibitors (Mansfield *et al.*, 1997). It seems obvious, therefore, that there are significant differences in the induction and execution of hypersensitive cell death in different plant-pathogen combinations. Although there may be some fundamental features common to all examples of the HR such as the involvement of cysteine proteases, there are sufficient differences between systems that a single model of HR induction (e.g. Jabs, 1999) may not apply to all.

Functions of the HR

With the experimental separation of hypersensitive cell death from defence gene induction and disease resistance (Jakobek and Lindgren, 1993; del Pozo and Lam, 1998; Heath, 1998a; Yu *et al.*, 1998) the necessity for cell death within the HR becomes more of an enigma. For pathogens that require living host cells for their survival, it is possible that cell death alone is a defence mechanism, but even this assumption is not always supported by experimental data (Richael and Gilchrist, 1999). Perhaps the reason why cell death is so consistently associated with disease resistance is related to the fact that localized plant cell death caused by such treatments as pathogen elicitors (Dorey *et al.*, 1997) or point-freezing (Rahe and Arnold, 1975) releases signals (endogenous elicitors) that cause defensive responses in surrounding cells. Hypersensitive cell death also has been suggested to release signals that condition adjacent cells to become responsive to pathogen elicitors (Graham and Graham, 1999) and that activate systemic resistance throughout the plant. Alvarez *et al.* (1998) have suggested that this systemic acquired resistance (SAR) in *Arabidopsis* depends on secondary oxidative bursts in distant tissues and the formation of 'micro-HRs'. Such data suggest that the cell death component of the HR may function more as a signalling system than as a direct defence mechanism (Figure 1).

Prevention of hypersensitive cell death by biotrophic pathogens

If localized, rapid cell death in association with the development of an antimicrobial environment is the normal response of plants to invading pathogens, then the question arises of why is it absent in disease-susceptible plants. For micro-organisms that can survive in dead cells, the production of toxins or cell wall-degrading enzymes, coupled to fast pathogen growth, may narcotize or kill cells before the HR can be mobilized. Interestingly, some pathogen toxins induce what appears to be a PCD (Gilchrist, 1998), which raises interesting questions as to how this 'susceptible' PCD differs from the programmed cell death of the HR. However, biotrophic pathogens that require living cells for their survival must prevent any type of cell death as well as preventing the inducible defence responses that living cells are able to muster. How they do this is still a mystery although *Phytophthora* species have been suggested to produce death-suppressing glucans (Doke *et al.*, 1998). Although not a biotroph, the fungus *Mycosphaerella pinodes* produces peptide suppressors of defence responses in its host, pea, that may interfere with signal transduction by inhibiting plasma membrane ATPase activity and polyphosphoinositide metabolism (Shiraishi *et al.*, 1994); perhaps biotrophs similarly can interfere with signal transduction pathways leading to defence responses and the HR.

Exploitation of the HR for disease control

Although the HR is commonly used in resistant host genotypes for disease control, its conventional use is often limited by the rapid evolution of parasite strains with modified *avr* gene products that are no longer recognized by host *R* genes (Heath, 1997). Therefore, a number of novel strategies are currently being devised to enhance the durability and usefulness of HR-controlled resistance. One is to use *avr/R* gene pairs as two-component sensor systems which could be introduced into crop plants with the *avr* gene under the control of a pathogen-inducible promoter so that infection by any pathogen will trigger a HR (Laugé and de Wit, 1998); a similar strategy is to fuse such a promoter to the gene coding for a non-specific death elicitor (Keller *et al.*, 1999). Other possibilities are to use such gene pairs to induce low levels of cell death that trigger SAR (Hammond-Kosack *et al.*, 1998), or to manipulate the signal transduction pathways that lead to the HR (Higgins *et al.*, 1998).

*Note added in proof

The direct interaction of resistance gene and avirulence gene products has recently been reported for the rice blast system (Jia, Y., McAdams, S.A., Bryan, G.T., Hershey, H.P., and Valent, B. 2000. EMBO Journal 19, pp. 4004–4014).

References

Aist, J.R. and Bushnell, W.R. 1991. Invasion of plants by powdery mildew fungi, and cellular mechanisms of resistance. In: G.T. Cole and H.C. Hoch (Eds.) The Fungal Spore and Disease Initiation in Plants and Animals, Plenum Press, New York/London, pp. 321–345.

Alfano, J.R. and Collmer, A. 1996. Bacterial pathogens in plants: life up against the wall. Plant Cell 8: 1683–1698.

Allen, L.J., MacGregor, K.B., Koop, R.S., Bruce, D.H., Karner, J. and Bown, A.W. 1999. The relationship between photosynthesis and a mastoparan-induced hypersensitive response in isolated mesophyll cells. Plant Phyiol 119: 1233–1241.

Alvarez, M.E., Pennell, R.I., Meijer, P.-J., Ishikawa, A., Dixon, R.A. and Lamb, C. 1998. Reactive oxygen intermediates mediate a systemic signal network in the establishment of plant immunity. Cell 92: 773–784.

Atkinson, M.M. and Baker, C.J. 1989. Role of the plasmalemma H^+-ATPase in *Pseudomonas syringae*-induced K^+/H^+ exchange in suspension-cultured tobacco cells. Plant Physiol 91: 298–303.

Baker, J.C. and Orlandi, E.W. 1995. Active oxygen in plant pathogenesis. Annu. Rev. Phytopath. 33: 299–321.

Baudouin, E., Charpenteau, M., Ranjeva, R. and Ranty, B. 1999. Involvement of active oxygen species in the regulation of a tobacco defence gene by phorbol ester. Plant Sci. 142: 67–72.

Bestwick, C.S., Bennett, M.H. and Mansfield, J.W. 1995. Hrp mutant of *pseudomonas syringae* pv. *phaseolicola* induced cell wall alterations but not membrane damage leading to the hypersensitive reaction in lettuce. Plant Physiol. 108: 503–516.

Bestwick, C.S., Brown, I.R. and Mansfield, J.W. 1998. Localized changes in peroxidase activity accompany hydrogen peroxide generation during the development of a nonhost hypersensitive reaction in lettuce. Plant Physiol. 118: 1067–1078.

Birch, P.R.J., Avrova, A.O., Duncan, J.M., Lyon, G.D. and Toth, R.L. 1999. Isolation of potato genes that are induced during an early stage of the hypersensitive response to *Phytophthora infestans*. Mol. Plant-Microbe Interact. 12: 356–361.

Blumwald, E., Aharon, G.S. and Lam, B.C.-H. 1998. Early signal transduction pathways in plant-pathogen interactions. Trends Plant Sci. 3: 342–346.

Bolwell, G.P. and Wojtaszek, P. 1997. Mechanisms for the generation of reactive oxygen species in plant defence: a broad perspective. Physiol. Mol. Plant Path. 51: 347–366.

Boyes, D.C., Nam, J. and Dangl, J.L. 1998. The *Arabidopsis thaliana RPM1* disease resistance gene product is a peripheral plasma membrane protein that is degraded coincident with the hypersensitive response. Proc. Natl. Acad. Sci. USA 95: 15849–15854.

Buchanan-Wollaston, V. 1997. The molecular biology of leaf senescence. J. Exp. Bot. 48: 181–199.

Chandra, S., Heinstein, P.F. and Low, P.S. 1996a. Activation of phospholipase A by plant defense elicitors. Plant Physiol. 110: 979–986.

Chandra, S., Martin, G.B. and Low, P.S. 1996b. The Pto kinase mediates a signalling pathway leading to the oxidative burst in tomato. Proc. Natl. Acad. Sci. USA 93: 13393–13397.

Chen, C.-Y. and Heath, M.C. 1992. Effect of stage of development of the cowpea rust fungus on the release of a cultivar-specific elicitor of necrosis. Physiol. Mol. Plant Path. 40: 23–30.

Chen, C.-Y. and Heath, M.C. 1994. Elicitors of necrosis in rust diseases. In: K. Kohmoto and O.C. Yoder (Eds.) Host-Specific Toxin: Biosynthesis, Receptor and Molecular Biology, Tottori University, Japan, pp. 73–82.

Chivasa, S. and Carr, J.P. 1998. Cyanide restores *N* gene-mediated resistance to tobacco mosaic virus in transgenic tobacco expressing salicylic acid hydroxylase. Plant Cell 10: 1489–1498.

Collmer, A. 1998. Determinants of pathogenicity and avirulence in plant pathogenic bacteria. Curr. Opin. Plant Biol. 1: 329–335.

Dangl, J.L., Dietrich, R.A. and Richberg, M.H. 1996. Death don't have no mercy: cell death programs in plant-microbe interactions. Plant Cell 8: 1793–1807.

Dawson, W.O. 1999. Tobacco mosaic virus virulence and avirulence. Phil. Trans. R. Soc. Lond. B 354: 645–651.

Delaney, T.P., Uknes, S., Vernooij, B., Friedrich, L., Weymann, K., Negrotto, D., Gaffne, T., Gut-Rella, M., Kessmann, H., Ward, E. and Ryals, J. 1994. A central role of salicylic acid in plant disease resistance. Science 266: 1247–1250

Delledonne, M., Xia, Y., Dixon, R.A. and Lamb, C. 1998. Nitric oxide functions as a signal in plant disease resistance. Nature 394: 585–588.

del Pozo, O. and Lam, E. 1998. Caspases and programmed cell death in the hypersensitive response of plants to pathogens. Curr. Biol. 8: 1129–1132.

Dion, M., Chamberland, H., St-Michel, C., Plante, M., Darveau, A., Lafontaine, J.G. and Brisson, L.F. 1997. Detection of a homologue of bcl-2 in plant cells. Biochem. Cell Biol. 75: 457–461.

Doke, N. 1983. Involvement of superoxide anion generation in the hypersensitive response of potato tuber tissue to infection with an incompatible race of *Phytophthora infestans* and to the hypha wall components. Physiol. Plant Path. 23: 345–357.

Doke, N., Sanchez, L.M., Yoshioka, H., Kawakita, K., Miura, Y. and Park, H.-J. 1998. In: K. Kohmoto and O.C. Yoder (Eds.) Molecular Genetics of Host-Specific Toxins in Plant Disease, Kluwer Academic Publishers, Dordrecht, Netherlands, pp. 331–341.

Dong, Y.-H., Zhan, X.-C., Kvarnheden, A., Atkinson, R.G., Morris, B.A. and Gardner, R.C. 1998. Expression of a cDNA from apple encoding a homologue of DAD1, an inhibitor of programmed cell death. Plant Sci. 139: 165–174.

Dorey, S., Baillieul, F., Pierrel, M.-A., Saindrenan, P., Fritig, B. and Kauffmann, S. 1997. Spatial and temporal induction of cell death, defense genes, and accumulation of salicylic acid in tobacco leaves reacting hypersensitively to a fungal glycoprotein elicitor. Mol. Plant-Microbe Interact. 10: 646–655.

D'Silva, I. and Heath, M.C. 1997. Purification and characterization of two novel hypersensitive response-inducing specific elicitors produced by the cowpea rust fungus. J. Biol. Chem. 272: 3924–3927.

D'Silva, I., Poirier, G.G. and Heath, M.C. 1998. Activation of cysteine proteases in cowpea plants during the hypersensitive response – a form of programmed cell death. Exp. Cell Res. 245: 389–399.

Durner, J., Wendehenne, D. and Klessig, D.F. 1998. Defense gene induction in tobacco by nitric oxide, cyclic GMP, and cyclic ADP-ribose. Proc. Natl. Acad. Sci. USA 95: 10328–10333.

Ebel, J. and Scheel, D. 1997. Signals in host-parasite interactions. In: G.C. Carroll and P. Tudzynski (Eds.) The Mycota Vol. V. Plant Relationships Part A, Springer-Verlag, Berlin/Heidelberg, pp. 85–105.

Falk, A., Feys, B.J., Frost, L.N., Jones, J.D.G., Daniels, M.J. and Parker, J.E. 1999. *EDS1*, an essential component of *R* gene-mediated disease resistance in *Arabidopsis* has homology to eukaryotic lipases. Proc. Natl. Acad. Sci. USA 96: 3292–3297.

Freytag, S., Arabatzis, N., Hahlbrock, K. and Schmelzer, E. 1994. Reversible cytoplasmic rearrangements precede wall apposition, hypersensitive cell death and defense-related gene activation in potato/*Phytophthora infestans* interactions. Planta 194: 123–135.

Gabriel, D.W. 1999. Why do pathogens carry avirulence genes? Physiol. Mol. Plant Path. 55: 205–214.

Gilchrist, D.G. 1998. Programmed cell death in plant disease: the purpose and promise of cellular suicide. Annu. Rev. Phytopath. 36: 393–414.

Glazener, J.A., Orlandi, E.W. and Baker, C.J. 1996. The active oxygen response of cell suspensions to incompatible bacteria is not sufficient to cause hypersensitive cell death. Plant Physiol. 110: 759–763.

Goodman, R.N. and Novacky, A.J. 1994. The Hypersensitive Reaction in Plants to Pathogens. APS Press, St. Paul, MN.

Gopalan, S., Wei, W. and He, S.Y. 1996. *hrp* gene-dependent induction of *hin 1*: a plant gene activated rapidly by both harpins and the *avrPto* gene-mediated signal. Plant J. 10: 591–600.

Graham, T.L. and Graham, M.Y. 1999. Role of hypersensitive cell death in conditioning elicitation competency and defense potentiation. Physiol. Mol. Plant Path. 55: 13–20.

Green, D.R. and Reed, J.C. 1998. Mitochondria and apoptosis. Science 281: 1309–1312.

Groover, A. and Jones, A.M. 1999. Tracheary element differentiation uses a novel mechanism coordinating programmed cell death and secondary wall synthesis. Plant Physiol. 119: 375–384.

Groover A., Dewitt, N., Heidel, A. and Jones, A. 1997. Programmed cell death of plant tracheary elements differentiating in vitro. Protoplasma 196: 197–211.

Gross, P., Julius, C., Schmelzer, E. and Hahlbrock, K. 1993. Translocation of cytoplasm and nucleus to fungal penetration sites is associated with depolymerisation of microtubules and defence gene activation in infected cultured parsley cells. EMBO J. 12: 1735–1744.

Gus-Mayer, S., Naton, B., Hahlbrock, K. and Schmelzer, E. 1998. Local mechanical stimulation induces components of the pathogen defense response in parsley. Proc. Natl. Acad. Sci. USA 95: 8398–8403.

Hammond-Kosack, K.E. and Jones, J.D.G. 1997. Plant disease resistance genes. Annu. Rev. Plant Physiol. Plant Mol. Biol. 48: 575–607.

Hammond-Kosack, K.E., Tang, S., Harrison, K. and Jones, J.D.G. 1998. The tomato *Cf-9* disease resistance gene functions in tobacco and potato to confer responsiveness to the fungal avirulence gene product Avr9. Plant Cell 10: 1251–1266.

He, S.Y. 1998. Type III protein secretion systems in plant and animal pathogenic bacteria. Annu. Rev. Phytopath. 36: 363–392.

Heath, M.C. 1982. The absence of active defense mechanisms in compatible host-pathogen interactions. In: R.K.S. Wood (Ed.) Active Defense Mechanisms in Plants, Plenum Press, New York, pp. 143–156.

Heath, M.C. 1991. The role of gene-for-gene interactions in the determination of host species specificity. Phytopathology 81: 127–130.

Heath, M.C. 1996. Plant resistance to fungi. Can. J. Plant Path. 18: 469–475.

Heath, M.C. 1997. Evolution of plant resistance and susceptibility to fungal parasites. In: G.C. Carroll and P. Tudzynski (Eds.) The Mycota Vol. V. Plant Relationships Part B, Springer-Verlag, Berlin/Heidelberg, pp. 257–276.

Heath, M.C. 1998a. Apoptosis, programmed cell death and the hypersensitive response. Eur. J. Plant Path. 104: 117–124.

Heath, M.C. 1998b. Involvement of reactive oxygen species in the response of resistant (hypersensitive) or susceptible cowpeas to the cowpea rust fungus. New Phytol. 138: 251–263.

Heath, M.C., Nimchuk, Z.L. and Xu, H. 1997. Plant nuclear migrations as indicators of critical interactions between resistant or susceptible cowpea epidermal cells and invasion hyphae of the cowpea rust fungus. New Phytol. 135: 689–700.

Higgins, V.J., Lu, H., Xing, T., Gellie, A. and Blumwald, E. 1998. The gene-for-gene concept and beyond: interactions and signals. Can. J. Plant Path. 20: 150–157.

Hu, G., Richter, T.E., Hulbert, S.H. and Pryor, T. 1996. Disease lesion mimicry caused by mutations in the rust resistance gene *rp1*. Plant Cell 8: 1367–1376.

Hu, G., Yalpani, N., Briggs, S.P. and Johal, G.S. 1998. A porphyrin pathway impairment is responsible for the phenotype of a dominant disease lesion mimic mutant of maize. Plant Cell 10: 1095–1105.

Hückelhoven, R. and Kogel, K.-H. 1998. Tissue-specific superoxide generation at interaction sites in resistant and susceptible near-isogenic barley lines attacked by the powdery mildew fungus (*Erysiphe graminis* f. sp. *hordei*). Mol. Plant-Microbe Interact. 11: 292–300.

Jabs, T. 1999. Reactive oxygen intermediates as mediators of programmed cell death in plants and animals. Biochem. Pharmacol. 57: 231–245.

Jakobek, J.L. and Lindgren, P.B. 1993. Generalized induction of defense responses in bean is not correlated with the induction of the hypersensitive reaction. Plant Cell 5: 49–56.

Ji, C., Boyd, C., Slaymaker, D., Okinaka, Y., Takeuchi, Y., Midland, S.L., Sims, J.J., Herman, E. and Keen, N. 1998. Characterization of a 34-kDa soybean binding protein for the syringolide elicitors. Proc. Natl. Acad. Sci. USA 95: 3306–3311.

Johal, G.S. Briggs, S.P. 1992. Reductase activity encoded by the *HM1* disease resistance gene in maize. Science 258: 985–987.

Kamoun, S., van West, P., Vleeshouwers, V.G.A.A., de Groot, K.E. and Govers, F. 1998. Resistance of *Nicotiana benthamiana* to *Phytophthora infestans* is mediated by the recognition of the elicitor protein INF1. Plant Cell 10: 1413–1425.

Kato, S. and Misawa, T. 1976. Lipid peroxidation during the appearance of hypersensitive reaction in cowpea leaves infected with cucumber mosaic virus. Ann. Phytopath. Soc. Japan 42: 472–480.

Kawakita, K. and Doke, N. 1994. Involvement of a GTP-binding protein in signal transduction in potato tubers treated with the fungal elicitor from *Phytophthora infestans*. Plant Sci. 96: 81–86.

Kazan, K., Murray, F.R., Goulter, K.C., Llewellyn, D.J. and Manners, J.M. 1998. Induction of cell death in transgenic plants expressing a fungal glucose oxidase. Mol. Plant-Microbe Interact. 11: 555–562.

Keller, H., Pamboukdjian, N., Ponchet, M., Poupet, A., Delon, R., Verrier, J.-L., Roby, D. and Ricci, P. 1999. Pathogen-induced elicitin production in transgenic tobacco generates a hypersensitive response in nonspecific disease resistance. Plant Cell 11: 223–235.

Keppler, L.D. and Novacky, A. 1987. The initiation of membrane lipid peroxidation during bacteria-induced hypersensitive reaction. Physiol. Mol. Plant Path. 30: 233–245.

Lamb, C. and Dixon, R.A. 1997. The oxidative burst in plant disease resistance. Annu. Rev. Plant Physiol. Plant Mol. Biol. 48: 251–275.

Laugé, R. and de Wit, P.J.G.M. 1998. Fungal avirulence genes: structure and possible functions. Fungal Genet. Biol. 24: 285–297.

Levine, A., Pennell, R.I., Alvarez, M.E., Palmer, R. and Lamb, C. 1996. Calcium-mediated apoptosis in a plant hypersensitive disease resistance response. Curr. Biol. 6: 427–437.

Ligterink, W., Kroj, T., zur Nieden, U., Hirt, H. and Scheel, D. 1997. Receptor-mediated activation of a MAP kinase in pathogen defense of plants. Science 276: 2054–2057.

Lu, H. and Higgins, V.J. 1998. Measurement of active oxygen species generated *in planta* in response to elicitor AVR9 of *Cladosporium fulvum*. Physiol. Mol. Plant Path. 52: 35–51.

Lu, H. and Higgins, V.J. 1999. The effect of hydrogen peroxide on the viability of tomato cells and of the fungal pathogen *Cladosporium fulvum*. Physiol. Mol. Plant Path. 54: 131–143.

Malhó, R., Moutinho, A., van der Luit, A. and Trewavas, A.J. 1998. Spatial characteristics of calcium signalling: the calcium wave as a basic unit in plant cell calcium signalling. Phil. Trans. R. Soc. Lond. B 353: 1463–1473.

Mansfield, J., Bennett, M., Bestwick, C. and Woods-Tör, A. 1997. Phenotypic expression of gene-for-gene interactions involving fungal and bacterial pathogens: variation from recognition to response. In: I.R. Crute, E.B. Holub and J.J. Burdon (Eds.) The Gene-for-Gene Relationship in Plant-Parasite Interactions, CAB International, Wallingford, UK/ New York, pp. 265–291.

Meyer, S.L.F. and Heath, M.C. 1988a. A comparison of the death induced by fungal invasion or toxic chemicals in cowpea epidermal cells. I. Cell death induced by heavy metal salts. Can. J. Bot. 66: 613–623.

Meyer, S.L.F. and Heath, M.C. 1988b. A comparison of the death induced by fungal invasion or toxic chemicals in cowpea epidermal cells. II. Responses induced by *Erysiphe cichoracearum*. Can. J. Bot. 66: 624–634.

Michelmore, R.W. and Meyers, B.C. 1998. Clusters of resistance genes in plants evolve by divergent selection and a birth-and-death process. Genome Res. 8: 1113–1130.

Mittler, R., Simon, L. and Lam, E. 1997a. Pathogen-induced programmed cell death in tobacco. J. Cell Sci. 110: 1333–1344.

Mittler, R., del Pozo, O., Meisel, L. and Lam, E. 1997b. Pathogen-induced programmed cell death in plants, a possible defense mechanism. Dev. Genet. 21: 279–289.

Molina, A., Volrath, S., Guyer, D., Maleck, K., Ryals, J. and Ward, E. 1999. Inhibition of protoporphyrinogen oxidase expression in *Arabidopsis* causes a lesion mimic phenotype that induced systemic acquired resistance. Plant J. 17: 667–678.

Morel, J.-B. and Dangl, J.L. 1997. The hypersensitive response and the induction of cell death in plants. Cell Death Different. 4: 671–683.

Morel, J.-B. and Dangl, J.L. 1999. Suppressors of the *Arabidopsis lsd5* cell death mutation identify genes involved in regulating disease resistance responses. Genetics 151: 305–319.

Mould, M.J.R. and Heath, M.C. 1999. Ultrastructural evidence of differential changes in transcription, translation, and cortical microtubules during *in planta* penetration of cells resistant or susceptible to rust infection. Physiol. Mol. Plant Path. 55: 225–236.

Naton, B., Hahlbrock, K. and Schmelzer, E. 1996. Correlation of rapid cell death with metabolic changes in fungus-infected, cultured parsley cells. Plant Physiol. 112: 433–444.

Nicholson, R.L. and Hammerschmidt, R. 1992. Phenolic compounds and their role in disease resistance. Annu. Rev. Phytopath. 30: 369–386.

Person, C. and Mayo, G.M.E. 1973. Genetic limitations on models of specific interactions between a host and its parasite. Can. J. Bot. 52: 1339–1347.

Peterhänsel, C., Freialdenhoven, A., Kurth, J., Kolsch, R. and Schulze-Lefert, P. 1997. Interaction analyses of genes required for resistance responses to powdery mildew in barley reveal distinct pathways leading to leaf cell death. Plant Cell 9: 1397–1409.

Pike, S.M., Ádám, A.L., Pu, X.-A., Hoyos, M.E., Laby, R., Beer, S.V. and Novacky, A. 1998. Effects of *Erwinia amylovora* harpin on tobacco leaf cell membranes are related to leaf necrosis and electrolyte leakage and distinct from perturbations caused by inoculated *E. amylovora*. Physiol. Mol. Plant Path. 53: 39–60.

Pontier, D., Tronchet, M., Rogowsky, P., Lam, E. and Roby, D. 1998. Activation of *hsr203*, a plant gene expressed during incompatible plant-pathogen interactions, is correlated with programmed cell death. Mol. Plant-Microbe Interact. 11: 544–554.

Pontier, D., Gan, S., Amasino, R.M., Roby, D. and Lam, E. 1999. Markers for hypersensitive response and senescence show distinct patterns of expression. Plant Mol. Biol. 39: 1243–1255.

Rahe, J.E. and Arnold, R.M. 1975. Injury-related phaseolin accumulation in *Phaseolus vulgaris* and its implications with regard to specificity of host-parasite interaction. Can. J. Bot. 53: 921–928.

Rathjen, J.P., Chang, J.H., Staskawicz, B.J. and Michelmore, R.W. 1999. Constitutively active *Pto* induced a *Prf*-dependent hypersensitive response in the absence of *avrPto*. EMBO J. 18: 3232–3240.

Rao, M.V. and Davis, K.R. 1999. Ozone induced cell death occurs via two distinct mechanisms in *Arabidopsis*: the role of salicylic acid. Plant J. 17: 603–614.

Reichheld, J.-P., Vernoux, T., Lardon, F., Van Montagu, M. and Inzé, D. 1999. Specific checkpoints regulate plant cell cycle progression in response to oxidative stress. Plant J. 17: 647–656.

Richael, C. and Gilchrist, D. 1999. The hypersensitive response: a case of hold or fold? Physiol. Mol. Plant Path. 55: 5–12.

Romeis, T., Peidras, P. Zhang, S., Klessig, D.F., Hirt, H. and Jones, J.D.G. 1999. Rapid Avr9- and Cf-9-dependent activation of MAP kinases in tobaccco cell cultures and leaves: convergence of resistance gene, elicitor, wound and salicylate responses. Plant Cell 11: 273–287.

Ryerson, D.E. and Heath, M.C. 1996. Cleavage of nuclear DNA into oligonucleosomal fragments during cell death induced by fungal infection or be abiotic treatments. Plant Cell 8: 393–402.

Schaller, A. and Oecking, C. 1999. Modulation of plasma membrane H^+-ATPase activity differentially activates wound and pathogen defense responses in tomato plants. Plant Cell 11: 263–272.

Seehaus, K. and Tenhaken, R. 1998. Cloning of genes by mRNA differential display induced during the hypersensitive reaction of soybean after inoculation with *Pseudomonas syringae* pv. *glycinea*. Plant Mol. Biol. 38: 1225–1234.

Shiraishi, T., Yamada, K., Toyoda, K., Kato, T., Kin, H.M., Ishinose, Y. and Oku, H. 1994. Regulation of ATPase and signal transduction for pea defense responses by the suppressor and elicitor from *Mycosphaerella pinodes*. In: K. Kohmoto and O.C. Yoder (Eds.) Host-Specific Toxin: Biosynthesis, Receptor and Molecular Biology, Tottori University, Japan, pp. 169–182.

Shirasu, K., Nakajima, H., Rajasekhar, V.K., Dixon, R.A. and Lamb, C. 1997. Salicylic acid potentiates an agonist-dependent gain control that amplifies pathogen signals in the activation of defense mechanisms. Plant Cell 9: 1–10.

Škalamera, D. and Heath, M.C. 1998. Changes in the cytoskeleton accompanying infection-induced nuclear movements and the hypersensitive response in plant cells invaded by rust fungi. Plant J. 16: 191–200.

Stakman, E.C. 1915. Relation between *Puccinia graminis* and plants highly resistant to its attack. J. Agric. Res. 4: 193–200.

Suzuki, K., Yano, A. and Shinshi, H. 1999. Slow and prolonged activation of the p47 protein kinase during hypersensitive cell death in a culture of tobacco cells. Plant Physiol. 119: 1465–1472.

Thordal-Christensen, H., Zhang, Z., Wei, Y. and Collinge, D.B. 1997. Subcellular localization of H_2O_2 in plants. H_2O_2 accumulation in papillae and hypersensitive response during the barley-powdery mildew interaction. Plant J. 11: 1187–1194.

Tomiyama, K., Sato, K., and Doke, N. 1982. Effect of cytochalasin B and colchicine on hypersensitive death of potato cells infected by incompatible race of *Phytophthora infestans*. Ann. Phytopath. Soc. Japan 48: 228–230.

van der Biezen, E.A. and Jones, J.D.G. 1998. The NB-ARC domain: a novel signalling motif shared by plant resistance gene products and regulators of cell death in animals. Curr. Biol. 8: R226–R227.

Warren, R.F., Henk, A., Mowery, P., Holub, E. and Innes, R.W. 1998. A mutation within the leucine-rich repeat domain of the *Arabidopsis* disease resistance gene *RPS5* partially suppresses multiple bacterial and downy mildew resistance genes. Plant Cell 10: 1439–1452.

Xie, Z. and Chen, Z. 1999. Salicylic acid induces rapid inhibition of mitochondrial electron transport and oxidative phosphorylation in tobacco cells. Plant Physiol. 120: 217–225.

Xu, H. and Heath, M.C. 1998. Role of calcium in signal transduction during the hypersensitive response caused by basidiospore-derived infection of the cowpea rust fungus. Plant Cell 10: 585–597.

Yano, A., Suzuki, K. and Shinshi, H. 1999. A signalling pathway, independent of the oxidative burst, that leads to hypersensitive cell death in cultured tobacco cells includes a serine protease. Plant J. 18: 105–109.

Yu, I.-c., Parker, J. and Bent, A.F. 1998. Gene-for-gene disease resistance without the hypersensitive response in *Arabidopsis dnd1* mutant. Proc. Natl. Acad. Sci. USA 95: 7819–7824.

Zeyen, R.J., Bushnell, W.R., Carver, T.L.W., Robbins, M.P., Clark, T.A., Boyles, D.A. and Vance, C.P. 1995. Inhibiting phenyl-

alanine ammonia lyase and cinnamyl-alcohol dehydrogenase suppresses *Mla1* (HR) but not *mlo5* (non-HR) barley powdery mildew resistances. Physiol. Mol. Plant Path. 47: 119–140.
Zhao, Y., Jiang, Z.-F., Sun, Y.-L. and Zhai, Z.-H. 1999. Apoptosis of mouse liver nuclei in the cytosol of carrot cells. FEBS Lett. 448: 197–200.
Zhou, J., Tang, X., Frederick, R. and Martin, G. 1998. Pathogen recognition and signal transduction by the Pto kinase. J. Plant Res. 111: 353–356.

Plant Molecular Biology **44:** 335–344, 2000.
E. Lam, H. Fukuda and J. Greenberg (Eds.), Programmed Cell Death in Higher Plants.

Transgene-induced lesion mimic

Ron Mittler[1,*] and Ludmila Rizhsky
Department of Plant Sciences, The Hebrew University of Jerusalem, Jerusalem 91904, Israel
[1]*Current address: Department of Biology, Technion-Israel Institute of Technology, Technion City, Haifa 32000, Israel (*author for correspondence; fax 972-2-6585093, e-mail: mittler@tx.technion.ac.il)*

Key words: biotechnology, lesion mimic, plant-pathogen interactions, programmed cell death

Abstract

Lesion mimic, i.e., the spontaneous formation of lesions resembling hypersensitive response (HR) lesions in the absence of a pathogen, is a dramatic phenotype occasionally found to accompany the expression of different, mostly unrelated, transgenes in plants. Recent studies indicated that transgene-induced lesion formation is not a simple case of necrosis, i.e., direct killing of cells by the transgene product, but results from the activation of a programmed cell death (PCD) pathway. Moreover, activation of HR-like cell death by transgene expression is viewed as an important evidence for the existence of a PCD pathway in plants. The study of lesion mimic transgenes is important to our understanding of PCD and the signals that control it in plants. PCD-inducing transgenes may provide clues regarding the different entry points into the cell death pathway, the relationships between the different branches of the pathway (e.g., developmental or environmental), or the different mechanisms involved in its induction or execution. Cell death-inducing transgenes may also be useful in biotechnology. Some lesion mimic transgenes were found to be induced in plants a state of systemic acquired resistance (SAR). These genes can be used in the development of pathogen-resistant crops. Other cell death-inducing transgenes may be used as specific cell ablation tools. Although mainly revealed unintentionally, and at times considered 'an adverse phenotype', lesion mimic transgenes should not be ignored because they may prove valuable for studying PCD as well as developing useful traits in different plants and crops.

Introduction

Programmed cell death and plant defense

The recognition of an invading pathogen by plant cells results in the induction of different antimicrobial defenses. These include production of reactive oxygen intermediates (ROI), strengthening of cell walls, synthesis of phytoalexins, and induction of pathogenesis-related (PR) proteins (Hammond-Kosack and Jones, 1996; Yang *et al.* 1997). Occasionally these responses are accompanied by a rapid death of cells at and around the site of infection (Dangl *et al.*, 1996). This response results in the formation of a zone of dead cells referred to as a 'lesion'. Since lesions may also form in plants during late stages of infection (i.e., as disease symptoms), the appearance of lesions at an early stage of infection makes it seem as if the plant hyper-reacts to the pathogen. This apparent phenomenology led to the coining of the term 'hypersensitive response' (HR), which is used to describe the early and rapid cell death response that accompanies pathogen resistance (Goodman and Novacky , 1994).

The exact cause of HR cell death is not clear. However, it is now established that this death is not directly caused by the pathogen. Thus, as opposed to cell death that occurs in a susceptible plant, late during infection, usually caused by pathogen proliferation or by different toxins produced and secreted by the pathogen, the early and rapid cell death response that occurs during the HR appears to be the outcome of a plant encoded mechanism for programmed cell death (PCD; Dangl *et al.,* 1996; Greenberg, 1996; Mittler and Lam, 1996; Pennel and Lamb, 1997). The term PCD is used to describe cell death which results from the activation of a cell suicide pathway encoded by the genome of the dying cell (Raff, 1992; Schwartzman and Cidlowski, 1993). This type of cell death is different from that which occurs during necrosis, i.e., a death not controlled or mediated by the cell (Mittler and Lam, 1996). Plants appear to activate a PCD pathway as part of their defense response against invading pathogens (i.e., the HR). It is thought that by killing of cells at and around the site of infection the plant generates a physical barrier composed of dead cells and limits the availability of nutrients to the pathogen due to the rapid dehydration that accompanies tissue death (Goodman and Novacky, 1994; Dangl *et al.*, 1996). As indicated above, it is not entirely clear whether the death of plant cells during the HR is a 'side-effect' caused by the activation of other defense responses such as increased ROI production and synthesis of phytoalexins, or is due to a specific cell death signal that may be similar to the signal(s) that activate PCD during developmental processes such as formation of tracheary elements (Dangl *et al.*, 1996; Fukuda, 1997).

Evidence supporting the assumption that HR cell death is a PCD includes: (1) the activation of the HR in a gene-dependent manner by different elicitors which are compounds produced or secreted by pathogens; (2) the inhibition of pathogen-induced HR cell death by different metabolic inhibitors such as cycloheximide or amanitin; (3) the spontaneous formation of HR lesions in the absence of pathogens in different mutants; and (4) the activation of HR cell death by expression of specific transgenes (Dangl *et al.*, 1996; Mittler and Lam, 1996). Of these the activation of HR cell death in transgenic plants or mutants in the absence of a pathogen (i.e., lesion mimics), is perhaps the strongest evidence for the existence of a cell suicide pathway that is activated by the plant upon pathogen recognition. In this review we will discuss the different examples of transgene-induced lesion mimics, their mode of action, and their possible applications.

Naturally occurring lesion mimic mutants

Key to our understanding of transgene-induced lesion mimics are studies of naturally occurring lesion mimic mutants. Reports of lesion mimics date as early as 1923 and 1948 (Emerson, 1923; Langford, 1948), and include different plants such as maize, rice, barley, tomato, and *Arabidopsis* (Walbot *et al.*, 1983; Greenberg and Ausubel, 1993; Wolter *et al.*, 1993). Mutants similar to the naturally occurring lesion mimics can also be isolated after mutagenesis (i.e., chemical-, radiation-, or transposon-induced; Greenberg and Ausubel, 1993; Dietrich *et al.*, 1994; Greenberg *et al.*, 1994). Lesion mimics were classified according to their appearance into two groups: initiation, and feedback or propagation mutants (Walbot *et al.*, 1983; Dietrich *et al.*, 1994). This classification is based upon the assumption that two different mechanisms are involved in coordinating the HR: (1) a pathway for the initiation of PCD, and (2) a mechanism for the suppression of PCD. Initiation mutants develop spontaneous lesions with a defined border. They are thought to be defective in regulating the activation of PCD, but not its inhibition at the boundary of the lesion. The abnormal activation of cell death in these mutants may result from the lack of a negative regulator of cell death initiation (a recessive initiation mutant) or from the constitutive activation of a cell death signal (a dominant initiation mutant). Propagation or feedback mutants form spontaneous or induced lesions that spread indeterminately. They are presumed to be defective in down-regulating PCD in cells surrounding a developing lesion (recessive mutations). In these mutants cell death which is initiated randomly, upon infection, or after a mechanical injury will propagate uncontrollably, eventually resulting in the complete death of the leaf. The majority of naturally occurring cell death mutants appear to belong to the dominant initiation class (e.g., 23 of 32 different lesion mimic mutants in maize are dominant gain-of-function mutations (Walbot *et al.*, 1983; Dangl *et al.*, 1996). This observation suggests that a variety of cellular signals, some of which may not be directly related to pathogens, can activate the HR-PCD pathway (explained below; Dangl *et al.*, 1996; Mittler and Lam, 1996).

Several cell death mutants express molecular and biochemical markers associated with the antimicrobial defense response of plants. These include enhanced expression of PR proteins, accumulation of salicylic acid (SA), deposition of callose or other cell wall-strengthening compounds, and synthesis of phytoalexins (Dietrich *et al.*, 1994; Greenberg *et al.*, 1994). The activation of these antimicrobial defenses in the absence of a pathogen further suggests that the PCD pathway activated in these mutants may be similar to that activated during the response of plants to incompatible pathogens (Dangl *et al.*, 1996).

Cell death mutants are powerful tools for the study of PCD in plants. The cloning of several lesion mimic genes was recently reported and it is believed that their analysis will unravel the molecular mechanisms involved in the regulation of PCD in plants (Shirasu and

Schulze-Lefert, this issue). In addition, by crossing of different mutants for complementation studies the order of cell death genes along the PCD pathway may be determined (Dangl *et al.*, 1996).

Transgene-induced lesion mimics

Spontaneous formation of HR-like lesions in the absence of a pathogen has been reported in a number of transgenic plants that express foreign or modified genes (Table 1). In some cases, the activation of cell death was accompanied by the induction of multiple defense mechanisms and the induction of enhanced resistance, similar to systemic acquired resistance (SAR; Dangl *et al.*, 1996; Mittler and Lam, 1996). Most of these examples resemble the dominant initiation class of lesion-mimic mutants. Transgene-induced lesion mimics may be classified into four different groups: pathogen-derived genes, signal transduction-inducing genes, general metabolism-perturbing genes, and killer genes (Figure 1, Table 1). These genes can be expressed in a constitutive manner or may be placed under the control of an inducible promoter. As their classification implies, they may have different modes of action.

Pathogen-derived transgenes such as AvrRpt2, elicitin, and Avr9 (Culver and Dawson, 1991; Hammond-Kosack *et al.*, 1994; Keller *et al.*, 1999) appear to function as elicitors, thereby possibly directly interacting with a plant receptor, or a resistant (R) gene product, and activating the HR in the same manner a pathogen would. Their action therefore depends on the genetic make-up of the plant, and requires the presence of an R gene for a proper 'gene-for-gene' interaction (Flor, 1956; Bent, 1996). They are usually expressed from an inducible promoter since they induce a massive cell death response that completely kills the transgenic plant. At least in one instance the activation of the HR response by such a gene (Avr9) was found to depend on the developmental stage of the plant since cell death was initiated only at day 13 of seed germination (using a constitutive promoter; Hammond-Kosack *et al.*, 1994). The action of pathogen-derived genes seems to be somewhat different from that of the dominant lesion mimic genes since instead of causing the sporadic appearance of lesions throughout the leaf, the pathogen-derived genes appear to cause complete death of all cells in the leaf (with the exception of the TMVcp gene in the N′ background; Culver and Dawson, 1991). Pathogen-derived genes are shown in Figure 1 (class A) as genes that mimic the presence of the pathogen and interact with the plant receptors for pathogen recognition.

Signal transduction-inducing genes are transgenes that may activate or effect different components of the signal transduction pathway involved in pathogen recognition or defense response activation (Figure 1). They may mimic the flux of protons across the plasma membrane during the early stages of the HR (i.e., the XR; Yang *et al.*, 1997), such as the bO gene (Mittler *et al.*, 1995), or directly effect different signaling events that take place during the HR, as may be the case with the expression of a small GTP-binding protein, or the cholera toxin (i.e., G-protein signaling; Sano *et al.*, 1994; Beffa *et al.*, 1995). Antisense constructs for the peroxide detoxifying enzymes catalase (CAT; Chamnongpol *et al.*, 1996, 1998; Takahashi *et al.*, 1997) and ascorbate peroxidase (APX; Orvar and Elli, 1997), and an antisense gene for protoporphyrinogen oxidase (PPO; Molina *et al.*, 1999) were also found to induce the formation of lesions, either spontaneously (PPO) or in response to different environmental conditions (CAT or APX), in a manner that resembles lesion mimic mutations. These transgenes may mimic the enhanced production of ROI that accompanies the HR (i.e., the oxidative burst; AOII; Levine *et al.*, 1994; Hammond-Kosack and Jones, 1996). Although they appear to directly effect the signal transduction pathway that is activated during a pathogen-induced HR, signal transduction-inducing transgenes may also activate the HR response via pathways that are not directly related to the HR, much like the metabolic perturbing genes (explained below). The majority of signal transduction-related transgenes were found to induce a lesion mimic phenotype similar to that induced in the dominant initiation mutants. Signal transduction-inducing genes are shown in Figure 1 (class B) as genes that mimic different signal transduction events that occur during the HR.

Expression of metabolism-perturbing transgenes in plants is thought to result in the alteration of cellular homeostasis and the generation of a signal which activates the PCD response. Uncontrolled expression of genes such as invertase or hexokinase may drastically alter the metabolic balance of cells due to changes in hexose transport or metabolism (Herbers *et al.*, 1996; D. Granot, personal communication). A different effect on cellular metabolism may be caused by expression of genes such as rPS14 and CaMV gVI that may affect protein translation in transgenic plants (Takahashi *et al.*, 1989; Karrer *et al.*, 1998). It is possible that metabolism-perturbing transgenes activate PCD

Table 1. Lesion mimic transgenes.

Transgene	Source	Function	Defense	Reference
Class A. Pathogen-derived genes				
TMVcp	TMV (N′)	Avr elicitor	NT	Culver and Dawson, 1991
Avr9	*C. fulvum*	Avr elicitor	Yes	Hammond-Kosack *et al.*, 1994
Elicitin	*P. cryptogea*	Avr elicitor	Yes	Keller *et al.*, 1999
AvrRpt2	*P. syringeae*	Avr elicitor	Yes	McNellis *et al.*, 1998
Class B. Signal transduction-inducing genes				
bO	*H. halobium*	proton pump	Yes	Mittler *et al.*, 1995
Cholera toxin	*V. cholerae*	Inhibit GTPase	Yes	Beffa *et al.*, 1995
sGTP-BP	plant	GTP-binding protein	Yes	Sano *et al.*, 1994
Antisense CAT	plant	removal of ROI	Yes	Chamnongpol *et al.*, 1996; Takahashi *et al.*, 1997
Antisense APX	plant	removal of ROI	NT	Orvar and Ellis, 1997
Antisense PPO	plant	heme biosynthesis	Yes	Molina *et al.*, 1999
Class C. General metabolism-perturbing genes				
Invertase	yeast	hexose transport	Yes	Herbers *et al.*, 1996
Hexokinase	plant	hexose metabolism	Yes	D. Granot, pers. commun.
CaMV gVI	CaMV	inclusion body protein	Yes	Takahashi *et al.*, 1989
rPS14	plant	ribosomal protein	NT	Karrer *et al.*, 1998
Class D. Killer genes				
Barnase	*B. amyloliquefaciens*	RNase	Yes	Strittmatter *et al.*, 1995
DTA	*D. pertussis*	inhibits translation	NT	Nilsson *et al.*, 1998
Protease-related (Class B/C/D)				
Ubiquitin	plant	protein degradation	Yes	Bachmair *et al.*, 1990
Kunitz-type trypsin inhibitor	plant	protein degradation	NT	Karrer *et al.*, 1998

Abbreviations: APX, ascorbate peroxidase; bO, bacterio-opsin; CaMV, cauliflower mosaic virus; CAT, catalase; DTA, diphtheria toxin A subunit; NT, not tested; PPO, protoporphyrinogen oxidase; ROI, reactive oxygen intermediates; sGTP-BP, small GTP-binding protein; TMVcp, tobacco mosaic virus coat protein.

via a pathway that is unrelated to pathogen attack. However, infection of plants with some pathogens such as viruses or bacteria may cause general alterations in the metabolic balance of cells, similar to the changes induced by hexokinase or the CaMV gVI gene. Such pathogen-dependent perturbation in cellular metabolism may in turn activate PCD (Dangl *et al.*, 1996; Mittler and Lam, 1996). Therefore, the activation of PCD by some of the metabolic-perturbing genes may occur via the same pathway that is activated during a 'bona fide' HR. In animals many perturbations in cellular metabolism were shown to activate PCD and the PCD pathway was suggested to act as a 'funnel' that 'channels in' many different signals (Raff, 1992; Schwartzman and Cidlowski, 1993). Since, at least in animals, PCD appears to be activated as part of a general defense mechanism that prevents the growth and proliferation of damaged, infected, or mutated cells (including cells with general metabolic alterations), we included these genes as a separate group. The existence of metabolism-perturbing transgenes may explain the large number of dominant lesion mimic mutants. Many of these may be mutations in general housekeeping genes that cause alterations in cellular metabolism and activation of PCD. Some alterations in cellular homeostasis are thought to result in the excess production of ROI. As shown in Figure 1, ROI produced by these alterations may act as triggers for the induction of PCD. Although of pathogen origin, we classified the CaMV gVI gene as a metabolism-perturbing transgene. The CaMV gVI protein was suggested to affect the translational apparatus of plants and may therefore change the cellular homeostasis of cells (De Tapia *et al.*, 1993). Metabolism-perturbing transgenes are shown in Figure 1 (class C) as genes that activate the HR via the generation of different cellular signals that may be channeled into the HR-PCD pathway.

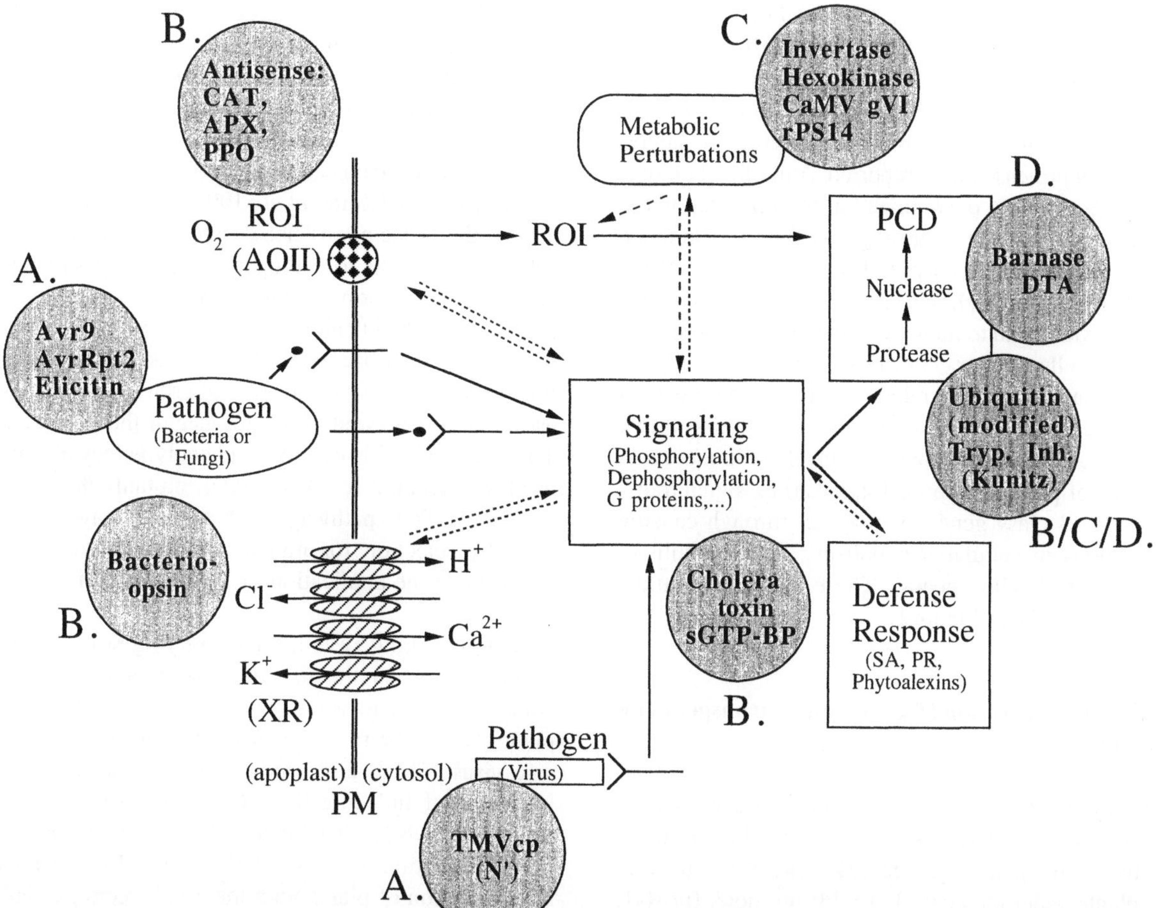

Figure 1. A model showing the different modes of PCD activation by lesion mimic transgenes, in relation to the HR pathway. HR-inducing transgenes (shaded circles; see Table 1 for abbreviations, description, and classification of the different transgenes) were classified into four groups (A to D) and placed at their possible sites of action along the pathogen-induced HR pathway (after Mittler *et al.*; Mittler and Lam, 1996). Abbreviations and symbols: AOII, oxidative burst; PCD, programmed cell death; PM, plasma membrane; PR, pathogen-related; ROI, reactive oxygen intermediates; SA, salicylic acid; XR, ion translocation response, >–, receptor.

Killer genes are genes thought to directly cause cell death. The expression of barnase, an RNase gene, in plants is one such example (Strittmatter *et al.*, 1995). Other genes that may function as killer genes are DNases, specific proteases, or subunit A of the diphtheria toxin (Nilsson *et al.*, 1998). Killer genes may mimic the action of different components of the PCD pathway, in particular cell death-executing genes. However, it should be noted that the expression of such genes may not be the direct cause of cell death and that they may simply be triggering a pre-existing PCD pathway by generating a cell death signal, similar to some examples in animals (Schwartzman and Cidlowski, 1993). It was suggested that in animals the PCD pathway is already present in cells in a 'ready-for-execution' mode which is not dependent on additional reactions of RNA or protein synthesis (Raff, 1992). If such a mechanism also functions in plant cells then a killer gene such as barnase or DTA may activate it and cause the execution of cell death in a manner that is independent of RNA metabolism or translation. Evidence for and against the existence of a 'ready-for-execution' cell death machinery in plants was reported (He *et al.*, 1994; Pennel and Lamb, 1997). Killer transgenes are shown in Figure 1 (class D) as genes that may mimic the action of plant genes involved in the final execution phase of the HR.

Proteases were recently suggested to be involved in the signal transduction pathway that leads to the activation and execution of PCD in plants and animals (Fraser and Evan, 1996; del Pozo and Lam, 1998). A cascade of proteases is thought to be activated during PCD and directly cause the death of cells by proteolytic cleavage (Fraser and Evan, 1996). At least two examples of cell death inducing transgenes involving proteases were reported. A modified ubiquitin gene unable to polymerize (an essential step in the protein degradation pathway) was found to induced a lesion mimic phenotype (Bachmair *et al.*, 1990; Becker *et al.*, 1993), and a 'Kunitz'-type trypsin inhibitor was found to induce cell death upon expression in plant cells (Karrer *et al.*, 1998). The action of these genes may be directly related to the protease execution machinery (i.e., Killer, D class genes), may effect the signal transduction pathway involved in the transmitting of pathogen-derived signals (i.e., signal transduction, B class gene), or may act through causing alterations in cellular metabolism (i.e., metabolism-perturbing, C class genes). These genes are marked as class B/C/D. in Figure 1.

Inducible expression of lesion mimic transgenes in plants

At least three different lesion mimic transgenes were expressed under the control of an inducible promoter in transgenic plants. The barnase gene was expressed in plants under the control of a PR promoter (prp1-1; Strittmatter *et al.*, 1995). It was found that upon infection of plants with a fungal pathogen the barnase gene was expressed and caused the induction of PCD. This induction enhanced the resistance of transgenic plants to pathogen attack, supporting the hypothesis that death of cells at and around the site of infection plays an important role in preventing pathogen proliferation. Ethylene was also found to induce the expression of the barnase gene and the activation of PCD (in the absence of a pathogen; Strittmatter *et al.*, 1995). Another pathogen-responsive promoter used to drive the expression of a lesion-inducing transgene is the *hrs203J* promoter (Keller *et al.*, 1999). This promoter was used to express the elicitor protein cryptogein (elicitin) in transgenic tobacco plants. Transgenic plants in which the elicitin gene was induced by a pathogen attack-activated PCD and enhanced the synthesis of PR proteins. These plants also displayed enhanced resistance to attack by different pathogens (Keller *et al.*, 1999).

McNellis *et al.* (1998) used a glucocorticoid-inducible promoter to drive the expression of avrRpt2 in the RPS2 genetic background. In this transgenic system the expression of avrRpt2 is controlled artificially by the addition of dexamethasone (DEX). Application of DEX caused the induction of PCD and the activation of defense mechanisms such as PR-1 gene expression. As opposed to the two inducible systems described above (Keller *et al.*, 1999; Strittmatter *et al.*, 1995), this system was not designed to enhance the tolerance of plants to pathogens. In place, it supplies an excellent platform for studying the PCD response in plants. Thus, PCD can be activated in a synchronous manner and the different events that take place during this response, such as activation of proteases or nucleases, can be followed in the absence of the pathogen. The DEX-inducible cell death phenotype may also be used in screening for *Arabidopsis* mutants that have a defective PCD pathway. Such mutants may be isolated by selection for mutated transgenic seedlings that survive the application of DEX (McNellis *et al.*, 1998).

An interesting system for identifying novel lesion mimic genes via the use of a virus-based expression vector was reported by Karrer *et al.* (1998). A plant cDNA library was constructed in tobacco mosaic virus (TMV) and used to infect tobacco plants incapable of inducing the HR in response to TMV infection. cDNAs that induced PCD were identified by the formation of lesions that resembled HR lesions after infection of plants with the TMV vector. cDNA inserts which encoded for cell death-inducing transgenes were cloned from the lesions by RT-PCR. In addition to a large number of unidentified novel genes, Karrer *et al.* isolated a ribosomal protein (rPS14), a 'Kunitz'-type protease inhibitor, with homology to the tobacco tumor-related protein miraculin, a glycine-rich protein, and a ubiquitin gene. The ubiquitin gene appeared to function in a manner similar to that of the ubiquitin mutant, which induced a lesion mimic phenotype in transgenic tobacco plants (Karrer *et al.*, 1998).

Use of transgene-induced lesion mimics in biotechnology

Two of the main uses for lesion mimic genes in biotechnology are enhancement of pathogen resistance (Strittmatter *et al.*, 1995; Mittler and Lam, 1996; Keller *et al.*, 1999) and tools for specific cell ablation (Nilsson *et al.*, 1998). Lesion mimic genes can be used to enhance pathogen resistance in two ways.

They can be expressed in a constitutive manner causing the activation of defense responses, appearance of lesions, and induction of a systemic state of resistance. This mode of expression results in a phenotype similar to the naturally occurring lesion mimic mutants that were used for many years by breeders to introduce resistance from wild-type cultivars into commercial plants (Langford, 1948). Alternatively, lesion mimic genes may be expressed from an inducible promoter (as described above; Keller *et al.*, 1999; Strittmatter *et al.*, 1995) that will limit their action to the site and time of infection. This method will not cause the constitutive induction of various defense mechanisms that may have adverse effects on the growth and yield of the plant due to high energetic requirements, the constitutive synthesis of various compounds such as phytoalexins and SA, and the presence of areas of dead cells (i.e., lesions). The choice between the two expression strategies may depend on the cultivar, its main disease agent, and the availability of a suitable promoter.

The productivity of some cultivars may not be drastically affected by the constitutive expression of a lesion mimic transgene. Thus, a constitutive expression system may be used in these plants to induce a high level of resistance without paying a high price in yield. The level of constitutive expression may also be controlled by the strength of the promoter; therefore, cultivars can be produced in which the lesion mimic gene is expressed at a low level that causes minimal adverse effects, but provides sufficient resistance against different pathogens (E. Lam, personal communication). Another crop-related consideration is that some crops, especially those grown in developing countries, may be under a continuous attack by a number of pathogens. In these cases an inducible promoter that will always stay active may be of no advantage over a strong constitutive promoter. The type of crop used, as well as its response to different levels of constitutive expression, and the area of the world in which it is grown may therefore greatly affect the choice of a promoter.

The disease agent(s) causing the most extensive damage to the crop of interest is also a key determinant in the choice of an expression strategy. Since the successful use of an inducible promoter is mainly dependent upon the timing of its induction, it may not be possible to use some of the existing inducible promoters to block the spread of a very aggressive pathogen. Developing resistance to these pathogens may require the use of a constitutive promoter. In addition, different pathogens may activate a specific promoter at different rates, so one inducible promoter may not be sufficient to combat a number of possible disease agents. Therefore, the choice of an expression system is also determined by the aggressiveness of the pathogen and its specific interactions with the plant.

The availability of suitable promoters appears to be the most limiting factor in developing lesion mimic-expressing crops that are pathogen-resistant. Especially critical is the availability of inducible promoters. An ideal inducible promoter is one that will have no basal level of expression and will be specifically and rapidly induced upon pathogen infection. The main problem with the currently available promoters is specificity. Some pathogen-responding promoters are also induced by factors that are not solely related to pathogen attack. For example, the prp1-1 promoter is induced by ethylene that may be produced in plants in the absence of a pathogen attack (Strittmatter *et al.*, 1995). In addition, some promoters of PR proteins are expressed in a development-dependent manner in the absence of pathogens, such as expression in flowers (Lotan *et al.*, 1989). In addition, the rapid induction of a PR promoter depends upon the recognition of the pathogen by the plant (i.e., the 'gene-for-gene' interaction; Flor, 1956; Bent, 1996). Therefore, some pathogens that may not be recognized by the plant may not cause a rapid enough induction of the PR promoter. In these specific cases the induction of the PR promoter will be a secondary event that occurs only late during infection. The 'gene-for-gene' recognition event may also be critical for choosing a lesion mimic gene. Some lesion mimic genes such as those belonging to class B, C or D (Figure 1) do not depend upon recognition and may function to induce PCD in many different plants; however, elicitor genes that belong to the class A genes (Figure 1) will require the presence of an R gene. These lesion mimic transgenes will depend on the genetic background of the plant and may not function in all plants. One solution to this problem is to introduce the R gene into the plant as well. This approach has successfully been tested (Hammond-Kosack *et al.*, 1998). At present it appears as if the successful use of lesion mimic genes in an inducible system will require the isolation of more inducible promoters, the molecular modification of currently available promoters, or the use of a combined strategy in which different constructs with different promoters will be used to transform the same plant.

Chemically inducible promoters such as DEX- or tetracycline-inducible promoters (Gatz, 1995) may also be used to drive the expression of lesion mimic transgenes. These promoters will be activated by the application of the chemical inducer only when disease symptoms are detected in the field or as part of a preventive maintenance program. As opposed to the other strategies described above the use of chemically inducible promoters will require continuous monitoring of fields, combined with chemical application.

Perhaps the most critical consideration with respect to the biotechnological use of lesion-inducing transgenes as resistance-enhancing tools is that they may not provide resistance against all pathogens. Since lesion mimic genes activate the plant's own defense mechanisms and induce PCD they may not be efficient against pathogens that the plant is incapable of resisting against even with all its defenses activated. Thus, against some pathogens a combined approach may be needed, that is, one that makes use of additional genes that encode for defenses which may not naturally occur in the plant, such as small lytic peptides. In addition to this consideration, there is always the consideration of evolutionary pressure on pathogens. Persistent use of a transgenic plant that activates all of its defense mechanisms continuously or in response to pathogen attack, for example by constitutive or inducible expression of a lesion mimic gene such as the bO gene, may exert an evolutionary pressure that will cause the development of resistance-breaking strains of pathogens. These are likely to be problematic since the plant will have no defenses left to use against them. In order to combat this problem refuge plots will have to be used.

Lesion-inducing transgenes can also be used as cell ablation tools. They can be expressed under the control of a tissue-specific promoter and used for the production of male-sterile plants, or seedless fruits. However, it was found that the action of some of these genes, such as bO or Avr9, is dependent on the developmental stage of the tissue. Therefore, in some tissues the activation of PCD will be prevented or suppressed by a particular developmental signal. In these examples other cell death-inducing genes may be used.

Conclusions and perspectives

The activation of HR cell death by transgene expression provides scientists with an excellent research tool to address different questions regarding the HR. For example, the various relationships between the pathogen and the plant, as well as the effect the pathogen may have on the activation and coordination of defense responses by the plant, may be addressed by comparing a transgene-induced HR that occurs in the absence of the pathogen to a pathogen-induced HR. This point may be specifically important for the understanding of complex plant-pathogen interactions such as those that occur between hemitrophic fungi and plants. In addition, transgenic plants that contain an inducible expression system for a cell death transgene may be used to study different biochemical, physiological, and molecular aspects of the PCD response, as well as for the isolation of mutants deficient in different steps along the PCD pathway (i.e., mutants that fail to die upon activation of the transgene; McNellis *et al.*, 1998).

HR cell death appears to be controlled in part by different developmental signals (Hammond-Kosack *et al.*, 1994; Mittler *et al.*, 1995). Lesion mimic transgenes expressed from different developmentally controlled promoters may unravel some of the basic relationships between development and the HR. Thus, we may find that plant cells undergoing a particular developmental program cannot enter the PCD pathway even though the HR-inducing transgene is expressed. Since we know of different lesion mimic transgenes that may function via different routes to induce PCD (i.e., classes A to D in Figure 1) we may test whether the developmental signal that inhibits, for example, class A or B of PCD-initiating genes will also inhibit class C or even D.

One interesting question that may be related to developmental signals is why does the activation of cell death in some of the lesion mimic transgenes or mutants occur sporadically throughout the leaf. What is the cause of cell death initiation in these particular areas of the leaf? This question is puzzling since the transgene or the mutated gene is present in all of the leaf cells. Is there a particular cellular parameter that may distinguish certain cells from others and make them more prone to the activation of PCD? This parameter/factor may be the level of a particular hormone, the cell cycle stage, or the presence or absence of other cellular or developmental signals. A dependence of PCD activation on a cellular parameter such as the cell cycle is known to occur in animals (Raff, 1992). An alternative explanation is that the cell death response may be randomly initiated in different cells of the leaf. The activation of PCD in these cells may cause the induction of a cell death repressor mechanism in neighboring cells thus resulting in the appearance of lesions only at and around the cells that randomly initiated the cell death response.

In addition to playing an important role in the study of PCD in plants, lesion mimic transgenes may be valuable for the biotechnological development of crops with enhanced resistance to pathogens. Transgenes that induce the lesion mimic phenotype may be incorporated into our arsenal of genes used to combat plant pathogens. They may be used alone or in combination with other defense strategies to create a 'super plant' with an enhanced resistance to attack by many different pathogens.

In the future, isolation and characterization of new lesion mimic transgenes will considerably advance our understanding of the PCD response and provide us with new tools to study and control this response as well as harness it to the development of crops with enhanced resistance traits.

Acknowledgements

We wish to thank Drs Leonora Reinhold, Robert Fluhr, and Alex Levine for critical comments. We gratefully acknowledge Drs David Granot and Eric Lam for sharing unpublished results. This work was supported by funding provided by the Israel Ministry of Agriculture, the Yigal Alon Fellowship, and the Hebrew University Intramural Research Fund Basic Project Awards.

References

Bachmair, A., Becker, F., Masterson, R.V. and Schell, J. 1990. Perturbation of the ubiquitin system causes leaf curling, vascular tissue alterations and necrotic lesions in higher plants. EMBO J. 9: 4543–4549.

Becker, F., Buschfeld, E., Schell, J. and Bachmair, A. 1993. Altered response to viral infection by tobacco plants perturbed in ubiquitin system. Plant J. 3: 875–881.

Beffa, R., Szell, M., Meuwly , P., Pay, A., Vogeli-Lange, R., Metraux, J.P., Neuhaus, G., Meins, F. Jr. and Nagy, F. 1995. Cholera toxin elevates pathogen resistance and induces pathogenesis-related gene expression in tobacco. EMBO J. 14: 5753–5761.

Bent, A.F. 1996. Plant disease resistance genes: function meets structure. Plant Cell 8: 1757–1771.

Chamnongpol, S., Willekens, H., Langebartels, C., Van Montagu, M., Inzé, D. and Van Camp, W. 1996. Transgenic tobacco with reduced catalase activity develops necrotic lesions and induces pathogenesis-related expression under high light. Plant J. 10: 491–503.

Chamnongpol, S., Willekens, H., Moeder, W., Langebartels, C., Sandermann, H., Van Montagu, M., Inzé, D. and Van Camp, W. 1998. Defense activation and enhanced pathogen tolerance induced by H_2O_2 in transgenic tobacco. Proc. Natl. Acad. Sci. USA 95: 5818–5823.

Culver, J.N. and Dawson, W.O. 1991. Tobacco mosaic virus elicitor coat protein genes produce a hypersensitive phenotype in transgenic *Nicotiana sylvestris* plants. Mol. Plant-Microbe Interact. 4: 458–463.

Dangl, J.L., Dietrich, R.A. and Richberg, M.H. 1996. Death don't have no mercy: cell death programs in plant-microbe interactions. Plant Cell 8: 1793–1807.

del Pozo, O. and Lam, E. 1998. Caspases and programmed cell death in the hypersensitive response of plants to pathogens. Curr. Biol. 8: 1129–1132.

De Tapia, M., Himmelbach, A. and Hohn, T. 1993. Molecular dissection of the cauliflower mosaic virus translation transactivator. EMBO J. 12: 3305–3314.

Dietrich, R.A., Delaney, T.P., Uknes, S.J., Ward, E.R., Ryals, J.A. and Dangl, J.L. 1994. *Arabidopsis* mutants simulating disease resistance response. Cell 77: 565–577.

Emerson, R.A. 1923. The inheritance of blotch leaf maize. Cornell Univ. Agric. Exp. Stn. Mem. 70: 3–16.

Flor, H.H. 1956. The complementary genic systems in flax and flax rust. Adv. Genet 8: 29–54.

Fraser, A. and Evan, G. 1996. A license to kill. Cell 85: 781–784.

Fukuda, H. 1997. Tracheary element differentiation. Plant Cell 9: 1147–1156.

Gatz, C. 1995. Novel inducible/repressible gene expression systems. Meth. Cell Biol. 50: 411–424.

Greenberg, J.T. 1996. Programmed cell death: a way of life for plants. Proc. Natl. Acad. Sci. USA 93: 12094–12097.

Greenberg, J.T. and Ausubel, F.M. 1993. Arabidopsis mutants compromised for the control of cellular damage during pathogenesis and aging. Plant J. 4: 327–342.

Greenberg, J.T., Ailan, G., Klessig, D.F. and Ausubel, F.M. 1994. Programmed cell death in plants: a pathogen-triggered response activated coordinately with multiple defence functions. Cell 77: 551–563.

Goodman, R.N. and Novacky, A.J. 1994. The Hypersensitive Response Reaction in Plants to Pathogens: A Resistance Phenomenon, American Phytopathological Society Press. St. Paul, MN.

Hammond-Kosack, K.E. and Jones, J.D.G. 1996. Resistance gene-dependent plant defense responses. Plant Cell 8: 1773–1791.

Hammond-Kosack, K.E., Harrison, K. and Jones, J.D.G. 1994. Developmentally regulated cell death on expression of the fungal avirulence gene Avr9 in tomato seedlings carrying the disease-resistance gene CF-9. Proc. Natl. Acad. Sci.USA 91: 10445–10449.

Hammond-Kosack, K.E., Tang, S., Harrison, K. and Jones, J.D.G. 1998. The tomato Cf-9 disease resistance gene functions in tobacco and potato to confer responsiveness to the fungal avirulance gene product avr9. Plant Cell 10: 1251–1266.

He, S.Y., Bauer, D.W., Collmer, A. and Beer, S.V. 1994. Hypersensitive response elicited by *Erwinia amylovora* harpin requires active plant metabolism. Mol. Plant-Microbe Interact. 7: 289–292.

Herbers, K., Meuwly, P., Frommer, W.B., Metraux, J.P. and Sonnewald, U. 1996. Systemic acquired resistance mediated by the ectopic expression of invertase: possible hexose sensing in the secretory pathway. Plant Cell 8: 793-803.

Karrer, E.E., Beachy, R.N. and Holt, C.A. 1998. Cloning of tobacco genes that elicit the hypersensitive response. Plant Mol. Biol. 36: 681–690.

Keller, H., Pamboukdjian, N., Ponchet, M., Poupet, A., Delon, R., Verrier, J.L., Roby, D. and Ricci, P. 1999. Pathogen-induced elicitin production in transgenic tobacco generates a hypersensitive response and nonspecific disease resistance. Plant Cell 11: 223–235.

Langford, A.N. 1948. Autogenous necrosis in tomatoes immune from *Cladosporium flavum* Cooke. Can. J. Res. 26: 35–64.

Levine, A., Tenhaken, R., Dixon, R. and Lamb, C. 1994. H_2O_2 from the oxidative burst orchestrates the plant hypersensitive disease resistance response. Cell 79: 583–593.

Lotan, T., Ori, N. and Fluhr, R. 1989. Pathogenesis-related proteins are developmentally regulated in tobacco flowers. Plant Cell 1: 881–887.

McNellis, T.W., Mudgett, M.B., Li, K., Aoyama, T., Horvath, D., Chua, N.H. and Staskawicz, B.J. 1998. Glucocorticoid-inducible expression of a bacterial avirulence gene in transgenic *Arabidopsis* induces hypersensitive cell death. Plant J. 14: 247–257.

Mittler, R. and Lam, E. 1996. Sacrifice in the face of foes: pathogen-induced programmed cell death in higher plants. Trends Microbiol. 4: 10–15.

Mittler, R., Shulaev, V. and Lam, E. 1995. Coordinated activation of programmed cell death and defense mechanisms in transgenic tobacco plants expressing a bacterial proton pump. Plant Cell 7: 29–42.

Molina, A., Volrath, S., Guyer, D., Maleck, K., Ryals, J. and Ward, E. 1999. Inhibition of protoporphyrinogen oxidase expression in *Arabidopsis* causes a lesion-mimic phenotype that induces systemic acquired resistance. Plant J. 17: 667–678.

Nilsson, O., Wu, E., Wolfe, D.S. and Weigel, D. 1998. Genetic ablation of flowers in transgenic *Arabidopsis*. Plant J. 15: 799–804.

Orvar, B.L. and Elli, B.E. 1997. Transgenic tobacco plants expressing antisense RNA for cytosolic ascorbate peroxidase show increased susceptibility to ozone injury. Plant J. 11: 1297–1305.

Pennel, R.I. and Lamb, C. 1997. Programmed cell death in plants. Plant Cell 9: 1157–1168.

Raff, C.M. 1992. Social controls on cell survival and cell death. Nature 356: 397–400.

Sano, H., Seo, S., Orudgev, E., Youssefian, S., Ishizuka, K. and Ohashi, Y. 1994. Expression of the gene for a small GTP binding protein in transgenic tobacco elevates endogenous cytokinin levels, abnormally induces salicylic acid in response to wounding and increases resistance to tobacco mosaic virus infection. Proc. Natl. Acad. Sci.USA 91: 10556–10560.

Schwartzman, R.A. and Cidlowski, J.A. 1993. Apoptosis: the biochemistry and molecular biology of programmed cell death. Endocrine Rev. 14: 133–151.

Strittmatter, G., Janssens, J., Opsomer, C. and Botterman, J. 1995. Inhibition of fungal disease development in plants by engineering controlled cell death. Bio/technology 13: 1085–1089.

Takahashi, H., Shimamoto, K. and Ehara, Y. 1989. Cauliflower mosaic virus gene VI causes growth suppression, development of necrotic spots and expression of defence-related genes in transgenic tobacco plants. Mol. Gen. Genet. 216: 188–194.

Takahashi, H., Chen, Z., Du, H., Liu, Y. and Klessig, D.F. 1997. Development of necrosis and activation of disease resistance in transgenic tobacco plants with severely reduced catalase levels. Plant J. 11: 993–1005.

Walbot, V., Hoisington, D.A. and Neuffer, M.G. 1983. Disease lesion mimics in maize. In: T. Kosuge and C. Meredith (Eds.) Genetic Engineering of Plants, Plenum, New York, pp. 431–442.

Wolter, M., Hollricher, K., Salamini, F. and Schulze-Lefert, P. 1993. The *mlo* resistance alleles to powdery mildew infection in barley trigger a developmentally controlled defence mimic phenotype. Mol. Gen. Genet. 239: 122–128.

Yang, Y., Shah, J. and Klessig, D.F. 1997. Signal perception and transduction in plant defense responses. Genet. Dev. 11: 1621–1639.

Plant Molecular Biology **44:** 345–358, 2000.
E. Lam, H. Fukuda and J. Greenberg (Eds.), Programmed Cell Death in Higher Plants.

Ozone: a tool for probing programmed cell death in plants

Mulpuri V. Rao[1], Jennifer R. Koch[2] and Keith R. Davis[3,*]
[1]*Department of Plant Biology and the Plant Biotechnology Center, 1060 Carmack Road, Ohio State University, Columbus, OH 43210, USA;* [2]*Department of Molecular Genetics, Ohio State University and USDA Forest Service, Delaware, USA;* [3]*Paradigm Genetics Inc., Building 2, PO Box 14528, 104 Alexander Drive, Research Triangle Park, NC 27709, USA (*author for correspondence; fax: 919-544-8094; e-mail: kdavis@paragen.com)*

Key words: cross-talk, ethylene, jasmonic acid, ozone, programmed cell death, salicylic acid

Introduction

Rapid increases in industrialization and other human activities during the twentieth century have contributed significant amounts of toxic gaseous pollutants to the tropospheric environment that pose a significant threat for the survival and productivity of native and cultivated ecosystems (Krupa and Kickert, 1989). Among various pollutants studied to date, the gaseous air pollutant ozone (O_3) has caused more damage to both natural and cultivated crop plants in industrialized nations than any other pollutant (Heagle, 1989; Krupa and Kickert, 1989). O_3 is produced both in the troposphere and in the stratosphere. Stratospheric O_3 is generated through photolysis of molecular oxygen by ultraviolet solar radiation (Chameides *et al.*, 1994). In the troposphere, O_3 is produced by oxygen/ozone equilibrium reactions involving NO_2/NO and light (Figure 1, reaction 1). In addition, reactions involving the photo-oxidation of carbon monoxide (Figure 1, reaction 2), unburned hydrocarbons such as methane (Figure 1, reaction 3), formaldehyde and other non-methane organic compounds (Figure 1, reaction 4; Fishman *et al.*, 1985) also contribute to tropospheric O_3.

Stratospheric O_3 shields biologically harmful UV radiation from reaching the earth's surface; however, somewhat paradoxically, tropospheric O_3 is toxic to biological organisms. The discovery of the phytotoxicity of O_3 during mid 1950s (Richards *et al.*, 1958) prompted widespread studies of the effects of O_3 on plant physiological processes under both laboratory and field conditions and led to the implementation of both national and international limits on ambient O_3 concentrations. However, in spite of these control measures, the concentrations of tropospheric O_3 in most urban areas and even in remote areas of most industrialized nations often exceed peak values of 0.1–0.2 ppm (Krupa *et al.*, 1995). The extensive studies conducted by the National Crop Loss Assessment Network (NCLAN) suggested that tropospheric O_3 pollution contributed significantly to forest decline (Preston and Tingey, 1988) and it is estimated that a 40% decrease in ambient O_3 would have the net benefit of increasing crop yield by 3 billion dollars per year (Adams *et al.*, 1990).

Mode of action

The effects of O_3 on biological organisms have been studied for at least 50 years, but the molecular basis of O_3 toxicity is still not well understood. O_3 is a three-atom allotrope of oxygen that reacts with plants in (1) solid phase (e.g., with cuticular components of plant leaves), (2) gas phase (e.g., reactions with the hydrocarbons emitted by plants) and (3) liquid phase, which includes the dissolution of O_3 in aqueous media followed by reaction with lipids, proteins and other cellular components. O_3 reactions in both solid and liquid phases are more relevant in plants; however, most studies have focused on the reactions of O_3 in the liquid phase because O_3 dissociation in the leaf extracellular spaces has the biggest effect on plants (Mudd, 1997).

Based on studies conducted by Weiss (1935), it was proposed that O_3 dissociates in aqueous solutions as a function of the endogenous hydroxyl ion concentration, generating active oxygen species (AOS) such as superoxide anion ($^{\bullet}O_2^-$) and hydrogen peroxide (H_2O_2, Figure 2). While O_3 reacts with lipid molecules generating stoichiometric amounts of aldehydes

$$NO_2 + O_2 + h\upsilon \Leftrightarrow NO + O_3 \quad (1)$$

$$CO + 2O_2 + h\upsilon \rightarrow CO_2 + O_3 \quad (2)$$

$$CH_4 + 4O_2 + 2h\upsilon \rightarrow HCHO + H_2O + 2O_3 \quad (3)$$

$$RH + 4O_2 + 2h\upsilon \rightarrow R'CHO + H_2O + 2O_3 \quad (4)$$

Figure 1. Chemical reactions of O_3 formation in the troposphere.

Weiss Reaction:	$O_3 + OH^- \Rightarrow {}^{\bullet}O_2^-$
Ozonolysis:	$O_3 \Rightarrow H_2O_2$
Dismutation:	${}^{\bullet}O_2^- + H^+ \Rightarrow HO_2^{\bullet}$ $HO_2^{\bullet} + HO_2^{\bullet} \Rightarrow H_2O_2 + O_2$ ${}^{\bullet}O_2^- + HO_2^{\bullet} + H_2O \Rightarrow H_2O_2 + OH^- + O_2$
Haber-Weiss Reaction:	$Fe^{3+} + {}^{\bullet}O_2^- \Rightarrow Fe^{2+} + O_2$ $Fe^{2+} + H_2O_2 \Rightarrow Fe^{3+} + OH^{\bullet} + OH^-$

Figure 2. Presumptive chemical reactions of O_3 dissociation in leaf extracellular spaces. O_3 enters the leaf through the stomata and dissociates to generate superoxide radical (${}^{\bullet}O_2^-$). In addition, O_3 also reacts with lipid molecules generating stoichiometric amounts of H_2O_2 and aldehydes. ${}^{\bullet}O_2^-$ is dismutated to H_2O_2 either spontaneously or enzymatically with the help of endogenous superoxide dismutases. The resulting ${}^{\bullet}O_2^-$ and H_2O_2 react with transition metals or other secondary phenolic compounds to generate hydroxyl radicals ($OH^{\bullet}$), a more lethal active oxygen species. Since plants are not capable of metabolizing $OH^{\bullet}$, it is believed that plants' ability to metabolize both ${}^{\bullet}O_2^-$ and H_2O_2 reduces the formation of $OH^{\bullet}$ and, therefore, the toxic reactions damaging to cellular membranes, proteins and organelles.

and H_2O_2 – a process referred to as ozonolysis –, both ${}^{\bullet}O_2^-$ and H_2O_2 react with transition metals generating hydroxyl radicals ($OH^{\bullet}$) via the Haber-Weiss reaction. Furthermore, O_3 reacts with thiol groups, amines and/or phenolic compounds such as caffeic acid and extracellular ascorbate, exacerbating the production of $OH^{\bullet}$ and singlet oxygen (${}^1O_2$; Kanofsky and Sima, 1995).

Initial site reactions

It has been clearly established that O_3 is a highly toxic agent responsible for the development of foliar necrotic lesions, but the sequence of reactions starting from the exposure of the plant and ending with the development of symptoms is not completely understood. Based on measurements of O_3 flux into leaves, Laisk *et al.* (1989) suggested that O_3 does not penetrate deep into intercellular spaces but rather decomposes at the cell wall and plasma membrane. Then how does O_3 penetrate into the cellular machinery and instigate a physiological response? The plant cell wall contains many phenolic groups, olefinic compounds and amide proteins. In addition, the adjacent plasma membrane contains many unsaturated lipids. It is likely that the first set of bio-molecules which can react with O_3 will be encountered within the cell wall regions just outside the plasma membrane and form highly toxic AOS (Heath, 1987). These O_3-derived AOS are believed to alter the physicochemical properties of plasma membrane by initiating lipid peroxidation (Pauls and Thompson, 1980) and altering Ca^{2+} and ion fluxes (Castillo and Heath, 1990; Clayton *et al.*, 1999) that together disrupt the cellular machinery causing a reduction in net photosynthesis (Reich and Amundson, 1985). It is possible that either the O_3-derived AOS or the intermediates generated due to the reaction of O_3 with cellular components are propagated throughout the cell causing a variety of biochemical changes.

Acute versus chronic exposures

Depending on the concentration of O_3 and the plant species, O_3 causes two different types of plant responses commonly referred to as acute and chronic. Acute exposures, which involve higher concentrations of O_3 (150–300 ppb) for relatively small periods of time (4–6 h), rapidly cause visible injury (necrotic) symptoms on the leaf surfaces. The necrotic lesions and plant responses induced by acute O_3 exposures are reminiscent of the hypersensitive response (HR) that occurs as a result of incompatible plant-pathogen interactions. Chronic exposures involve low concentrations of O_3 ($\leq$100 ppb) for relatively longer periods (days to months). The chronic injury is more subtle and, depending on plant species, may include symptoms such as chlorosis and premature senescence (Pell *et al.*, 1997). While acute O_3 exposure initiates cell death prior to any detectable changes in plant productivity, chronic exposures diminishes whole plant productivity prior to the occurrence, if any, of lesion formation (Heagle, 1989; Heath and Taylor, 1997). However, studies conducted with crop plants grown under simulated field conditions revealed that acute levels of O_3 may be important in the induction of plant responses that lead to reduced crop yield (Heagle, 1989). Since the main thrust of this review is to establish the usefulness of O_3 to study programmed cell death (PCD), we have deliberately excluded chronic exposures and their influence on various plant physiological processes. For exhaustive information on the

O_3 effects on various plant physiological processes, the reader is referred to reviews by Darrall (1989), Heath and Taylor (1997) and Pell *et al.* (1997).

Programmed cell death (PCD): the ins and outs

PCD is a physiological process that selectively targets and eliminates unwanted cells in both animal and plant systems. In animals cells, PCD occurs during elimination of cells that have served temporary functions and is involved in several disorders such as AIDS, Alzheimer's disease, Huntington disease, Parkinson's disease and Lou Gehrig's disease (Jacobson *et al.*, 1997; Jabs, 1999). Similarly in plants, PCD is involved in the reproductive and vegetative stages of development including senescence, formation of vessels and tracheids, fruit ripening and aleurone development. PCD also plays a role in influencing plant responses to a wide variety of biotic and abiotic stimuli (Greenberg, 1997; Kukolova *et al.*, 1997; Alvarez *et al.*, 1998; Danon and Gallios, 1998).

PCD in animal cells has been extensively studied during the past few decades and the mechanisms that regulate PCD are well documented. However, as discussed in other contributions to this issue, much less is known about how plants regulate cell death. In recent years, cell death activated during plant-pathogen interactions and other stress responses has received significant attention. Plants under the influence of microbial invaders and other stress factors use a variety of constitutive and inducible defense systems. One such rapid induced defense system is a hypersensitive response (HR) that often precedes systemic acquired resistance (SAR) and occurs at the site of pathogen entry. HR is characterized by the induction of various cellular protectant genes, host cell death and restriction of pathogen growth and spread (Hammond-Kosack and Jones, 1996; Greenberg, 1997; Lamb and Dixon, 1997). Hypersensitive cell death is often activated as part of an efficient defense response, but some key questions remain. First, why do plants activate a cell death program that decreases the functional leaf area index and affects plant productivity? Second, what are the factors that regulate PCD in plants and are the mechanisms underlying cell death in plants similar to those in animals? Third, do plants possess anti-cell death pathways similar to animals? At the moment, it is not possible to offer satisfactory explanations for most of the above questions; however, studies aimed at investigating the role of AOS, Ca^{2+}, cysteine proteases and poly(ADP-ribose) polymerases indicate that at least some of the mechanisms that influence PCD are conserved in plants and animals (Levine *et al.*, 1994; D'Silva and Heath, 1998; Amor *et al.*, 1998; Solomon *et al.*, 1999). Furthermore, the existence of lesion mimic mutants and the identification of putative animal apoptosis suppressor gene (*DAD1*) from *Arabidopsis* support the idea that plants possess anti-cell death pathways (Greenberg, 1997; Gallois *et al.*, 1997).

AOS, the regulators of PCD in plants

Plants, as a consequence of normal cellular metabolism, are continuously producing AOS such as $^{\bullet}O_2^-$, H_2O_2 and $^1O_2$. Under normal conditions, plants rapidly metabolize these AOS with the help of antioxidant enzymes and/or metabolites (Scandalios, 1997). However, various environmental perturbations can cause excess AOS production, overwhelming the system and necessitating additional defenses. Unless efficiently metabolized, these AOS rapidly oxidize membrane lipids, proteins and other cellular organelles leading to their dysfunction and ultimately inducing cell death that is manifested by the appearance of necrotic lesions.

Based on the assumption that AOS are highly reactive and lead to cellular and organelle dysfunction, their generation is frequently considered deleterious and harmful. However, recent studies indicate that AOS may be important components of signal transduction pathways that influence plant defense responses and PCD to a wide variety of stimuli, including pathogens (Lamb and Dixon, 1997; Jabs, 1999). Evidence for this comes from the observations that, firstly, a number of environmental stress factors such as O_3 (Schraudner *et al.*, 1998; Rao and Davis, 1999), cold (Prasad *et al.*, 1996), high light (Karpinski *et al.*, 1999), heavy metals (Richards *et al.*, 1998), mechanical and physical stresses (Orozco-Cardenas and Ryan, 1999), drought (Scandalios, 1997), UV radiation (Rao *et al.*, 1996; Surplus *et al.*, 1998) and pathogens (Doke, 1997; Alvarez *et al.*, 1998) act by stimulating the production of AOS. Secondly, several stimuli that generate oxidative stress induce the expression of the same sub-set of defense genes such as *GST1*, *GPX*, and *PR1* and activate a PCD pathway that was also shown to be induced by $^{\bullet}O_2^-$ or H_2O_2 treatment alone (Levine *et al.*, 1994; Jabs *et al.*, 1996, 1997; Kukolova *et al.*, 1997; Shirasu *et al.*, 1997; Alvarez *et al.*, 1998, Danon and Gallios, 1998; Karpinski *et al.*, 1999; Rao and Davis, 1999; Solomon *et al.*, 1999; Aguilar, Sharma and Davis, unpublished results).

Although data supporting a role for AOS in influencing PCD are compelling, questions such as which of the AOS species, $^{\bullet}O_2^-$ or H_2O_2, are physiologically relevant and whether AOS alone are sufficient to regulate cell death are under strong debate (Delledonne *et al.*, 1998; Durner *et al.*, 1998; Jabs, 1999). Recent studies suggest that $^{\bullet}O_2^-$, but not H_2O_2, induce cellular defenses and cell death in the *Arabidopsis lsd1* mutant and in parsley cell cultures treated with different elicitors (Jabs *et al.*, 1996, 1997). Treatment of whole plants with a glucose/glucose oxidase mixture that continuously generates H_2O_2-induced HR in primary as well as in secondary leaves (Alvarez *et al.*, 1998; Rao and Davis, 1999), while only high physiologically irrelevant concentrations of neat H_2O_2 (>100 mM) induce cell death (Rao and Davis, unpublished results). Since AOS are rapidly removed by the presence of endogenous detoxifying antioxidant enzymes/metabolites, it is not known whether the increased production of AOS alone is sufficient to orchestrate cell death directly by killing the cells and/or indirectly by signaling further cellular responses (Delledonne *et al.*, 1998). Sub-lethal levels of H_2O_2 were sufficient to induce defense gene activation in a catalase-deficient tobacco plant, while complete induction of defense genes and cell death required additional signal molecules such as salicylic acid (SA) to amplify the AOS signal (Chamnongpol *et al.*, 1998; Rao and Davis, 1999). In addition, recent studies suggest that nitric oxide (NO) can act in concert with AOS to orchestrate a HR (Delledonne *et al.*, 1998; Durner *et al.*, 1998; Bolwell, 1999).

Development of a model

The studies presented above exemplify more than ever that AOS alone may not be sufficient to induce PCD at the whole-plant level and that AOS acts in concert with other signaling components to trigger cell death. What are the components that might interact with AOS at the whole-plant level? To answer this question, an overall understanding of the essential players that can act together with AOS to influence cellular defenses and cell death is required.

The ability of plants to withstand changes in their environment depends on the speed and the efficiency at which they recognize the stress, generate signal molecules and initiate a physiological response that adjusts its metabolism to compensate for the external changes. Therefore, a typical plant response to a given stress is determined by three major steps: (1) recognition of the stress, (2) induction of signaling molecules and/or amplification of signaling response that trigger changes in gene expression and protein activities to govern appropriate physiological response and (3) maintenance and/or termination of the signaling events (Figure 3).

After the recognition of a pathogen or other stress factors, a number of downstream signaling events are activated to initiate a physiological response. These include changes in the physicochemical properties of plasmamembrane, alterations in ion and Ca^{2+} fluxes leading to alkalinization, changes in phospholipase and protein kinase activities and increased production of AOS, which together and/or independently regulate the biosynthesis of various secondary signaling molecules such as SA, jasmonic acid (JA) and ethylene that can transduce and/or amplify the signaling response (Figure 3). For example, a leaky membrane may influence the kinetics of ion and Ca^{2+} fluxes (Castillo and Heath, 1990; Clayton *et al.*, 1999) that might activate phospholipase, NADPH-oxidase and GTPases resulting in the induction of nitric oxide synthase (NOS), MAP kinases, AOS such as $^{\bullet}O_2^-$ and H_2O_2 and SA. Several studies have demonstrated that SA potentiates the feed-back amplification loop resulting in the production of excess AOS and SA and activating a cell death pathway (Draper, 1997; Rao *et al.*, 1997; Shirasu *et al.*, 1997). Likewise, NO generated by NOS reacts with $^{\bullet}O_2^-$ generating a potential oxidizing species, peroxynitrite, that might cause havoc to cellular membranes and proteins. One might also speculate that the AOS- or O_3-induced changes in plasma membrane lipid peroxidation (Pauls and Thompson, 1980) might lead to the synthesis of JA or methyl jasmonate (MeJA), alkanes, C_{6-12} volatiles and a number of other lipid-based signaling molecules (Creelman and Mullet, 1997; Bolwell, 1999; Heiden *et al.*, 1999). Likewise, O_3-induced changes in the kinetics of ion and Ca^{2+} fluxes might favor the biosynthesis of ethylene (Kieber, 1997). Although the sequence of these events is poorly understood, the overall order seems to reflect the activation of multiple signaling pathways.

Most of the studies that have explored the mechanisms of signal transduction mechanisms have utilized the classic cell culture model systems (Levine *et al.*, 1994; Shirasu *et al.*, 1997; Desikan *et al.*, 1998; Solomon *et al.*, 1999). Although these studies have significantly advanced the subject, very often the data obtained from such studies were difficult to extrapolate to the situation occurring at the whole-plant level. Furthermore, cell cultures often underestimate the role of the photosynthetic machinery in influencing oxidative burst (Allen *et al.*, 1999). Since it is increasingly realized that the final response of the plant is dependent on not one, but several interacting and interdependent signaling pathways, cell culture model systems may not always be appropriate to study such an interaction. One direct approach to study the interaction between different signaling pathways at the whole-plant level is by infecting plants with pathogens. However, pathogen-induced cell death is dependent on complex interactions between the plant and the pathogen and often occurs slowly (24–48 h after treatment) masking or underestimating the possible interaction between different signaling pathways. To circumvent this problem, it would be useful to have an elicitor that could act at the whole-plant level. Do we have an effective elicitor that can be used at the whole-plant level? During the remaining part of this review, we will present several examples on how O_3 exposure mimics plant pathogen infection in several aspects and will summarize recent data demonstrating that O_3 can be used to increase our understanding of the complex network of interacting signaling pathways involved in regulating the HR.

Acute O_3, a pathological agent

The fact that O_3 spontaneously generates AOS in leaf extracellular spaces, combined with the resemblance between the physiognomy of necrotic lesions induced by acute O_3 exposures and those resulting from a HR, lead to the suggestion that a significant overlap may exist between O_3-, pathogen- and other stress-induced responses (Kangasjarvi *et al.*, 1994; Sharma and Davis, 1994; Sandermann *et al.*, 1998). The interesting observation made in a number of studies with different plant species is that O_3 induces (1) the expression of diverse defense-related genes such as PR proteins, *GST*, *GPX* and *SOD*, (2) the emission of ethylene and other volatile organic compounds such as isoprene, mono- and sesquiterpenes, including methyl salicylate, (3) the biosynthesis of signaling molecules such as SA, JA, ethylene and ABA and (4) DNA fragmentation and nuclear chromatin condensation, both hallmark features of PCD, which are also shown to be induced by other stress factors such as cold, UV radiation and pathogen exposure (Ernst *et al.*, 1992; Croft *et al.*, 1993; Kangasjarvi *et al.*, 1994; Sharma and Davis, 1994; Wellburn and Wellburn, 1996; Rao *et al.*, 1996; Sharma *et al.*, 1996; Koukalova *et al.*, 1997; Pare and Tumlinson, 1997; Shulaev *et al.*, 1997; Tuomainen *et al.*, 1997; Alvarez *et al.*, 1998; Danon and Gallois, 1998; Sandermann *et al.*, 1998; Schraudner *et al.*, 1998; Surplus *et al.*, 1998; Clayton *et al.*, 1999; Heiden *et al.*, 1999; Rao and Davis, 2000; Rao *et al.*, 2000; Aguilar, Sharma and Davis, unpublished results; Rao and Davis, unpublished results; T. Jabs, personal communication; J. Kangasjarvi, personal communication). This phenomenon is termed 'cross-induction' (Sharma and Davis, 1997; Sandermann *et al.*, 1998) and suggests that distinct stresses may activate the same, or at least overlapping, signal transduction pathways.

The other important aspect that illustrates the similarities between the mechanisms underlying the PCD activated by O_3 and other stress factors is the mitochondrial permeability transition. Recent evidence suggests that signals derived from mitochondria play an important role in influencing PCD in both plant and animal systems (Naton *et al.*, 1996; Jabs, 1999). Attempts to localize O_3-induced H_2O_2 levels indicated that mitochondria accumulate H_2O_2 prior to the disintegration of mitochondrial matrix and the appearance of visible lesions (Pellinen *et al.*, 1999). Furthermore, O_3 exposure induced a mitochondrial phosphate translocator that might stabilize the cellular redox state and compromise cell death (Kiiskinen *et al.*, 1997). A list that summarizes the similarities between O_3- and pathogen-induced changes at the morphological, physiological and molecular levels is presented in Table 1.

Based on these similarities, we and others have suggested that acute O_3-induced responses resemble those of pathogen-induced responses not only at the morphological level, but also at the physiological and molecular levels (Kangasjarvi *et al.*, 1994; Sharma and Davis, 1997; Sandermann *et al.*, 1998). However, this remained a hypothesis until our laboratory showed that pre-exposure of plant to O_3 induces resistance to subsequent pathogen infection. Since SA is thought to be an important component of disease resistance conferred by HR and SAR, we used the *Arabidopsis* lines NahG and *npr1*, which are deficient in SA accumulation and SA-mediated signal transduction, respectively, to demonstrate that the O_3-induced disease resistance is similarly dependent on SA (Sharma *et al.*, 1996).

Table 1. A summary of the similarities in plant responses at the morphological, physiological and molecular level to O_3 and pathogen exposure.[a]

Morphological responses	Physiological responses	Antioxidant or other defense responses	Signaling molecules
chlorotic lesions (+)	ion fluxes (+)	ascorbate, glutathione (+/−)	salicylic acid (+)
necrotic lesions (+)	photosynthetic pigments (−)	APX, MDHR, DHAR (+/−)	jasmonic acid (+)
	photorespiration (+)	GR, GST, GPX, POX (+/−)	ethylene (+)
	photosynthesis (−)	phenolics, AS, NOS (+)	$^{\bullet}O_2^-$ and H_2O_2 (+)
	photoinhibition (+)	LOX, AS, lignin (+)	NO (+)
	lipid peroxidation (+)	PAL, CAD, phytoalexins (+)	Ca^{2+} fluxes (+)
	ATP depletion (+)	PR proteins, AtOZ1(+)	calmodulin (+)
	programmed cell death (+)	STS, lignin, callose (+)	ABA, MeJA (+)
		LOX, NOS (+)	C-6 volatiles (+)
		polyamines (+)	sesquiterpenes (+)
		GTPases, MAP kinases (+)	
		phospholipase-C and D (+)	
		ACC synthase, ACC oxidase (+)	

(+), increase; (−), decrease; (+/−), increase or decrease depending on plant species; ACC oxidase, 1-aminocyclopropane-1-carboxylate oxidase; ACC synthase, 1-aminocyclopropane-1-carboxylate synthase; APX, ascorbate peroxidase; MDHR, monodehydroascorbate reductase; DHAR, dehydroascorbate reductase; GR, glutathione reductase; GST, glutathione *S*-transferase; GPX, glutathione peroxidase; POX, peroxidases; PAL, phenylalanine ammonia-lyase; CAD, cinnamyl alcohol dehydrogenase; PR, pathogenesis-related proteins, basic and acidic; STS, stilbene synthase; AS, allene oxide synthase; LOX, lipoxygenase; NOS, nitric oxide synthase; AtOZ1, *Arabidopsis thaliana* ozone-induced protein; MAP kinases, mitogen-activated protein kinases; $^{\bullet}O_2^-$, superoxide radical; H_2O_2, hydrogen peroxide; NO, nitric oxide.

[a]The data presented are a summary of several articles (for a list, the reader is referred to Schraudner *et al.*, 1997; Sharma and Davis, 1997; Pell *et al.*, 1997; Sandermann *et al.*, 1998 and references therein). We apologize for not being able to cite individual references due to space limitations.

O_3-induced oxidative burst

Given the fact that O_3 induces the expression of PR proteins, a molecular marker for pathogen-induced HR and SAR in a variety of plant systems, we proposed that O_3 mimics pathogen responses by serving as an abiotic inducer of the oxidative burst, a primary signaling step during the recognition of avirulent pathogens. By using various cell permeable dyes that distinguish $^{\bullet}O_2^-$ and H_2O_2, it was shown that O_3 exposure triggers an oxidative burst that is physiologically similar to that induced during plant-pathogen interactions prior to lesion formation in the O_3-sensitive plants (Overmeyer *et al.*, 1998; Schraudner *et al.*, 1998; Rao and Davis, 1999). The fact that plants exhibit increased $^{\bullet}O_2^-$ production even after the termination of O_3 exposure and diphenylene iodonium, a suicidal inhibitor of membrane-localized oxidases, blocked O_3-induced $^{\bullet}O_2^-$ radicals suggest that an endogenous oxidative burst is initiated and maintained in O_3-sensitive plants, exacerbating the O_3-induced responses (Rao and Davis, 1999). Cytochemical localization of H_2O_2 accumulation in O_3-exposed plants indicated that H_2O_2 accumulate primarily at the plasmamembrane and cell wall during the O_3 exposure, while the cytosol, mitochondria and peroxisomes accumulated H_2O_2 only during post-fumigation periods (Pellinen *et al.*, 1999).

Although other studies have presented evidence for increased production of $^{\bullet}O_2^-$ in O_3-exposed plants (Runeckles and Vaarnou, 1997), these studies elegantly demonstrated that O_3-sensitive plants trigger an endogenous source of oxidative burst in the absence of a pathogen and provided substantial credibility for the hypothesis that acute O_3 mimics a pathological agent. However, one might argue that O_3-induced cell death might simply result from the toxicity of AOS, not being PCD. While the O_3 sensitivity in the *Arabidopsis* mutant (*vtc1*) that exhibited increased sensitivity to a wide variety of other oxidative stress factors (Conklin *et al.*, 1996) as well supports this assumption, neither Cvi-0 nor a O_3-sensitive mutant (*ozs2*) are more sensitive to other oxidative stress factors such as paraquat, ultraviolet-B radiation and SO_2 compared to Col-0 (H.

Saji, personal communication; J. Kangasjarvi, personal communication). These studies clearly indicated the specificity of O_3 in triggering an oxidative burst and initiating a PCD pathway that is similar to HR during plant-pathogen interactions. Since the studies described above overwhelmingly suggest that O_3 mimics a oxidative burst that is known to be necessary and sufficient to influence plant resistance to pathogen infection, our subsequent studies have focused on using O_3 to understand the interacting signaling pathways that influence cellular defenses and cell death (Figure 3). However, it should be remembered that O_3-induced cell death is associated with O_3 sensitivity, while cell death activated during plant pathogen interactions is considered a resistance response because of its linkage to disease resistance.

Salicylic acid

SA is a major phenylpropanoid compound that influences plant resistance to pathogens and probably to other stress factors (Sharma *et al.*, 1996; Durner *et al.*, 1997; Surplus *et al.*, 1998). Although the biochemistry of SA biosynthesis is not completely understood, recent studies suggest that both NOS and/or AOS such as H_2O_2 regulate SA biosynthesis in pathogen-infected plants (Draper, 1997; Durner *et al.*, 1998). SA influences cellular redox state (Durner *et al.*, 1997), cellular defenses (Thulke and Conrath, 1997), potentiates AOS production and triggers PCD (Draper, 1997; Durner *et al.*, 1997; Shirasu *et al.*, 1997). Identification of structural similarities between NPR1/NIM1 that regulates SA-mediated HR and human IκB (Cao *et al.*, 1997; Ryals *et al.*, 1997) suggested a role for SA in regulating cellular redox state and PCD. Furthermore, the fact that several of the cell death/lesion mimic mutants constitutively accumulated high levels of SA (Greenberg, 1997; Weymann *et al.*, 1995) in the absence of a pathogen indicated a relationship between SA and cell death. In addition, the *lls1* gene from maize was shown to encode an aromatic-ring hydroxylation dehydrogenase suggesting that a phenolic compound like SA may likely mediate cell death (Gray *et al.*, 1997). Taken together, these studies convincingly suggest SA or SA-mediated signaling pathways to play an important role in regulating cell death.

Similar to the biphasic production of AOS and SA during plant-pathogen interactions (Draper, 1997), O_3 exposure induced a biphasic production of AOS and SA in O_3-sensitive tobacco cv. Bel W3 (H. Sandermann, unpublished results). Likewise, a lesion-mimic mutant (*lsd1*) accumulated high levels of SA prior to lesion formation in response to an acute O_3 exposure (T. Jabs, personal communication) suggesting a general relation between SA and cell death. Having shown that O_3-induced HR and SAR is dependent on SA (Sharma *et al.*, 1996), our initial studies focused on SA-mediated signaling pathways that regulate cellular defenses and cell death. By using different genotypes that accumulate different SA levels in response to O_3 exposure, we have shown that O_3-induced cell death occurs via two distinct mechanisms (Rao and Davis, 1999). Compromising SA accumulation (as seen in NahG plants) resulted in the weak induction of defense responses and the accumulation of toxic metabolites that cause toxic cell death, while high levels of SA (as seen in Cvi-0) potentiated the feed-back amplification loop resulting in the production of excess AOS and triggered a cell death pathway. These studies suggested that optimal concentrations of SA are required to achieve maximal induction of defense responses and to minimize the adverse influence of O_3 without activation of PCD.

Our other studies involving hybrid poplar clones that are differentially sensitive to O_3 (O_3-tolerant NE-245 and O_3-sensitive NE-388) further confirmed the involvement of SA in influencing cell death in different species (Koch *et al.*, 1998). O_3-sensitive NE-388 was found to lack the ability to perceive SA and, unlike O_3-tolerant NE-245, failed to exhibit DNA fragmentation (a hallmark feature of PCD) in response to both O_3 and pathogen exposure (Koch *et al.*, 2000). The tolerant clone underwent SA-dependent HR activating SA-mediated defense gene expression, while in the sensitive clone lesion formation appears to be due to toxic cell death, similar to NahG plants. Thus, O_3-induced lesions can be caused either by the activation of a PCD or due to the toxicity of toxic metabolites that can occur when antioxidant defense mechanisms are not induced.

Our ongoing studies aimed at elucidating the factors that influence SA biosynthesis in O_3-exposed plants indicated a role for NOS and we are currently attempting to identify the molecular components that control NOS activity and their influence on O_3-induced cell death. To date, we have identified a putative *Arabidopsis* homologue of an animal NOS regulator that is rapidly down-regulated by O_3 in a O_3-sensitive Cvi-0 (Rao, Dias and Davis, unpublished results). Based on these studies, we propose a

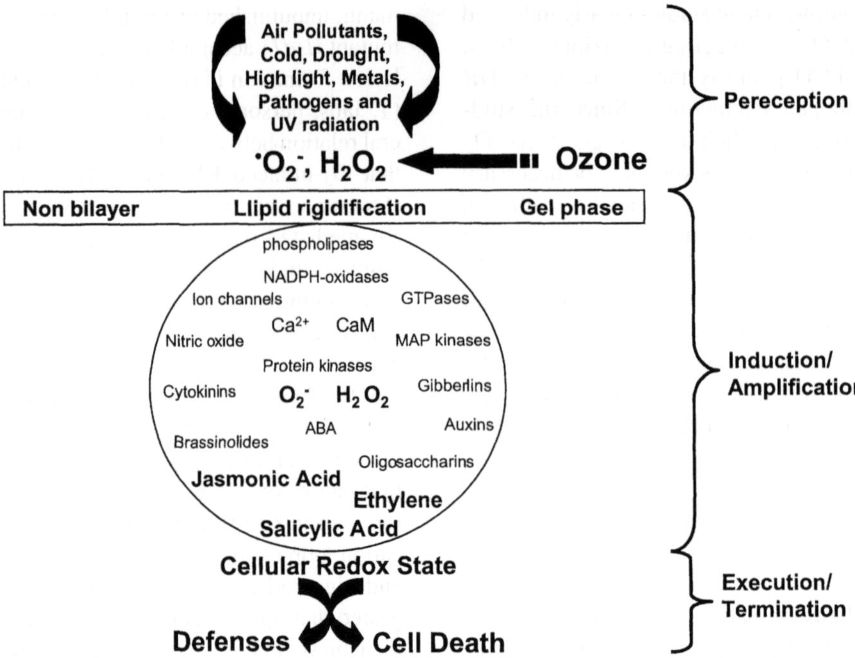

Figure 3. A simplified model identifying various signaling components that may act either in concert or antagonistically to influence plant defense responses and PCD in response to a wide variety of stimuli that generate active oxygen species. Various stress factors are known to act by generating active oxygen species such as $^\bullet O_2^-$ and H_2O_2 which oxidize membrane lipids causing distinct gel phase domains, a characteristic feature of leaky membranes. These AOS-induced changes in the physicochemical properties of plasma membrane are believed to modify phospholipase activity by altering the ion fluxes, Ca^{2+} channels, GTPases, MAP kinases, etc. that, in turn, influence the NOS activity generating NO. These changes are believed to influence the biosynthesis of signaling molecules such as salicylic acid (SA), jasmonic acid (JA), ethylene and/or other phytohormones. These signal molecules act together to amplify the signal and influence the cellular redox state which, in turn, influences the transcription of defense genes and induction of cell death. Since O_3 is believed to spontaneously dissociate in leaf extracellular spaces generating active oxygen species such as $^\bullet O_2^-$ and H_2O_2 (Figure 2) we proposed that O_3 be used as an elicitor at the whole-plant level to exclusively focus on the downstream components of AOS. $^\bullet O_2^-$, superoxide radical; H_2O_2, hydrogen peroxide; UV-B, ultraviolet-B radiation; NO, nitric oxide; ABA, abscisic acid; MAP kinases, mitogen-activated protein kinases; CaM, calmodulin.

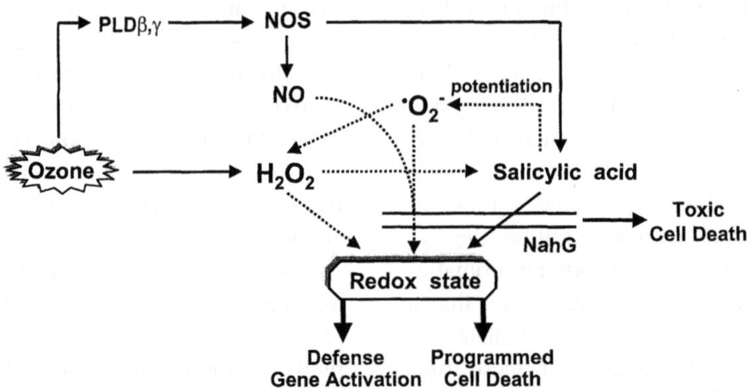

Figure 4. A schematic model illustrating the central role of SA in O_3-induced defenses and cell death responses. O_3-induced NOS and/or H_2O_2 trigger SA biosynthesis, which is required to maintain the cellular redox state and activate defense gene expression. Compromising SA accumulation (NahG) greatly reduced defense responses and accumulated toxic metabolites causing toxic cell death. High levels of SA (Cvi-0) potentiates the feed-back amplification loop resulting in the production of excess AOS and activating a PCD pathway (represented by dotted lines). $^\bullet O_2^-$, superoxide radical; H_2O_2, hydrogen peroxide; PLDβ, γ, phospholipase D β or γ isoform; NOS, nitric oxide synthase; NO, nitric oxide.

hypothetical model to illustrate the central role of SA in O_3-induced HR and cell death (Figure 4). O_3, by activating phospholipase-mediated NOS, triggers SA biosynthesis that is required to maintain the cellular redox state and to induce and potentiate the activation of cellular defense responses necessary to minimize the O_3-induced oxidative stress. Plants compromised for SA accumulation (e.g., NahG plants) or having a reduced ability to perceive SA (e.g., hybrid poplar clone NE-388) result in weak induction of antioxidant defenses that can lead to toxic cell death (necrosis), while high levels of SA as seen in Cvi-0 potentiate AOS production and subsequent activation of a runaway-hypersensitive cell death pathway. As SA has been shown to possess a dual role in inducing as well as in potentiating defense genes (Thulke and Conrath, 1998), it is likely that distinct threshold concentrations of SA are required to achieve maximal induction of defense responses without inducing PCD. A higher threshold level of SA is thus required to activate PCD. Thus, it is likely that high levels of NOS activity could lead to more $^{\bullet}O_2^-$ and/or H_2O_2 resulting in an amplifying cycle producing more SA and NO. Therefore, it appears that, once a sufficient response is reached, plants' ability to desensitize the amplification cycle would reduce the cell death.

Understanding the complex network of interacting signaling pathways

The outcome of a plant interaction with a given stress is governed by several factors including the genotype, the stress factor and the complex exchange of signals triggered by other environmental factors. Analysis of SAR in a wide range of plant species suggested that plant defense responses and cell death are regulated by a complex signaling network that involves many factors affecting various aspects of plant resistance (Dong, 1998). Although earlier studies suggested that SA is the primary regulator of disease resistance responses, recent studies indicate the existence of alternative pathways that activate HR and induced systemic resistance (ISR) which are SA-independent but JA- and/or ethylene-dependent (Penninckx *et al.*, 1996; Thomma *et al.*, 1998). Therefore, it appears that many defense reactions are dependent on the plant's ability to respond to SA, JA and ethylene and the nature and the final response depend on the extent of cross-talk that exists between these molecules (Dong, 1998; Niki *et al.*, 1998; Shah *et al.*, 1999).

Interaction between SA and JA signaling pathways

O_3 is primarily believed to react with the plasma membrane, altering the lipid composition and increasing the production of linoleic acid (Mudd, 1997), a precursor of JA biosynthesis. JA is a cyclopentanone compound that plays a major role in influencing plant growth and development, as well as plant responses to different stress factors including pathogens and wounding (Creelman and Mullet, 1997; Vijayan *et al.*, 1998). Considering the overwhelming similarities between O_3- and pathogen-induced responses, it was not surprising to learn that acute O_3 treatment results in an induction of not only SA levels, but JA levels as well in both *Arabidopsis* and hybrid poplar (Rao *et al.*, 2000; Koch *et al.*, 2000). A potential role of JA in O_3-induced responses is supported by studies showing that wounding or JA application prior to O_3 exposure reduced O_3-induced necrotic lesions (Orvar *et al.*, 1997) and H_2O_2 levels (Schraudner *et al.*, 1998) in O_3-sensitive tobacco.

Recently, our laboratory obtained evidence suggesting that JA-signaling pathways desensitize the O_3-induced, SA-mediated hypersensitive cell death pathway (Rao *et al.*, 2000). Based on the analysis of molecular markers for JA responsiveness, the ozone-sensitive ecotype Cvi-0 was found to have a greatly reduced ability to perceive JA. However, Cvi-0 is capable of responding to high concentrations of MeJA (200 μM), which attenuated O_3-induced oxidative burst and SA accumulation and completely abolished O_3-induced cell death. The role of JA in regulating O_3-induced cell death was found to be specific for PCD, as MeJA treatment failed to attenuate toxic cell death exhibited by NahG plants exposed to O_3.

Additional evidence implicating the role of JA in modulating the magnitude of O_3-induced, SA-dependent PCD was obtained by exploiting the JA-insensitive mutants (*jar1* and *coi1*) and the triple mutant, *fad3/7/8*, blocked in JA biosynthesis in Col-0 background. Our earlier studies have suggested that Col-0 is relatively resistant to O_3 and does not develop lesions even in response to prolonged O_3 exposure (4–5 days; Sharma and Davis, 1994). However, both of the JA mutants rapidly developed lesions in response to a O_3 exposure, suggesting a role for JA in regulating O_3-induced PCD. Detailed analysis of the responses of *jar1* and *coi1* suggested that the JA perception is essential to control O_3-induced oxidative burst, SA content and hypersensitive cell death pathway (Rao *et al.*, 2000). Additional support for the hypothesis that

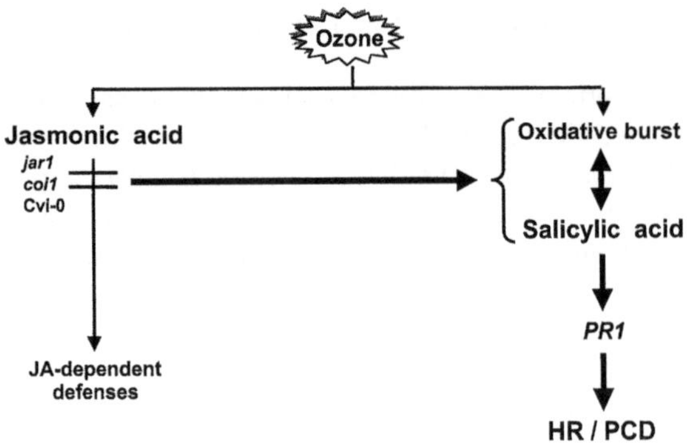

Figure 5. A hypothetical model illustrating the role of JA in influencing the magnitude of O_3-induced, SA-dependent HR/PCD. O_3 exposure induces both SA- and JA-dependent signal pathways that are known to act antagonistically. The interaction between these pathways results in sufficient SA production to trigger defense gene expression sufficient to confer resistance, but not high enough to trigger large-scale lesion formation. However, JA insensitivity increased the magnitude of O_3-induced oxidative burst accumulating high levels of SA, that, in turn, potentiated the feed-back amplification loop resulting in the runaway cycle of hypersensitive cell death (represented by thick lines). HR, hypersensitive response; PR1, pathogenesis-related protein; PCD, programmed cell death.

JA-signaling pathways regulate the magnitude of SA-mediated cell death in other plant systems is provided by the observations that both O_3-sensitive poplar clone NE-388 and the *Arabidopsis* mutant *ozs2* were also defective in the JA-dependent signaling pathways (Koch *et al.*, 2000; J. Kangasjarvi, personal communication).

Taken together, these studies suggest that JA perception is essential to desensitize the O_3-induced oxidative burst and SA-mediated amplification loop that results in the production of excess AOS and hypersensitive cell death (Figure 5). In the case of the O_3-tolerant Col-0 and hybrid poplar clone NE-245, O_3 induces both SA and JA pathways that then act antagonistically. The interaction between these pathways results in sufficient SA production to trigger defense gene expression sufficient to confer resistance, but not high enough to trigger large-scale lesion formation. At the moment it is not clear whether JA perception primarily regulates O_3-induced oxidative burst and/or SA biosynthesis, but our preliminary studies indicate that JA perception may modulate the magnitude of O_3-induced oxidative burst. Thus, JA-mediated attenuation of cell death appears to shift the balance toward life processes, which would be beneficial in situations where cell death has no apparent benefit for the survival and plant growth and development.

Interaction between SA and ethylene signaling pathways

Ethylene is a simple gaseous hormone that is known to influence various physiological processes including plant growth and development as well as stress responses (Kieber, 1997). For quite some time, plants have been known to produce ethylene in response to a variety of stress factors including wounding, hypoxia, heavy metals, pathogen infection and O_3 exposure (Mehlhorn *et al.*, 1991; Bent *et al.*, 1992; Kieber, 1997; Tuomainen *et al.*, 1997; Lund *et al.*, 1998; Vahala *et al.*, 1998; Kettunen *et al.*, 1999). However, in general, the precise role of ethylene in influencing plant response to various stress factors is rather poorly understood.

A potential role of ethylene in influencing PCD during senescence (Orzaez and Garnell, 1997), hypoxia (He *et al.*, 1996) and in maize endosperm development (Young *et al.*, 1997) was documented in several plant species. Ethylene has been shown to mediate pathogen-induced damage by modulating the spread of cell death in *Arabidopsis* (Bent *et al.*, 1992) and was proposed to play a role in disease resistance and susceptibility in tomato (Lund *et al.*, 1998). Similarly, a recent survey of various plant species indicated a relationship between O_3 sensitivity and O_3-induced stress ethylene production (Wellburn and Wellburn, 1996). O_3-induced ethylene biosynthesis is believed to induce premature senescence (Pell *et al.*, 1997; Vahala *et al.*, 1998) and/or react with ethylene to form damaging AOS (Mehlhorn *et al.*, 1991). Additional evidence for the role of ethylene in influencing O_3-induced cell death comes from studies that have shown that (1) inhibition of ethylene biosynthesis reduced O_3-induced damage in tomato (Tuomainen *et al.*, 1997), (2) treatment of plants with norbornadiene, an inhibitor of ethylene perception, blocked O_3-induced cell death (Bae *et al.*, 1996) and (3) O_3-induced cell death is dependent on *ein2*, a known ethylene receptor (J. Kangasjarvi, personal communication). We have begun to extend these studies by examining the response of specific *Arabidopsis thaliana* ethylene mutants to acute O_3 exposure. Preliminary results indicate that ethylene is a major regulator of O_3-induced defense responses and cell death and that ethylene signaling pathways interact with both SA- and JA-dependent signaling pathways (Rao, Koch and Davis, unpublished results).

Epistasis analysis

The studies described above indicated that the final response that we visualize as cell death is dependent on the nature and the extent of interaction between AOS, SA, JA and probably ethylene. Then, one question remains: how can we understand the contribution of each of these individual compounds at the whole-plant level? It might be difficult to assess the overall contribution of each of these individual compounds due to their involvement in overlapping and synergistic pathways (Dong, 1998). However, one approach our laboratory was making is to use the well-characterized mutants of SA-, JA- and ethylene-signaling pathways and perform epistasis analysis. We have generated double mutants of JA-insensitive, ethylene-insensitive and ethylene over-producer mutants expressing either NahG or *npr1* to understand the new resistance-related phenotypes and perform more in-depth analyses. Our preliminary studies with these double mutants indicated that compromising SA accumulation in JA and ethylene mutants reduced the magnitude of O_3-induced cell death (Rao and Davis, unpublished results). This further supports the hypothesis that cell death triggered by AOS is influenced by the nature and the extent of cross-talk between SA, JA and ethylene signaling. Our ongoing detailed analyses of these double mutants are expected to provide information on the complex resistance pathways that influence plant resistance responses.

Conclusion

It is an undeniable fact that O_3 is a major pollutant that affects plant growth and productivity. By exploiting the ability of O_3 to generate AOS, we have made significant progress in understanding the wide spectrum of plant defense responses including PCD that are modulated by signaling molecules such as AOS, SA, JA and ethylene. Cell death in plants is a unique process that drives several functions and the role it plays in plant defense responses has received significant attention from pathologists and stress physiologists alike. In this review we have provided several examples to illustrate how O_3 is facilitating our understanding of the intricacies of AOS-dependent biological disorders. The unique ability of O_3 to mimic several pathogen-induced responses simultaneously is helpful in the study of interacting signaling pathways that influence PCD – work that would have been difficult with model cell culture systems. No doubt, further studies using O_3 as an abiotic elicitor will provide more detailed information on the interaction of different signaling pathways during oxidative stress.

Acknowledgements

We acknowledge Drs Thorsten Jabs, Jakko Kangasjarvi, Eva Pell, Hikaru Saji and H. Sandermann for making pre-prints available prior to publication. We thank Dr Grotewold, Ohio State University, Dr Saji, NIES, Japan and Prof. Doug Ormrod, University of Victoria, Canada, for critical reading of the manuscript. We apologize to many of our colleagues for not being able to cite many exceptional articles due to space limitations. USDA co-operative State Research Service (No. 96-35100-3214) supported the work described in the author's laboratory.

References

Adams, R.M., Rosenzweig, C., Peart, R.M., Ritchie, J.T., McCarl, B.A., Glyer, J.D., Curry, R.B., Jones, J.W., Boote, K.J. and Allen, L.H. Jr. 1990. Global climate change and US agriculture. Nature 345: 219–224.

Allen, L.J., MacGregor, K.B., Koop, R.S., Bruce, D.H., Karner, J. and Brown, A.W. 1999. The relationship between photosynthesis and a mastoparan induced hypersensitive response in isolated mesophyll cells. Plant Physiol. 119: 2133–1241.

Alvarez, M.E., Pennell, R.I., Meijer, P.J., Ishikawa, A., Dixon, R.A. and Lamb, C. 1998. Reactive oxygen intermediates mediate a systemic signal network in the establishment of plant immunity. Cell 92: 773–784.

Amor, Y., Babiychuk, E., Inzé, D. and Levine, A. 1998. The involvement of poly(ADP-ribose) polymerase in the oxidative stress responses in plants. FEBS Lett. 440: 1–7.

Bae, G.Y., Nakajima, N., Ishizuka, K. and Kondo, N. 1996. The role in ozone phytotoxicity of the evolution of ethylene upon induction of 1-aminocyclopropane-1-carboxylic acid synthase by ozone fumigation in tomato plants. Plant Cell Physiol. 37: 129–134.

Bent, A.F., Innes, R.W., Ecker, J.R. and Staskawicz, B.J. 1992. Disease development in ethylene-insensitive *Arabidopsis thaliana* infected with virulent and avirulent *Pseudomonas* and *Xanthomonas* pathogens. Mol. Plant-Microbe Interact. 5: 372–378.

Bolwell, G.P. 1999. Role of active oxygen species and nitric oxide in plant defense responses. Curr. Opin. Plant Biol. 2: 287–294.

Cao, H., Glazebrook, J., Clarke, J.D., Voko, S. and Dong, X. 1997. The *Arabidopsis NPR1* gene that controls systemic acquired resistance encodes a novel protein containing ankyrin repeats. Cell 88: 57–63.

Castillo, F.J. and Heath, R.L. 1990. Ca^{2+} transport in membrane vesicles from pinto bean leaves and its alteration after ozone exposure. Plant Physiol. 94: 788–795.

Chameides, W.L., Kasibhata, P.S., Yienger, J. and Levy, H. 1994. Growth of continental-scale metroagro-plexes, regional ozone pollution and world food production. Science 264: 74–77.

Chamnongpol, S., Willekens, H., Moeder, W., Langebartels, C., Sandermann Jr., H., Van Montagu, M., Inzé, D. and Van Camp, W. 1998. Defense activation and enhanced pathogen tolerance induced by H_2O_2 in transgenic tobacco. Proc. Natl. Acad. Sci. USA 95: 5818–5823.

Chapmann, K.D. 1998. Phospholipase activity during plant growth and development and in response to environmental stress. Trends Plant Sci. 11: 419–426.

Clayton, H., Knight, M.R., Knight, H., McAinsh, M.R. and Hetherington, A.M. 1999. Dissection of ozone-induced calcium signature. Plant J. 17: 575–579.

Conklin, P.L., Williams, E.H. and Last, R.L. 1996 Environmental stress sensitivity of an ascorbic acid-deficient *Arabidopsis* mutant. Proc. Natl. Acad. Sci. USA. 93: 9970–9974.

Creelman, R.A. and Mullet, J.E. 1997. Biosynthesis and action of jasmonates in plants. Annu. Rev. Plant Physiol. Plant Mol. Biol. 48: 355–381.

Croft, P.C., Juttner, F. and Slusarenko, A.J. 1993. Volatile products of the lipoxygenase pathway evolved from *Phaseolus vulgaris* L. leaves inoculated with *Pseudomonas syringae* pv. *phaseolicola*. Plant Physiol. 101: 13–24.

D'Silva, I., Poirier, G.G. and Heath, M.C. 1998. Activation of cysteine proteases in cowpea plants during the hypersensitive response: a form of programmed cell death. Exp. Cell Res. 245: 389–399.

Danon, A. and Gallois, P. 1998. UV-C radiation induces apoptotic-like changes in *Arabidopsis thaliana*. FEBS Lett. 437: 131–136.

Darrall, N.M. 1989. The effect of air pollutants on physiological processes in plants. Plant Cell Envir. 12: 1–39.

Delledonne, M.A., Xia, Y., Dixon, R.A. and Lamb, C.J. 1998. Nitric oxide functions as a signal in plant disease resistance. Nature 394: 585–588.

Desikan, R., Reynolds, A., Hancock, J.T. and Neil, S.J. 1998. Harpin and hydrogen peroxide both initiate programmed cell death but have differential effects on defense gene expression in *Arabidopsis* suspension cultures. Biochem. J. 330: 115–120.

Doke, N. 1997. The oxidative burst: roles in signal transduction and plant stress. In. J.G. Scandalios (Ed.) Oxidative Stress and the Molecular Biology of Antioxidant Defenses, Cold Spring Harbor Laboratory Press, Plainview, NY, pp. 785–814.

Dong, X. 1998. SA, JA, ethylene and disease resistance in plants. Curr. Opin. Plant Biol. 1: 316–323.

Draper, J. 1997. Salicylate, superoxide synthesis and cell suicide in plant defense. Trends Plant Sci. 2: 162–165.

Durner, J., Shah, J. and Klessig, D.F. 1997. Salicylic acid and disease resistance in plants. Trends Plant Sci. 2: 266–274.

Durner, J., Wendehenne, D. and Klessig, D.F. 1998. Defense gene induction in tobacco by nitric oxide, cyclic GMP and cyclic ADP-ribose. Proc. Natl. Acad. Sci. USA 95: 10328–10333.

Ernst, D., Schraudner, M., Langebartels, C. and Sandermann, H. 1992. Ozone-induced changes of mRNA levels of β-1,3-glucanase, chitinase and 'pathogenesis-related' protein 1b in tobacco. Plant Mol. Biol. 20: 673–682.

Fishman, J., Vukovich, F.M. and Browell, E.V. 1985. The photochemistry of synoptic-scale ozone synthesis: implications for the global ozone budget. J. Atm. Chem. 3: 299–320.

Gallois, P., Makishima, T., Hecht, V., Despress, B., Laudie, M., Nishimotot, T. and Cooke, R. 1997. An *Arabidopsis thaliana* cDNA complementing a hamster apoptosis suppressor mutant. Plant J. 11: 1325–1331.

Gray, J., Close, P.S., Briggs, S.P. and Johal, G.S. 1997. A novel suppressor of cell death in plants encoded by the *Lls1* gene of maize. Cell 89: 25–31.

Greenberg, J.T. 1997. Programmed cell death in plant-pathogen interactions. Annu. Rev. Plant Physiol. Plant Mol. Biol. 48: 525–545.

Hammond-Kossack, K.E. and Jones, J.D.G. 1996. Resistance gene-dependent plant defense responses. Plant Cell 8: 1773–1791.

He, C.-J., Morgan, P.W. and Drew, M.C. 1996. Transduction of an ethylene signal is required for cell death and lysis in the root cortex of maize during aerenchyma formation induced by hypoxia. Plant Physiol. 112: 463–472.

Heagle, A.S. 1989. Ozone and crop yield. Annu. Rev. Phytopathol. 27: 397–423.

Heath, R.L. 1987. The biochemistry of ozone attack on the plasma membrane of plant cells. Rec. Adv. Phytochem. 21: 29–54.

Heath, R.L. and Taylor, G.E. 1997. Physiological processes and plant responses to ozone exposure. In: H. Sandermann, A. Wellburn and R.L. Heath (Eds.) Forest Decline and Ozone: A Comparison of Controlled Chamber and Field Experiments. Ecological Studies Vol. 127, Springer-Verlag, Berlin, pp. 317–368.

Heiden, A.C., Hoffmann, T., Kahl, J., Kley, D., Klockow, D., Langebartels, C., Mehlhorn, H., Sandermann H. Jr., Schraudner, M., Schuh, G. and Wildt, J. 1999. Emission of volatile organic compounds from ozone-exposed plants. Ecol. Appl. (in press).

Jabs, T. 1999. Reactive oxygen intermediates as mediators of programmed cell death in plants and animals. Biochem. Pharmacol. 57: 231–245.

Jabs, T., Dietrich, R.A. and Dangl, J. 1996. Initiation of runaway cell death in an *Arabidopsis* mutant by extracellular superoxide. Science 273: 1853–1856.

Jabs, T., Tschope, M., Colling, C., Hahlbrock, K. and Schell, D. 1997. Elicitor-stimulated ion fluxes and O_2^- from the oxidative burst are essential components in triggering defense gene activation and phytoalexin synthesis in parsley. Proc. Natl. Acad. Sci. USA 94: 4800–4805.

Jacobson, M.D., Weil, M. and Raff, M.C. 1997. Programmed cell death in animal development. Cell 88: 347–354.

Kangasjarvi, J., Talvinen, J., Utriainen, M. and Karjalainen, R. 1994. Plant defense systems induced by ozone. Plant Cell Envir. 17: 783–794.

Karpinski, S., Reynolds, H., Karpinska, B., Wingsle, G., Creissen, G. and Mullineaux, P. 1999. Systemic signaling and acclimation in response to excess excitation energy in *Arabidopsis*. Science 284: 654–657.

Kanofsky, G. and Sima, H. 1995. Singlet oxygen generation from the reaction of O_3 with plant leaves. J. Biol. Chem. 270: 7850–7852.

Kettunen, P., Overmyer, K. and Kangasjarvi, J. 1999. The role of ethylene in the formation of cell damage during ozone stress. Does ozone induced cell death require concomitant AOS and ethylene production. In: A. Kanellis, C. Chang, H. Klee, A.B. Bleecker, J.C. Pech and D. Grierson (Eds.) Biology and Biotechnology of the Plant Hormone Ethylene vol. II, Kluwer Academic Publishers, Dordrecht, Netherlands (in press).

Kieber, J.J. 1997. The ethylene response pathway in *Arabidopsis*. Annu. Rev. Plant Physiol. Plant Mol. Biol. 48: 277–296.

Kiiskinen, M., Korhonen, M. and Kangasjarvi, J. 1997. Isolation and characterization of a cDNA for a plant mitochondrial phosphate translocator *Mpt1*: O_3 stress induces *MPT1* mRNA accumulation in birch *Betula pendula* Roth. Plant Mol. Biol. 35: 271–279.

Koch, J.R., Scherzer, A.J., Eshita, S.M. and Davis, K.R. 1998. Ozone sensitivity in hybrid poplar is correlated with the lack of defense gene activation. Plant Physiol. 118: 1243–1252.

Koch, J.R., Creelman, J.A., Eshita, S.M., Seskar, M., Mullet, J. and Davis, K.R. 2000. Ozone sensitivity in hybrid poplar correlates with insensitivity to both salicylic acid and jasmonic acid: the role of programmed cell death in lesion formation. Plant Physiol. 123, 1–10.

Koukalova, B., Kovarik, A., Fajkus, J. and Siroky, J. 1997. Chromatin fragmentation associated with apoptotic changes in tobacco cells exposed to cold stress. FEBS Lett. 414: 289–292.

Krupa, S.V. and Kickert, R.N. 1989. The greenhouse effect: the impacts of carbon dioxide CO_2, ultraviolet-B UV-B radiation and ozone O_3 on vegetation. Envir. Pollut. 61: 263–293.

Krupa, S.V., Grunhage, L., Jager, H.-J., Nosal, M., Manning, W.J., Legge, H. and Hanewald, K. 1995. Ambient ozone O_3 and adverse crop response: a unified view of cause and effect. Envir. Pollut. 87: 119–126.

Laisk, A., Kull, O. and Moldau, H. 1989. Ozone concentration in leaf intracellular spaces is close to zero. Plant Physiol. 90: 1163–1167.

Lamb, C. and Dixon, R.A. 1997. The oxidative burst in plant disease resistance. Annu. Rev. Plant Physiol. Plant Mol. Biol. 48: 251–275.

Levine, A., Tenhaken, R., Dixon, R. and Lamb, C. 1994. H_2O_2 from the oxidative burst orchestrates the plant hypersensitive disease resistance response. Cell 79: 583–593.

Lund, S., Stall, R. and Klee, H. 1998. Ethylene regulates the susceptible response to pathogen infection in tomato. Plant Cell 10: 371–382.

Mehlhorn, H., O'Shea, J.M. and Wellburn, A.R. 1991. Atmospheric ozone interacts with stress ethylene formation by plants to cause visible plant injury. Plant Cell Envir. 13: 971–976.

Mudd, J.B. 1997. Biochemical basis for the toxicity of ozone. In: M. Yunus and M. Iqbal (Eds.) Plant Response to Air Pollution, Wiley, New York, pp. 267–284.

Naton, B., Hahlbrock, K. and Schmelzer, F. 1996. Correlation of rapid cell death with metabolic changes in fungus-infected, cultured parsley cells. Plant Physiol. 112: 433–444.

Niki, T., Mitsuhara, I., Seo, S., Ohtsubo, N., Ohashi, Y. 1998. Antagonistic effect of salicylic acid and jasmonic acid on the expression of pathogenesis-related PR protein genes in wounded mature tobacco leaves. Plant Cell Physiol. 39: 500–507.

Orozco-Cardenas, M. and Ryan, C.A. 1999. Hydrogen peroxide is generated systematically in plant leaves by wounding and systemin via the octadecanoid pathway. Proc. Natl. Acad. Sci. USA 96: 6553–6557.

Orvar, B.L., McPherson, J. and Ellis, B.E. 1997. Pre-activating wounding response in tobacco prior to high-level ozone exposure prevents necrotic injury. Plant J. 11: 203–212.

Orzaez, D. and Grannel, A. 1997. DNA fragmentation is regulated by ethylene during carpel senescence in *Pisum sativum*. Plant J. 11: 137–144.

Overmyer, K., Kangasjarvi, J., Kuittinen, T. and Saarma, M. 1998. Gene expression and cell death in ozone-exposed plants: is programmed cell death involved in ozone damage in ozone-sensitive *Arabidopsis* mutants. In:. L.J. DeKok and I. Stulen (Eds.) Responses of Plant Metabolism to Air Pollution and Global Change, Backhuys Publishers, Leiden, Netherlands, pp. 403–406.

Pare, P.W. and Tumilson, J.H. 1997. Induced synthesis of plant volatiles. Nature 385: 30–31.

Pauls, K.P. and Thompson, J.E. 1980. *In vitro* simulation of senescence-related membrane damage by ozone-induced lipid peroxidation. Nature 283: 504–506.

Pell, E.J., Schlagnhaufer, C.D. and Arteca, R.N. 1997. Ozone-induced oxidative stress: mechanisms of action and reaction. Physiol. Plant. 100: 264–273.

Pellinen R., Palva T. and Kangasjarvi J. 1999. Subcellular localization of ozone-induced hydrogen peroxide production in birch (*Betula pendula*) leaf cells. Plant J. 20: 349–356.

Penninckx, I.A.M.A., Eggermont, K., Terras, F.R.G., Thomma, B.P.H.J., De Samblanx, G.W., Buchala, A., Metraux, J.P., Manners, J.M. and Broekaert, W.F. 1996. Pathogen-induced systemic activation of a plant defensin gene in *Arabidopsis* follows a salicylic acid-independent pathway. Plant Cell 8: 2309–2323.

Prasad, T.K. 1996. Mechanisms of chilling-induced oxidative stress injury and tolerance in developing maize seedlings: changes in antioxidant system, oxidation of proteins and lipids and protease activities. Plant J. 10: 1017–1026.

Preston, E.M. and Tingey, D.T. 1988. The NCLAN program for crop loss assessment. In: W.W. Heck (Ed.), Assessment of crop loss from air pollution, Elsevier Applied Science Publishers, London, pp. 45–62

Rao, M.V. and Davis, K.R. 1999. Ozone-induced cell death occurs via two distinct mechanisms. The role of salicylic acid. Plant J. 16: 603–614.

Rao, M.V., Paliyath, G. and Ormrod, D.P. 1996. Ultraviolet-B- and ozone-induced biochemical changes in antioxidant enzymes of *Arabidopsis thaliana*. Plant Physiol. 110: 125–136.

Rao, M.V., Paliyath, G., Ormrod, D.P., Murr, D.P. and Watkins, C.B. 1997. Influence of salicylic acid on H_2O_2 production, oxidative stress and H_2O_2 metabolizing enzymes. Salicylic acid-mediated oxidative damage requires H_2O_2. Plant Physiol. 115: 137–149.

Rao, M.V., Lee, H.-iL., Creelman, R.A., Mullet, J.E. and Davis, K.R. 2000. Jasmonic acid signaling modulates ozone-induced hypersensitive cell death. Plant Cell, 12 (in press).

Reich, P.B. and Amundson, R.G. 1985. Ambient levels of ozone reduce net photosynthesis in tree and crop species. Science 230: 566–570.

Richards, B.L., Middleton, J.T. and Hewitt, W.B. 1958. Air pollution with relation to agronomic crops. V. Oxidant stipple of grape. Agron. J. 50: 599–561.

Richards, K.D., Schott, E.J., Sharma, Y.K., Davis, K.R. and Gardner, R.C. 1998. Aluminum induces oxidative stress genes in *Arabidopsis thaliana*. Plant Physiol. 116: 409–418.

Runeckles, V.C. and Vaartnou, M. 1997. EPR evidence for superoxide anion formation in leaves during exposure to low levels of ozone. Plant Cell Envir. 20: 306–314.

Ryals, J., Weyman, K., Lawton, K., Friedrich, L., Ellis, D., Steiner, H.-Y., Johnson, J., Delaney, T.P., Jesse, T., Vos, P. and Uknes, S. 1997. The *Arabidopsis* NIM1 protein shows homology to the mammalian transcription factor inhibitor IκB. Plant Cell 9: 425–439.

Sandermann, H. Jr., Ernst, D., Heller, W. and Langebartles, C. 1998. Ozone: an abiotic elicitor of plant defense reactions. Trends Plant Sci. 3: 47–50.

Scandalios, J.G. 1997. Oxidative Stress and the Molecular Biology of Antioxidant Defenses. Cold Spring Harbor Laboratory Press, Plainview, NY.

Schraudner, M., Moeder, W., Wiese, C., Van Camp, W., Inzé, D., Langebartels, C. and Sandermann, H. Jr. 1998. Ozone-induced oxidative burst in the ozone biomonitor plant, tobacco Bel W3. Plant J. 16: 235–245.

Shah, J., Kachroo, P. and Klessig, D.F. 1999. The *Arabidopsis ssi1* mutation restores pathogenesis-related gene expression in *npr1* plants and renders defensin gene expression salicylic acid dependent. Plant Cell 11: 191–206.

Sharma, Y.K. and Davis, K.R. 1994. Ozone-induced expression of stress-related genes in *Arabidopsis thaliana*. Plant Physiol. 105: 1089–1096.

Sharma, Y.K. and Davis, K.R. 1997. The effects of ozone on antioxidant responses in plants. Free Rad. Biol. Med. 23: 480–488.

Sharma, Y.K., Leon, J., Raskin, I. and Davis, K.R. 1996. Ozone-induced expression of stress-related genes in *Arabidopsis thaliana*: the role of salicylic acid in the accumulation of defense-related transcripts and induced resistance. Proc. Natl. Acad. Sci. USA 93: 5099–5104.

Shirasu, K., Nakajima, H., Rajashekar, K., Dixon, R.A. and Lamb, C. 1997. Salicylic acid potentiates an agonist-dependent gain control that amplifies pathogen signals in the activation of defense mechanisms. Plant Cell 9: 261–270.

Shulaev, V., Silverman, P. and Raskin, I. 1997. Airborne signaling by methyl salicylate in plant pathogen resistance. Nature 385: 718–721.

Solomon, M., Belenghi, B., Delledonne, M., Menachem, E. and Levine, A. 1999. The involvement of cysteine proteases and protease inhibitor gene in the regulation of programmed cell death in plants. Plant Cell 11: 431–443.

Surplus, S.L., Jordan, B.R., Murphy, A.M., Carr, J.P., Thomas, B. and Mackerness, S.A.H. 1998. Ultraviolet-B-induced responses in *Arabidopsis thaliana*: role of salicylic acid and reactive oxygen species in the regulation of transcripts encoding photosynthetic and acidic pathogenesis-related proteins. Plant Cell Envir. 21: 685–694.

Thomma, B.P.H.J., Eggermont, K., Penninckx, I.A.M.A., Mauch-Mani, B., Vogelsang, R., Cammue, B.P.A. and Broekaert, W.F. 1998. Separate jasmonate-dependent and salicylate-dependent defense-response pathways in *Arabidopsis* are essential for resistance to distinct microbial pathogens. Proc. Natl. Acad. Sci. USA 95: 15107–15111.

Thulke, O. and Conrath, U. 1998. Salicylic acid has a dual role in the activation of defense related genes in parsley. Plant J. 14: 35–42.

Tuomainen, J., Betz, C., Kangasjarvi, J., Ernst, D., Yin, Z.H., Langebartles, C. and Sandermann, H. Jr. 1997. Ozone induction of ethylene emission in tomato plants: regulation by differential transcript accumulation for the biosynthetic enzymes. Plant J. 33: 1151–1162.

Vahala, J., Schlagnhaufer, C.D. and Pell, E.J. 1998. Induction of an ACC synthase cDNA by ozone in light-grown *Arabidopsis thaliana* leaves. Physiol. Plant. 103: 45–50.

Vijayan, P., Shockey, J., Levesque, C.A., Cook, R.J. and Browse, J. 1998. A role for jasmonate in pathogen defense of *Arabidopsis.* Proc. Natl. Acad. Sci. USA 95: 7209–7214.

Weiss, J. 1935. Investigations on the radical HO_2 in solution. Trans. Faraday Soc. 31: 668–681.

Wellburn, F.A.M. and Wellburn, A.R. 1996. Variable patterns of antioxidant protection but similar ethene emission differences in several ozone-sensitive and ozone-tolerant plant selections. Plant Cell Envir. 19: 754–760.

Weymann, K., Hunt, M., Uknes, U., Neuenschwander, U., Lawton, K. and Ryals, J. 1995. Suppression and restoration of lesions formation in *Arabidopsis lsd* mutants. Plant Cell 7: 2013–2022.

Young, T.E., Gallie, D.R. and DeMason, D.A. 1997. Ethylene-mediated programmed cell death during maize endosperm development of wild type and *shrunken2* genotypes. Plant Physiol. 115: 737–751.

Plant Molecular Biology **44:** 359–368, 2000.
E. Lam, H. Fukuda and J. Greenberg (Eds.), Programmed Cell Death in Higher Plants.

Programmed cell death in cell cultures

Paul F. McCabe* and Christopher J. Leaver
*Department of Plant Sciences, University of Oxford, South Parks Road, Oxford, OX1 3RB, UK (fax: +44 1865 275144; e-mail: chris.leaver@plants.ox.ac.uk) *Present address: Department of Botany, University College Dublin, Belfield, Dublin 4, Ireland (e-mail: paul.mccabe @ ucd.ie)*

Key words: Arabidopsis, PCD induction, programmed cell death, protoplasts, suspension cell cultures

Abstract

In plants most instances of programmed cell death (PCD) occur in a number of related, or neighbouring, cells in specific tissues. However, recent research with plant cell cultures has demonstrated that PCD can be induced in single cells. The uniformity, accessibility and reduced complexity of cell cultures make them ideal research tools to investigate the regulation of PCD in plants. PCD has now been induced in cell cultures from a wide range of species including many of the so-called model species. We will discuss the establishment of cell cultures, the fractionation of single cells and isolation of protoplasts, and consider the characteristic features of PCD in cultured cells. We will review the wide range of methods to induce cell death in cell cultures ranging from abiotic stress, absence of survival signals, manipulation of signal pathway intermediates, through the induction of defence-related PCD and developmentally induced cell death.

Introduction

Programmed cell death (PCD) is an indispensable facet of plant development, defence responses and architecture (Pennell and Lamb, 1997), and is a highly organised, regulated process which we are just beginning to understand. Studying the mechanism of PCD in whole plants can however be difficult as it often occurs in a small group of inaccessible cells buried in a bulk of surrounding healthy cells. Tissue cultures have been used to study a wide range of physiologically important cell processes and in many instances it would be convenient to be able to study aspects of the PCD process *in vitro*. The relevance of studying PCD in cell cultures depends on how comparable the mechanism of death in a cultured cell is to PCD *in vivo*. *In vivo* and *in vitro* the process of PCD consists of three clearly identifiable phases, an induction phase – where the cell perceives a death signal and initiates cell death, an effector phase – where the cell commits irrevocably to death and the cell death machinery is activated–, and a degradation phase – where the action of the death machinery results in the morphological and biochemical features that are hallmarks of the cell death process (Green and Kroemer, 1998). During the induction phase a wide range of extra- or intracellular stimuli can influence whether or not a cell activates PCD (Vaux and Strasser, 1996). For example, PCD can be activated as a consequence of abiotic or biotic stress, pathogen attack, developmental cues or cell-cell signalling. It seems probable that in plant cells during the effector phase, as with animal cells, once the cell death signal has been perceived the disparate signalling pathways feed into a common (or limited number of) PCD pathway(s) with a shared intrinsic death machinery. Activation of the shared PCD machinery culminates in the degradation phase where a number of hallmark features arise that can be used to identify active cell death. These features include shrinkage of the protoplast and nucleus and activation of endonucleases that cleave DNA. As PCD appears to utilise a core mechanism, death activated in a cell culture will share fundamental similarities with a whole-plant developmental or defence-activated cell death. Therefore, the dissection of the death program *in vitro* is directly relevant to understanding the regulation and mechanism of plant cell death programs during growth and development.

There are a number of reasons that make cell cultures an attractive system with which to study the mechanism of PCD, for example:

- *Uniformity.* Cell cultures can be utilised to generate rapidly dividing, relatively homogeneous, collection of cells. Large quantities of cell material can be maintained or quickly generated from stocks.
- *Accessibility.* It is relatively simple to examine quantitatively the growth or death of cells in a rapidly dividing homogenous culture using simple vital stains. Compounds can be easily added or removed from cultures, and the cells are relatively accessible which facilitates the evaluation of the role of hormones or signal pathway intermediates in the regulation of PCD. Easy access to cultured cells also means one can track cells (either by sampling at time intervals or by following individual cells in real time) under the microscope for morphological changes associated with cell death.
- *Reduced complexity.* Suspension cultures are generally regarded as collections of undifferentiated cells, and this is indeed often true. Absence of cellular differentiation can be a distinct advantage when one wants to study the basal response of cells during cell death programs. However, suspension cultures are not always undifferentiated and it is also possible to use cell cultures to study developmentally regulated PCD associated with, for example, embryogenesis, xylogenesis or senescence. Additionally, defence responses can be activated by adding avirulent pathogens or biotic elicitors of the hypersensitive response (HR) directly to cell cultures. The developmental cell deaths that occur in culture are often more amenable to investigation than their *in vivo* counterparts.

While the dissection of the mechanism of PCD can be difficult in whole plants and is often easier in cell cultures it is, in several instances, even easier with single cells. It appears that every cell has an intrinsic cell death program with constitutive expression of the PCD machinery (Weil *et al.*, 1996; McCabe *et al.*, 1997a) and it is technically straightforward to establish cultures consisting of single cells or protoplasts which further reduce the complexity of the cellular environment. Single cells and protoplasts can be used to achieve maximum cell homogeneity or used to study the social controls and survival signals that regulate PCD.

This review discusses aspects of how plant cell cultures can be used to recognise, induce and ultimately study PCD in cell cultures. PCD can be activated in several of the so-called model species, including *Arabidopsis*, carrot, tobacco and *Zinnia* cultures, as well as in species such as cucumber, soybean, spruce and parsley (Table 1). Similarly, a diverse range of chemicals and compounds which induce or inhibit PCD in cell suspension cultures, for example, hypersensitive response elicitors, H_2O_2, signalling pathways intermediates (e.g. Ca^{2+}), environmental extremes (e.g. temperature stress, UV radiation) and developmental cues (embryogenesis, xylogenesis and senescence) (Table 1), are all discussed. Where appropriate, brief details are given of the use of *Arabidopsis* cultures to study PCD; more detailed technical information on using *Arabidopsis* cultures or cultures from other species can be obtained from the appropriate literature cited in the text.

Establishing cell suspension cultures

Suspension cultures of most plant species are relatively straightforward to establish and there are a large number of published protocols (Ammarito, 1984; Dudits *et al.*, 1991). Where possible a culture regime should be selected which results in a large proportion of small cell aggregates and single cells. We routinely use an *Arabidopsis thaliana* suspension culture which was initiated from stem explants and whose growth conditions have been previously defined (May and Leaver, 1993). Briefly, cells are grown in Murashige and Skoog medium containing 0.5 mg/l naphthalene acetic acid, 0.05 mg/l kinetin, 3% w/v sucrose, on a rotary shaker at 110 rpm and maintained at 24 °C under a 16 h photoperiod. Sub-culturing is carried out by pipetting 30 ml of suspension into 170 ml of fresh medium in a 500 ml conical flask every 7 days. This subculture routine produces a culture with a high percentage of small cell aggregates and single cells which is useful for microscopic analysis of cell morphology, cell fluorescence or single-cell dilution experiments.

Fractionating cell cultures

Isolation of single cells and dilution studies

For some applications, such as cell-cell signalling experiments, it may be necessary to segregate single cells from the culture. Single cells from *Arabidopsis* and carrot cell cultures can be isolated quite easily by repeatedly sieving the cell suspension culture through

Table 1. Examples of cell cultures used to study PCD in plants.

Species	PCD induced by	Reference
Arabidopsis thaliana	UV-C radiation	Danon and Gallois, 1998
	HR elicitor	Desikan *et al.*, 1998
	H_2O_2	Desikan *et al.*, 1998
Daucus carota	heat shock	McCabe *et al.*, 1997a
	low density	McCabe *et al.*, 1997a
	H_2O_2	McCabe *et al.*, 1997a
	calcium ionophore	McCabe *et al.*, 1997a
	embryogenesis cues	McCabe *et al.*, 1997a
Cucumis sativus	heat shock	Delorme *et al.*, 2000
Glycine max	H_2O_2	Levine *et al.*, 1996
	pathogen	Solomon *et al.*, 1999
Nicotiana plumbaginifolia	senescence cues	O'Brien *et al.*, 1998
	calcium ionophore	O'Brien *et al.*, 1998
	H_2O_2	O'Brien *et al.*, 1998
Nicotiana tabacum	cryptogein	Levine *et al.*, 1996
	cold stress	Koukalová *et al.*, 1997
	HR elicitor	Yano *et al.*, 1998
Picea abies	embryogenesis cues	Havel and Durzan, 1996
Petroselinum crispum	fungal infection	Naton *et al.*, 1996
Zinnia elegans	tracheary element formation	Fukuda, 1997; McCann, 1997

successively smaller-sized nylon meshes. Cell cultures should be passed through a 125 μm mesh and then serially through 50 μm, 38 μm, 30 μm, 22 μm and 10 μm meshes. The single-cell fraction should pass through the 22 μm mesh and not the 10 μm mesh and can therefore be collected from the surface of the 10 μm mesh. This series of meshes work well for a number of other species but the single cell size should be ascertained for any given cell culture and mesh sizes adjusted accordingly. The separated fraction containing the single cells should be washed in fresh medium and the cell density determined with a haemocytometer. The density can then be adjusted to the required titre by appropriate dilution into fresh medium.

Isolation of protoplasts

Instead of sieving single cells it may sometimes be more convenient to isolate protoplasts. This can be especially useful if the chosen cell culture does not contain appreciable numbers of single cells. Again there are numerous published protocols related to the isolation of protoplasts from a wide range of species. To isolate protoplasts from *A. thaliana* we use a variation of the protocol described in Altmann *et al.* (1992). Briefly, 10 ml of suspension culture is removed aseptically from a 7-day old culture and dispensed into a sterile centrifuge tube which is then centrifuged at 300 $\times$ *g* for 3 min, the liquid medium removed and replaced with 10 ml of fresh medium containing MS salts with 0.4 M sucrose, 8 mM $CaCl_2$, 1% w/v Cellulase Onozuka R-10 and 0.25% w/v Macerozyme R10. Tubes are slowly rotated overnight at 24 °C in the dark. The protoplasts are separated from undigested cells by filtering through a 60 μm nylon mesh into a clean centrifuge tube. Then 1 ml of W5 solution (150 mM NaCl, 125 mM $CaCl_2$, 5 mM KCl, 5 mM glucose) is carefully added to the surface of the enzyme solution. The tube is then centrifuged at 300 $\times$ *g* for 3 min. Intact protoplasts will float to the surface of the enzyme solution and can be removed together with the W5 solution to a clean centrifuge tube. The remaining enzyme solution is removed by adding 9 ml of W5 and recentrifugation to sediment the protoplasts. Finally, the W5 solution is replaced with culture medium (MS salts, 0.4 M sucrose, pH 5.6) at a density of 10^6 protoplasts per ml for use in PCD experiments.

How to identify PCD in cultured cells

Morphology

It is relatively easy to ascertain any visible morphological changes that occur in suspension-cultured cells induced to undergo PCD. The most obvious PCD-related, morphological change is the condensation of the protoplast away from the cell wall. This retraction leaves a gap between the wall and plasma membrane, which is visible under the light microscope. A wide range of PCD-inducing treatments result in this morphological change. For example, HR elicitors induce the condensed morphology in soybean and tobacco cells (Levine *et al.*, 1996; Yano *et al.*, 1998). McCabe *et al.* (1997a) observed the condensed morphology in cells in which PCD had been induced by a number of abiotic insults such as H_2O_2, ethanol and staurosporine. The morphology also occurs in cells that die in embryogenic culture (the cell wall also appears to thicken) and in cells deprived of survival signals (the cell walls also accumulate amorphous material) (Havel and Durzan 1996; McCabe *et al.*, 1997b). There appears to be a calcium signalling aspect to this morphological condensation as the condensation does not occur when PCD is induced in cells in the presence in $LaCl_3$, a calcium channel blocker (McCabe *et al.*, 1997a).

Viability checks

Fluorescein diacetate (FDA) is a hydrophobic molecule which enters the cell passively and is then cleaved by cytoplasmic enzymes. It becomes fluorescent and charged after cleavage, which prevents it from leaving the cell. Under ultraviolet light a bright green fluorescence is observed in viable cells with an intact plasma membrane. A 0.1% w/v stock of FDA can be made up in acetone and the stock stored at −20 °C. The stock can be diluted into fresh culture medium at a 1:50 ratio and an equal volume of the diluted FDA and cell culture added to a microscope slide. Epifluorescence is observed through a FITC (fluorescein isothiocyanate) filter. Cell viability can also be assessed with Evans blue staining (Levine *et al.*, 1996) as lethally damaged cells are unable to exclude the dye. A 0.025% w/v stock of Evans blue should be made up in liquid medium and added 1:1 to the cell suspension sample.

Nuclear morphology

The nucleus often condenses during PCD. Danon and Gallois (1998) observed three types of nuclear morphology upon UV irradiation of protoplasts: (1) normal round nuclei, (2) elongated, crescent-shaped nuclei and (3) fragmented nuclei. They suggested these three types of nuclei may constitute a nuclear morphological time course. Nuclear condensation has also been observed in senescing *Nicotiana plumbaginifolia* cells (O'Brien *et al.*, 1998) and tobacco cells treated with fungal HR elicitors (Yano *et al.*, 1998). We have observed similar nuclear condensation morphologies after induction of PCD by heat treatment in carrot and *Arabidopsis* cells. In our experience, however, nuclear condensation is a much more variable morphological change than protoplast condensation and, while a useful indicator of PCD in plants, it should not be considered as reliable a hallmark feature as it is in apoptotic animal cells.

4′,6′-diamidino-2-phenylindole (DAPI) is a DNA stain used to follow any changes in nuclear morphology. A stock of 0.0001% w/v DAPI in cell culture medium can be stored at −20 °C. Nuclear DNA is stained by adding DAPI solution to cells and transferring to a glass slide with cover slip (McCabe and Pennell, 1996). Cellular DNA is viewed under the microscope with the cells illuminated by UV light. Propidium iodide is an alternative DNA stain and can be added to cells at a final concentration of 10 μM. Propidium iodide can only enter cells whose plasma membrane have been compromised; once it enters the cell it preferentially stains nucleic acids which fluorescent red when viewed under a fluorescent microscope (Yano *et al.*, 1998). Propidium iodide staining also results in a level of fluorescence which is indicative of the extent of chromatin condensation and can therefore be used to discriminate early and late stages of PCD. Cells can then be sorted into early- and late-stage PCD on the basis of flow cytometry. This technique has been developed in tobacco cells by O'Brien *et al.* (1998) who used it to show that the early stages of plant PCD are reversible.

DNA cleavage

One of the hallmark features of PCD is the cleavage of DNA at specific chromosomal sites by endonucleases (Schwartzman and Cidlowski, 1993). Simple breakage of DNA can be detected with terminal deoxynucleotidyl transferase-mediated dUTP nick-end labelling (TUNEL) which reveals accumulated DNA 3′-OH groups (Gravrieli *et al.*, 1992). Cells are fixed in 2% v/v formaldehyde in phosphate-buffered saline (PBS) buffer pH 7.2 for 2 h, then washed in PBS, immobilised in 4% w/v low-melting-point agarose and embedded in Paraplast in a graded ethanol series with tertiary butanol as the carrier. Sections (10 μm) are cut with a microtome and the sections dried onto glass microscope slides and labelled with a TUNEL kit according to the manufacturer's instructions (Promega).

The death of cells upon abiotic stress can be necrotic or programmed. Necrotic cell death is an uncontrolled event after irreversible failure of cell membranes, while programmed cell death is a well organised cell death which results in a series of characteristic morphological changes and degradation of nuclear DNA by endonucleases. When cells are killed by a necrotic heat treatment the DNA is not cleaved and does not run far in an agarose gel during electrophoresis, or, alternatively, at very high temperatures, is destroyed by the heat rather than endonucleases and is visualised as a time-independent smear in an agarose gel (McCabe *et al.*, 1997a). If, however, the heat treatment results in activation of PCD then the DNA will only be cleaved after nucleases have been activated and, therefore, as DNA cleavage is a late feature of PCD, fragmentation should not be detectable for some time (5–12 h) after the PCD-inducing heat treatment has occurred. 5 h after heat treatment cells that have undergone PCD or necrosis both label with TUNEL. However, a time-course of DNA samples show that necrotic DNA breakage is instant while PCD endonuclease cleavage is only detectable several hours after heat treatment (McCabe *et al.*, 1997a). Also, after application of heat to the cell cultures, samples can be removed at regular intervals and the cells pelleted at 300 × *g* for 3 min and as much medium removed as possible. Cells can be ground to a fine powder in liquid nitrogen and the DNA extracted by standard procedures (Dellaporta *et al.*, 1983). DNA samples can be digested with DNase-free RNase for 1 h at 37 °C and the DNA content determined by measuring the absorption at 260 nm. DNA can then be fractionated by electrophoresis in a 1.5% w/v agarose gel or staining DNA-containing agarose gels with ethidium bromide will confirm whether the DNA fragmentation is time-dependent or not (McCabe *et al.*, 1997a).

Often PCD-activated nucleases cleave the DNA at linker sites between nucleosomes resulting in DNA fragments which are multimers of ca. 180 bp and run as a 'ladder' pattern when separated by electrophoresis in agarose gels (Mittler and Lam, 1996; Koukalová *et al*, 1997; Danon and Gallois, 1998). When the DNA has been cleaved into the ca. 180 multimers the 'laddering' pattern can sometimes be difficult to detect with ethidium bromide staining alone, but can be resolved by hybridising a Southern blot of the gel with a total DNA probe that has been randomly digested into fragments and labelled with ^{32}P. After a 10 min/55 °C heat treatment we have observed endonuclease degradation and laddering of DNA in suspension cultures of carrot, cucumber and *Arabidopsis*.

Induction of PCD in cell suspension cultures

As previously discussed, cells can die via PCD pathways or via uncontrolled necrosis. Many chemicals kill cells in a dose-dependent manner with low doses not being fatal, higher doses triggering PCD and higher doses still causing uncontrolled necrotic cell death (Lennon *et al.*, 1991). It is useful to be able to recognise the characteristics of necrotic versus PCD death in a cell culture before proceeding with further cell death experiments for a number of reasons: firstly, to determine the level of treatment which will trigger maximum PCD; secondly, cells dying necrotically can be a useful control in PCD experiments; and thirdly, different cultures of cells may differ in their sensitivity to noxious treatments with a level that causes widespread PCD in one culture causing necrosis in another. Being able to discriminate between PCD and necrosis by morphological differences allows easy microscopic assessment of how cells have died after treatment with a death-inducing agent. PCD can be differentiated from necrosis by simply applying an incremental range of a particular stress condition to the cell culture. Below we explain how to do this with a heat treatment but this experiment has been repeated with various different dose-dependent insults with the same PCD or necrotic cell death response (McCabe *et al.*, 1997a).

Stress-induced PCD

PCD may be triggered by extremes of many environmental stimuli, including temperature, water stress and UV-irradiation. Exposing cell cultures to predetermined temperatures is a convenient method of inducing PCD in a synchronous manner. Exposure of cells to different temperatures also allows an incremental response of cells from survival, through

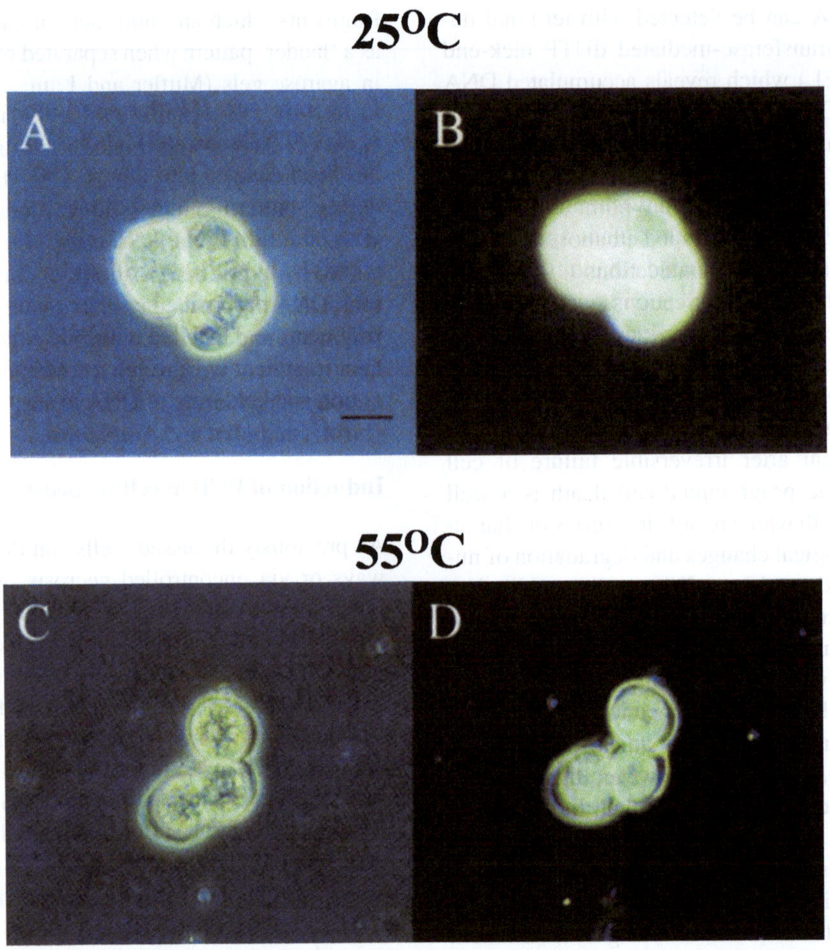

Figure 1. Cell death morphology in *Arabidopsis* cell cultures. *Arabidopsis* cell cultures 7 days old were subjected to different heat treatments (either 25 or 55 °C) for a 10 min period and returned to 25 °C for 24 h. Cells were then removed from the cultures and viewed under the microscope. Cells treated at 25 °C (A, B) have a healthy appearance 24 h after treatment while cells treated at 55 °C have undergone PCD (confirmed with FDA) and the protoplast has condensed and left a gap between plasma membrane and cell wall (C, D). Cells were photographed under phase contrast (A, C) or dark field (B, D). Bar is 10 μm for A–D.

PCD and ultimately to necrotic cell death. We have successfully triggered PCD in carrot, cucumber and *Aradidopsis* cell cultures with heat. McCabe *et al.* (1997a) subjected carrot cell cultures to 55 °C for 10 min and triggered PCD in over 90% of the cells. Morphologically the protoplasts condense away from the cell walls leaving a large gap between plasma membrane and wall, the nucleus also becomes condensed and activated nucleases cleave the DNA in a time-dependent fashion. This morphological appearance is the most obvious and easily scorable difference between cells that have undergone heat-induced PCD or necrosis. No condensation occurs in cells that have died necrotically and they are subsequently more difficult to distinguish from living cells under the microscope without vital stains. We now routinely observe this condensed morphology when PCD is induced in *Arabidopsis* cell cultures. About 10 ml of 7-day old culture is removed and the cells recovered by low-speed centrifugation. The medium is removed and replaced with fresh medium that has been heated to the desired temperature and the cells immersed in a water bath at the same temperature for 10 min. Cells are then removed from the heat treatment and returned to normal culture conditions at a constant temperature in the growth room. Morphological changes are normally first observed 1–3 h after heat treatment but it is easier to discern PCD after maximal protoplast condensation so it is usually best to wait 12–24 h to observe and score cells under the light microscope (Figure 1). After

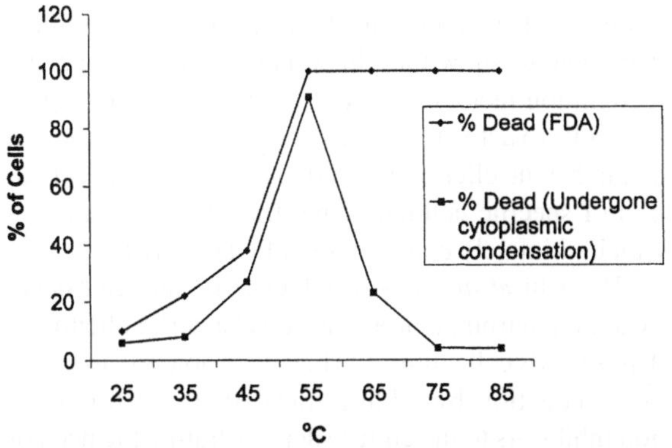

Figure 2. Induction of PCD or necrosis by heat treatment. *Arabidopsis* cell cultures 7 days old were subjected to different heat treatments for 10 min and returned to 25 °C for 24 h. Cells were then removed from the cultures and viewed under the microscope, for viability using FDA, and for PCD using cell morphology. Up to 55 °C the majority of the dead cells also showed the condensation of the protoplast which is a characteristic feature of PCD in plants. Above this temperature cells were unable to activate PCD and die necrotically and there was no evidence of protoplast condensation in dead cells.

a 10 min heat treatment at temperatures below 45 °C most cells survive, above this temperature the incidence of PCD increases to a peak of over 90% of the cells after exposure to 55 °C, above this temperature cells still die but the number that do so by PCD falls and the cells begin to die necrotically (Figure 2). Cells are sampled at regular intervals after heat shock to check for DNA laddering. Initial PCD-induced cleavage generally begins 5–7 h after heat treatment with laddering becoming clearly detectable after 12–24 h by staining with ethidium bromide (Figure 3).

Other environmental stresses have been shown to trigger PCD in cell cultures; for example, Koukalová *et al.* (1997) triggered PCD with cold stress (cells were grown at 5–6 °C in a cold room) in tobacco cell cultures. Morphologically the protoplast and nucleus were shown to have condensed and the DNA was cleaved into ca. 180 bp oligonucleosomal multiples which ran as a 'ladder' on agarose gels. Danon and Gallois (1998) subjected *Arabidopsis* protoplasts to UV radiation to trigger PCD and showed DNA laddering associated with nuclear condensation and fragmentation.

Manipulation of signal pathway intermediates

PCD in plants and animals involves cell signalling events such as an elevation in cytosolic calcium concentrations and/or an increase in reactive oxygen species. These cellular signalling events can be in-

Figure 3. Cleavage of DNA by endonucleases after activation of PCD. *Arabidopsis* cell cultures 7 days old were subjected to different heat treatments (either 25 or 55 °C) for 10 min and returned to 25 °C for 48 h. Cells were frozen in liquid nitrogen at various time points over the 48 h time period after heat treatment prior to extraction of genomic DNA. A 500 ng portion of genomic DNA were fractionated in a 1.5% w/v agarose gel. DNA was detected by staining the gel with ethidium bromide.

duced easily and rapidly in cell cultures. For example, gating calcium channels with the calcium ionophore A23187 generates an artificial calcium influx which induces widespread PCD in soybean (Levine *et al.*, 1996), carrot (McCabe *et al.*, 1997a) and tobacco (O'Brien *et al.*, 1998) cell cultures. Similarly blocking calcium influxes with calcium channel blockers modifies PCD activation (Levine *et al.*, 1996; McCabe *et al.*, 1997a; Yano *et al.*, 1998). Oxidative bursts and reactive oxygen species are also often involved in triggering PCD and this can be reproduced by adding H_2O_2 directly to cell culture medium (Levine *et al.* 1996; McCabe *et al.*, 1997a; O'Brien *et al.*, 1998). Desikan *et al.* (1998) induced cell death when they added H_2O_2 to *Arabidopsis* cell cultures and were able to show, by adding catalase which removes H_2O_2 from the culture medium, that the point of irreversibility of PCD was 60 min after addition of H_2O_2.

Defence response-induced PCD

A hypersensitive response (HR) occurs when a plant cell recognises, directly or indirectly, the product of an invading pathogen's avirulence gene. This recognition event activates a signalling cascade that initiates the up-regulation of a host of plant defence responses. Ultimately PCD is induced in challenged cells, as the HR signalling cascade appears to converge on and trigger the cell death pathway. PCD could be a vital component of a plant's defence arsenal as invasion of avirulent plant pathogens is blocked due to the death of the challenged cells. The HR can be induced in cell cultures by adding the pathogen to the suspension medium. Naton *et al.* (1996) studied the activation of plant defence responses by triggering the HR in parsley cells treated with the phytopathogenic fungus *Phytophthora infestans*. Using their model system the authors induced 60 % cell death of infected single cells within 6 h of co-cultivation. Interestingly, they also demonstrated that the age of the culture was important to activation of cell death with many more cells dying in the first 6 h from 5-day old cultures compared to 7-day old ones. They also showed that reactive oxygen species (ROS) were involved in regulating PCD as incubating cells with salicylic acid (an inhibitor of catalase) promoted PCD, while incubating cells with *n*-propyl gallate (a scavenger of ROS) inhibited PCD. Death of the cell reduced the number of secondary hyphae thereby terminating the growth of the fungal pathogen.

Elicitors are molecules produced by a pathogen that cause HR in their hosts. Several research groups have added pathogen elicitors directly to cell cultures activating the HR and often cell death. Levine *et al.* (1996) induced the HR and cell death in tobacco cell cultures by adding the fungal peptide elicitor cryptogein. Morphologically the protoplast was seen to condense leaving 'corpses' with the typical cell death morphology. The same morphology was evident in soybean cell cultures which were induced to undergo PCD by adding avirulent *Pseudomonas syringae* pv. *glycinea* (Solomon *et al.* 1999).

Yano *et al.* (1998, 1999) added xylanase to suspension cultures of tobacco. Xylanase is a proteinase elicitor of the HR obtained from the fungus *Trichoderma viride* and had previously been shown to induce the defence response in leaves of tobacco. When it was added directly to cell suspension cultures it induced cell death accompanied by shrinkage of the cytoplasm and nuclear condensation. Up to 70% of the cells died within 24 h of addition of the elicitor. The authors were able to show that the xylanase required specific recognition factors to be present on the plant suspension cells to activate the HR, and demonstrated the cell death that the elicitor induced to be the result of activation of specific cellular signal transduction pathways which ultimately converge on a PCD pathway.

Desikan *et al.* (1998) induced HR and cell death by adding harpin, a proteinaceous bacterial elicitor, to *Arabidopsis* cell cultures. They were able to block this PCD induction by adding translation and transcription inhibitors to the culture demonstrating the need of protein synthesis for induction of PCD in their system.

Absence of survival signals

Raff (1992) proposed that PCD in animal cells occurs by default unless suppressed by signals from neighbouring cells. Evidence to support this proposal has come from several subsequent studies. For example, Barres *et al.* (1992) showed that 50% of oligodendrocytes die in the developing rat optic nerve and suggested this was due to competition for limiting amounts of survival signals. Increasing the amount of survival signals present in the developing optic nerve greatly reduced normal oligodendrocyte death. Ishizaki *et al.* (1993) showed cultures of rat neonatal lens cells died when cultured at low cell density. Culture medium from high-density cultures promotes the survival of cells grown at low density suggesting that lens cells secrete survival signals for other lens cells.

Similarly, plant cells do not proliferate, and subsequently die, when cultured below a critical threshold density. However, cells can survive at sub-optimal densities if they are cultured in 'conditioned medium'. Conditioned medium is medium in which cells have previously been actively growing and contains factors that sustain cells at low densities. McCabe *et al.* (1997a) demonstrated that the death of carrot cells at low cell densities was in fact PCD and, furthermore, that this PCD could be suppressed by culturing the cells in conditioned medium which contained cell-released 'survival signals'. Schröder and Knoop (1995) purified a survival signal from tobacco cell culture conditioned medium and showed it to be a branched oligosaccharide. It is not clear yet if this molecule is representative of all survival molecules, whether it is a member of a specialized family of molecules, or if it is one of a vast plethora of molecules that can suppress PCD in plant cells. However, cell cultures should be well-suited to answering this question.

Developmental cell death in embryogenic cultures

Certain single cells in carrot embryogenic cell cultures can develop into somatic embryos. These cells initially undergo an asymmetric division which results in a vacuolate 'suspensor' type cell and a cytoplasm-rich 'embryo proper' cell. While the cytoplasmic cell begins to divide and differentiate into the embryo the suspensor cell dies (McCabe *et al.*, 1997b). The cell death results in protoplast condensation and the activation of endonucleases resulting in DNA cleavage. PCD morphology and endonuclease activation has also been shown to occur in suspensor cells in embryogenic cultures of *Picea abies* (Havel and Durzan, 1996).

Developmental cell death in Zinnia *cultures*

In *Zinnia elegans* cell cultures, single mesophyll cells can transdifferentiate into tracheary elements (TE) in response to plant growth regulators (Fukuda, 1997 and this issue; McCann, 1997). Homogeneity of this single cell culture, along with the high percentage of cells that can be synchronously induced to transdifferentiate directly into TE cells, has resulted in this system being extremely useful in studying TE differentiation. Towards the end of TE differentiation, cell death occurs and this model system can be used to study the biochemical and molecular activation of TE PCD (McCann, 1997).

Conclusions

Programmed cell death is a basic cellular process that is a vital component of plant growth and development. As well as being an important basic cellular process early in development, the involvement of PCD in later events such as senescence, fertility, pathogen defence and xylogenesis means an improved understanding of PCD at the molecular and biochemical level will potentially have important agronomic implications. In this review we have provided a range of examples as to how cell cultures can be used to investigate the regulation and mechanism of PCD. The examples we have provided are not exhaustive, and furthermore, as the research field of plant PCD is relatively young there are many opportunities to study PCD in cell culture which have not yet been explored. In many instances suspension cultures are an effective tool in studying the regulation of PCD and will be instrumental in uncovering at least some of the answers as to how cells die.

Acknowledgements

Our studies described in this review have been funded in part by grants from the Biotechnology and Biological Sciences Research Council.

References

Altmann, T., Damm, B., Halfter, U., Willmitzer, L. and Morris, P.-C. 1992. Protoplast transformation and methods to create specific mutants in *Arabidopsis thaliana*. Theor. Appl. Genet. 84: 371–383.

Ammarito, P.V. 1984. Induction, maintenance, and manipulation of development in embryogenic cell suspension cultures. In: I.K. Vasil (Ed.) Cell Culture and Somatic Cell Genetics of Plants: Laboratory Procedures and their Applications, Academic Press, Orlando, FL, pp. 139–151.

Barres, B.A., Hart, I.K., Coles, H.S.R., Burne, J.F., Voyvodic, J.T., Richardson, W.D. and Raff, M.C. 1992. Cell death and control of cell survival in the oligodendrocyte lineage. Cell 70: 31–46.

Danon, A. and Gallois, P. 1998. UV-C radiation induces apoptotic-like changes in *Arabidopsis thaliana*. FEBS Lett. 437: 131–136.

Dellaporta, S.L., Wood, J. and Hicks, J.B. 1983. A plant DNA minipreparation: version II. Plant Mol. Biol. Rep. 1: 19–21.

Delorme, V.G.R., Mc Cabe, P.F., Kim, D.-J. and Leaver, C.J. 2000. A matrix metalloproteinase gene is expressed at the boundary of senescence and programmed cell death in cucumber. Plant Physiol. 123: 917–927.

Desikan, R., Reynolds, A., Hancock, J.T. and Neill, S.J. 1998. Harpin and hydrogen peroxide both initiate programmed cell death but have differential effects on defence gene expression in *Arabidopsis* suspension cultures. Biochem. J. 330: 115–120.

Dudits, D., Börge, L. and Györgyev, J. 1991. Molecular and cellular approaches to the analysis of plant embryo development from somatic cells *in vitro*. J. Cell Sci. 99: 475–484.

Fukuda, H. 1997. Tracheary element differentiation. Plant Cell 9: 1147–1156.

Gavrieli, Y., Sherman, Y. and Ben-Sasson, S.A. 1992. Identification of programmed cell death *in situ* via specific labeling of nuclear DNA fragmentation. J. Cell Biol. 119: 493–501.

Green, D and Kroemer, G. 1998. The central executioners of apoptosis: caspases or mitochondria? Trends Cell Biol. 8: 267–271.

Havel, L. and Durzan, D.J. 1996. Apoptosis during diploid parthenogenesis and early somatic embryogenesis of Norway spruce. Int. J. Plant Sci. 157: 8–16.

Ishizaki, Y., Voyvodic, J.T., Burne, J.F. and Raff, M.C. 1993. Control of lens epithelial cell survival. J. Cell Sci. 121: 899–908.

Koukalová, B., Kovarík, A., Fajkus, J. and Siroký, J. 1997. Chromatin fragmentation associated with apoptotic changes in tobacco cells exposed to cold stress. FEBS Lett. 414: 289–292.

Lennon, S.V., Martin, S.J. and Cotter, T.G. 1991. Dose-dependent induction of apoptosis in human tumour cell lines by widely diverging stimuli. Cell Prolif. 24: 203–214.

Levine, A., Pennell, R.I., Alvarez, M.E., Palmer, R. and Lamb, C. 1996. Calcium-mediated apoptosis in a plant hypersensitive disease response. Curr. Biol. 6: 427–437.

May, M.J. and Leaver, C.J. 1993. Oxidative stimulation of glutathione synthesis in *Arabidopsis thaliana* suspension cultures. Plant Physiol. 103: 621–627.

McCabe, P.F. and Pennell, R.I. 1996. Apoptosis in plant cells *in vitro*. In: T.G. Cotter and S.J. Martin (Eds.) Techniques in Apoptosis, Portland Press, London, pp. 301–326.

McCabe, P.F., Levine, A., Meijer, P.-J., Tapon, N.A. and Pennell, R.I. 1997a. A programmed cell death pathway activated in carrot cells cultured at low cell density. Plant J. 12: 267–280.

McCabe, P.F., Valentine, T.A., Forsberg, S. and Pennell, R.I. 1997b. Soluble signals from cells identified at the cell wall establish a developmental pathway in carrot. Plant Cell 12: 2225–2241.

McCann, M.C. 1997. Tracheary element formation: building up to a dead end. Trends Plant Sci. 2: 333–338.

Mittler, R. and Lam, E. 1996. Sacrifice in the face of foes: pathogen-induced programmed cell death in plants. Trends Microbiol. 4: 10–15.

Naton, B., Hahlbrock, K. and Schmelzer, E. 1996. Correlation of rapid cell death with metabolic changes in fungus-infected, cultured parsley cells. Plant Physiol. 112: 433–444.

O'Brien, I.E.W., Baguley, B.C., Murray, B.G., Morris, B.A.M. and Ferguson, I.B. 1998. Early stages of the apoptotic pathway in plant cells are reversible. Plant J. 13: 803–814.

Pennell, R.I. and Lamb, C. 1997. Programmed cell death in plants. Plant Cell 9: 1157–1168.

Raff, M.C. 1992. Social controls on cell survival and cell death. Nature 356: 397–400.

Schröder, R. and Knoop, B. 1995. An oligosaccaride growth factor in plant suspension cultures: a proposed structure. J. Plant Physiol. 146: 139–147.

Schwartzman, R.A. and Cidlowski, J.A. 1993 Apoptosis: the biochemistry and molecular biology of programmed cell death. Endocrine Rev. 14: 133–151.

Solomon, M., Belenghi, B., Delledonne, M., Menachem, E. and Levine, A. 1999. The involvement of cysteine proteases and protease inhibitor genes in the regulation of programmed cell death in plants. Plant Cell 11: 431–443.

Vaux, D.L. and Strasser, A. 1996. The molecular biology of apoptosis. Proc. Natl. Acad. Sci. USA 93: 2239–2244.

Weil, M., Jacobsen, M.D., Coles, H.R.S., Davies, T.J., Gardner, R.L., Raff, K.D. and Raff, M.C. 1996. Constitutive expression of the machinery for programmed cell death. J. Cell Biol. 133: 1053–1059.

Yano, A., Suzuki, K., Uchimiya, H. and Shinshi, H. 1998. Induction of hypersensitive cell death by a fungal protein in cultures of tobacco cells. Mol. Plant-Microbe Interact. 11: 115–123.

Yano, A., Suzuki, K. and Shinshi, H. 1999. A signaling pathway, independent of the oxidative burst, that leads to hypersensitive cell death in cultured tobacco cells includes a serine protease. Plant J. 18: 105–109.

SECTION 3

MACHINE PARTS AND REGULATORS OF THE DEATH ENGINE IN PLANTS

Plant Molecular Biology **44:** 371–385, 2000.
E. Lam, H. Fukuda and J. Greenberg (Eds.), Programmed Cell Death in Higher Plants.

Regulators of cell death in disease resistance

Ken Shirasu and Paul Schulze-Lefert*
*The Sainsbury Laboratory, John Innes Centre, Colney Lane, Norwich, NR4 7UH, UK (*author for correspondence; current address: Max Planck Institut für Züchtungsforschung, D-50829 Köln, Carl-von-Linne-Weg 10, Germany; e-mail: schlef@mpiz-koeln.mpg.de)*

Key words: hypersensitive response, lesion mimic, programmed cell death, R gene, reactive oxygen intermediates

Abstract

Cell death and disease resistance are intimately connected in plants. Plant disease resistance genes (*R* genes) are key components in pathogen perception and have a potential to activate cell death pathways. Analysis of R proteins suggests common molecular mechanisms for pathogen recognition and signal emission whereas the subsequent signalling unexpectedly involves a network of pathways of parallel, branching and converging action. Disease resistance signalling mutants have revealed novel types of regulatory proteins whose biochemical functions are still unknown. Accumulation of small molecules such as salicylic acid, reactive oxygen intermediates, and nitric oxide amplifies resistance responses and directs cells to initiate cell death programs. Genetic analyses of lesion mimic mutants provide a glimpse of how cell death thresholds are set via an interplay of positive and negative regulatory components.

Abbreviations: CWAs, cell wall appositions; GPCR, G-protein-coupled receptors; HR, hypersensitive response; LRR, leucine-rich repeat; LZ, leucine zipper; NB, nucleotide binding; NO, nitric oxide; NOS, nitric oxide synthase; PCD, programmed cell death; ROI, reactive oxygen intermediates; SA, salicylic acid; TIR, Toll and mammalian interleukin-1 receptors; TM, transmembrane

Introduction

The abnormally rapid death of host plant cells in response to pathogen attack has been recognized as 'hypersensitiveness' as early as the beginning of this century (Stakman, 1915). Later, the term 'hypersensitive reaction' (HR) was introduced to recapitulate the defence reaction in plants which is accompanied by rapid tissue necrosis and often accumulation of small antimicrobial molecules (phytoalexins) in the incompatible plant-pathogen interaction (Klement, 1963, 1982). This early definition of HR (also often termed 'hypersensitive response') has led to considerable confusion. Does the necrotic lesion caused by a high concentration of inoculum in laboratory experiments reflect natural conditions of pathogen infection? Is the necrosis the same as the one caused by toxic metabolites of pathogens? Is HR cause or consequence of resistance?

Since pathogens have evolved different strategies to invade hosts, it is not surprising that plants also possess a variety of defence mechanisms. It is therefore unlikely that host cell death is the only component conferring growth arrest of pathogens and there is ample evidence suggesting that cell death reinforces or stimulates the induction of plant defence. Thus, we would like to define HR in this article as a whole process of plant defence responses including recognition of pathogens, host cell death, accumulation of antimicrobial compounds and inhibition of pathogen growth. Only recently genetic and biochemical data accumulated suggesting that HR cell death in plants is under intrinsic genetic control or, in other words, a type of programmed cell death (PCD). Then new questions arose. Are there molecular commonalities between PCD in HR and apoptosis in metazoan organisms? If not, how do plants control and execute cellular suicide in HR? Is the mechanism of PCD during HR the same as those seen in other physiological forms of plant PCD? We review here recent data obtained by genetic, biochemical and cytological techniques that provide a first glimpse into the molecular control of PCD in plant disease resistance.

Resistance (R)-gene-dependent activation of PCD

Conserved structures of R *gene products and their relation to cell death*

R-gene-dependent resistance has been genetically described in interactions of plants with all major classes of plant pathogens. A single gene in the host (the *R* gene) confers resistance only to those pathogen isolates containing a cognate avirulence gene (*Avr*) (Flor, 1971). This 'gene-for-gene' type resistance is generally interpreted by an elicitor-receptor model: the plant R protein recognizes directly or indirectly the pathogen-derived *Avr* product. Most of *R* gene-triggered resistance is associated with a HR. Rapid host cell death is indeed an attractive means to explain resistance to different classes of plant pathogens, including biotrophs and necrotrophs. HR can potentially deplete nutrient supply to biotrophs and may arrest ingress of necrotrophs via release of toxic host molecules. Conceptually, it seems plausible that speed of death execution is critical for efficient pathogen arrest whereas strict spatial confinement to attempted infection sites will be necessary to minimize tissue damage. Here we critically assess whether recent substantial progress in the molecular characterization of *R* genes and *R*-gene-dependent signalling components warrants the idea of an intrinsic death path that is activated upon pathogen perception.

The past few years have seen the molecular isolation of *R* genes specifying resistance to viruses, bacteria, fungi, nematodes and insects (Baker *et al.*, 1997; Ellis and Jones, 1998). An important observation from sequence alignments of the encoded proteins is a modular R protein architecture. The modules of R proteins are leucine-rich repeats (LRR), a nucleotide-binding domain (NB) and a serine/threonine kinase unit. The type and order of the modules present in R proteins forms the basis for R protein classes. The predominant class has a central NB domain and a carboxy-terminal LRR. This NB-LRR class is further subdivided based on sequence similarities close to the N-termini. One subclass contains similarities to the cytoplasmic domains of the *Drosophila* Toll and mammalian interleukin-1 receptors, the TIR-NB-LRR proteins. The second subclass has amino-terminal sequences with the potential to form coiled-coil structures and exhibit some similarities to a leucine-zipper domain, the LZ-NB-LRR proteins. Both the TIR-NB-LRR and the LZ-NB-LRR proteins are predicted to reside in the cytoplasm. The second R protein class is thought to be membrane-anchored by a single transmembrane helix. In this class, the LRR domain is exposed extracellularly and can be linked, as in the case of rice *Xa21*, to a cytoplasmic serine/threonine kinase module (Song *et al.*, 1995).

On the assumption that R proteins are directly involved in pathogen perception, one must predict the existence of at least two subfunctions, recognition of pathogen determinants and signal emission to downstream components. Recognition is primarily determined by LRR modules in each the membrane-anchored and cytoplasmic classes (Thomas *et al.*, 1997; Ellis *et al.*, 1999). Sequence comparison of *Cf-9* homologues in tomato and *L* alleles from flax identified hypervariable amino acid positions in a predicted β-strand/β-turn region of the LRRs (Parniske *et al.*, 1997; Ellis *et al.*, 1999). Structural studies of a crystallized LRR-containing protein, ribonuclease inhibitor (RI) and its interactor RNaseA revealed that the interacting point resides in the β-strand/β-turn region of the LRRs (Kobe and Deisenhofer, 1995). Indeed, a single amino acid substitution of a plant LRR-containing protein, polygalacturonase-inhibiting protein (PGIP), has been shown to confer a new ligand recognition capability (Leckie *et al.*, 1999). Thus, LRRs in R proteins appear to provide both a structural backbone and, via hypervariable solvent-exposed residues, a recognition surface for pathogen determinants. It should be noted, however, that, despite intensive efforts, direct biochemical evidence for this simple model is lacking. Even less is known about the mechanistic basis of signal emission by R proteins. This could be mediated by the LZ and TIR domain in NB-LRR type proteins (see below). The cytoplasmic kinase domain in Xa21 may serve such a role in the membrane-anchored class of R proteins.

Molecular characterization of tomato *Pto*, specifying bacterial resistance to *Pseudomonas syringae* containing *AvrPto*, has been particularly instructive for our understanding of plant *R* gene functions. Different from the multi-domain architecture of the above-mentioned two R protein classes, *Pto* was shown to encode a serine/threonine kinase only (Martin *et al.*, 1993). Interestingly, Pto requires for its function a LZ-

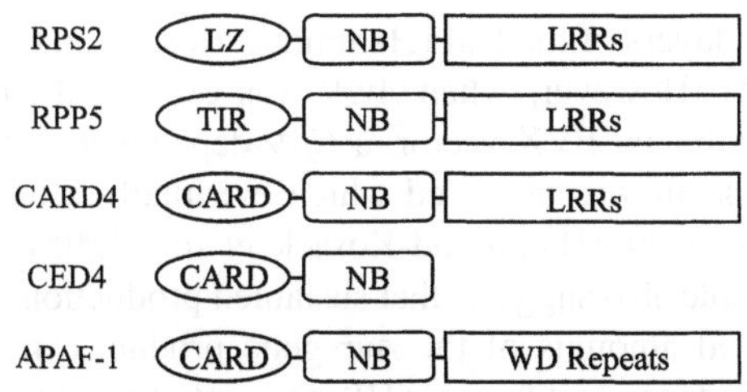

Figure 1. Comparative domain structure of R proteins, CARD4, CED4, and APAF-1. LZ, leucine zipper motif; NB, nucleotide-binding motif; LRR, leucine-rich repeat; TIR, Toll/interleukin-1 receptor homology domain; CARD, caspase recruitment domain.

NB-LRR protein, termed Prf (Salmeron *et al.*, 1994, 1996). Pto is the only R protein so far shown to interact directly with its cognate avirulence gene product AvrPto (Scofield *et al.*, 1996; Tang *et al.*, 1996). One possibility is that the LZ-NB-LRR protein Prf recognizes the AvrPto-Pto complex, thereby activating the HR response. This scenario is consistent with recent reports describing the use of constitutive active *Pto* mutants to probe the molecular basis of HR induction (Rathjen *et al.*, 1999). Activation of the HR by constitutive active Pto variants is still strictly dependent on the presence of *Prf* and requires Pto kinase activity. However, these variants have lost the capacity to bind AvrPto, thus they overcome the ligand dependence of the wild-type protein to induce HR. Binding of AvrPto to Pto in wild-type plants may lead to a conformational change and activation of Pto kinase, thereby providing a critical conformational substrate for the *Prf*-triggered HR response. In this model, the LZ-NB-LRR protein Prf may not directly interact with any pathogen determinant but may indirectly 'recognize' the product of AvrPto activity through the conformational change imposed on Pto. Future experiments will show whether this model is also applicable to other members of the NB-LRR R protein family.

Do the conserved modules of R proteins provide hints for PCD activation upon pathogen perception? One possible link has been exposed by sequence alignments of R proteins and domains known to be involved in animal PCD (Chinnaiyan *et al.*, 1997; van der Biezen and Jones, 1998; Aravind *et al.*, 1999). Five distinct motifs of NB-LRR type R proteins, encompassing the NB domain, share significant sequence similarities and collinear arrangement with *C. elegans* CED-4 and human APAF-1 (Figure 1). CED-4 and its human equivalent APAF-1 represent one component of a ternary complex, termed the 'apoptosome'. This inactive ternary complex becomes activated by extracellular death-inducing stimuli that promote the release of CED-4/APAF-1, in turn leading to an association and activation of caspases to initiate the death program (Thornberry and Lazebnik, 1998). The five motifs shared between NB-LRR proteins and CED-4/APAF-1 present a characteristic signature, termed NB-ARC or apoptotic Ap-ATPase domain (Aravind *et al.*, 1999, van der Biezen and Jones, 1998), that distinguishes it from other ATP or GTP binding domains in many proteins. Recently, a novel CED-4/APAF-1 family member (CARD4) was identified which shares striking similarity with plant NB-LRR R proteins (Bertin *et al.*, 1999) (Figure 1). Conceptually, one could envisage R proteins as components of a plant cell death machinery that is activated upon recognition of a pathogen-derived ligand. The Ap-ATPase domain is also found in a number of prokaryotic multi-domain proteins (Aravind *et al.*, 1999), suggesting that plant and animal Ap-ATPases have originated from a common ancestral prokaryotic domain. Thus, it is possible that plants and animals adapted the Ap-ATPase domain to perform different tasks. Although the overall modular architecture of NB-LRR proteins in plants and CED-4/APAF-1/CARD4 in animals is similar, the N-terminal caspase recruitment domain (CARD), which is present in the latter, has not been found in plant R proteins to date (Figure 1). Thus, biochemical evidence is needed to determine if and to what extent these structural similarities reflect functional similarities.

Proteins similar to other core components of the animal apototic machinery (caspases and Bcl2 homologues) have not been identified in plants. The recent report of caspase-independent PCD in *Dictyostelium* underlines the notion that a different cell death program may have emerged earlier in evolution (Olie *et al.*, 1998). Since plants diverged from primitive eukaryotes earlier than insects, nematodes and mammals, it remains a possibility that they have developed novel death programs.

Does R-gene-dependent resistance require cell death?

Most but not all *R*-gene-triggered resistance is accompanied by an HR cell death. For example, *Rx*-dependent resistance in potato against potato virus X (PVX) is not associated with host cell death (Bendahmane *et al.*, 1999). This has been designated extreme resistance or ER. *Rx*-dependent resistance is functional in isolated leaf protoplasts as indicated by suppression of viral replication (Adams *et al.*, 1986;

Köhm *et al.*, 1993). This apparent cell-autonomous *Rx* function contrasts with conventional HR-type resistance that is believed to involve cell-to-cell communication (Lamb and Dixon, 1997). Surprisingly, *Rx* encodes a LZ-NB-LRR type protein, similar to *RPS2* and *RPM1*, which are associated with HR cell death in *Arabidopsis* upon *Pseudomonas syringae* challenge. Transgenic *Nicotiana tabacum* and *N. benthamiana* containing *Rx* exhibit the same rapid arrest of PVX accumulation without HR, thus HR-free resistance is not a particular characteristic of potato. Somehow Rx must rapidly block PVX accumulation by interfering with an unknown step(s) in the viral replication cycle. Does this mean that Rx induces a resistance pathway that is distinct from the usual *R*-gene-triggered HR cell death pathway? Interestingly, when the PVX coat protein elicitor was constitutively expressed by an *Agrobacterium*-mediated expression system in leaf cells, HR cell death was triggered even in the *Rx* background. These data may indicate that an increased amount or a sustained expression of the coat protein elicitor can force the ER mode into HR-type resistance. Conversely, *Rx*-triggered arrest of viral replication can even override the induction of HR by another *R* gene. For these experiments *Rx* and *N*, encoding a TIR-NB-LRR type protein and associated with HR-type arrest of viral spread, were co-expressed in transgenic tobacco (Whitham *et al.*, 1994, Bendahmane *et al.*, 1999). To test for possible interference of the resistance responses, a recombinant tobacco mosaic virus (TMV) was used, expressing both the PVX elicitor for *Rx* and the TMV elicitor for *N* function. The recombinant TMV induced HR-type resistance on tobacco expressing the *N* gene only, but resistance without cell death in the presence of both *N* and *Rx* genes. Thus, *Rx*-mediated arrest of viral replication is epistatic to *N*-triggered resistance. Taken together, the case of *Rx* demonstrates that the activity of R proteins cannot be considered as a component of a death machinery but rather in providing signals that can be forced into a cell death pathway.

A second example of HR-independent resistance is *Cf-9*-mediated resistance in tomato to the fungus *Cladosporium fulvum*. *Cf-9* encodes a member of the membrane-anchored class of *R* genes with extracellular LRR units (Jones *et al.*, 1994). The cognate fungal avirulence gene *Avr9* encodes a small secreted peptide which in its active form has a length of only 28 amino acids (van den Ackerveken *et al.*, 1992). Incompatibility in plants homozygous for *Cf-9* is manifested as a gradual cessation of hyphal growth in the mesophyll layers of the leaf (Hammond-Kosack and Jones, 1994). However, when *Avr9* was expressed from a recombinant PVX vector, a *Cf-9*-dependent systemic cell death was observed which eventually killed the entire plant (Hammond-Kosack *et al.*, 1995). This example also suggests that sustained production or increased amounts of the *Avr* gene product can drive HR-independent towards HR-associated resistance.

Further evidence that *R*-gene-mediated pathogen growth arrest can be uncoupled from HR comes from studies of plants that are homo- or heterozygous for a particular *R* gene. Barley plants homozygous for the powdery mildew resistance gene *Mlg* exhibit a characteristic single-cell HR. Heterozygotes for *Mlg* retain fungal arrest but have lost the capacity to trigger HR (Görg *et al.*, 1993). Plants heterozygous for *Cf-9* are less effective in containing fungal growth than homozygotes (Hammond-Kosack and Jones, 1994). Since *Cf-9* confers resistance without HR, this suggests that *R* gene dose can modify resistance as well as HR. Taken together, these observations provide strong evidence that R proteins emit signals that feed into cell death pathways. It appears, however, that the amount of R and Avr protein or the duration of Avr expression, defined by each *R/Avr* combination, can determine whether resistance occurs with or without HR. This implies the existence of a threshold responding to the combined duration and amplitude of all the *R*-gene-derived signals, above which a cell commits itself to suicide.

A mutational approach was used to shed further light on the relationship between HR cell death and pathogen growth arrest (Yu *et al.*, 1998). If host cell death occurs downstream of growth arrest or is the result of the activation of a parallel pathway activated by an *R* gene, then it should be possible to find mutants that uncouple pathogen resistance from HR. This was found to be the case in *Arabidopsis* when using the gene pair *avrRpt2/RPS2*, specifying HR-type growth arrest to *P. syringae*. The identified *dnd1* mutation uncouples HR from resistance since *RPS2dnd1* plants exhibit full resistance to *Pseudomonas* expressing *avrRpt2* but fail to mount the HR seen in the wild-type interaction (Yu *et al.*, 1998). However, *dnd1* plants also exhibit partial resistance to an isogenic *Pseudomonas* strain lacking *avrRpt2*, and other virulent pathogens. In a *dnd1* genotype, salicylic acid levels are at least 10-fold higher than in wild-type plants in the absence of pathogens and *PR* genes are constitutively expressed. Administered physiological levels of SA can potentiate *R* gene-dependent defence responses in tissue-cultured soybean cells (Shirasu *et al.*, 1997). Thus, it is possible that *dndl* restricts bacterial growth and *Avr* gene product amounts to levels insufficient for the activation of *RPS2*-mediated resistance and HR. Alternatively, the *dndl* mutation may amplify signalling events downstream from the *RPS2*-derived signal, leading to an extremely rapid growth arrest without cell death. In this model, a short-lived *RPS2* signal would be insufficient to activate HR. It is then interesting to see if a sustained production of AvrRpt2 *in planta* by *Agrobacterium*-mediated expression could override HR suppression in *dndl*.

Signal components in R-gene-dependent PCD

Signalling molecules co-ordinating HR cell death

The accumulation of reactive oxygen intermediates (ROI) is a rapid event upon pathogen attack that precedes cell death in incompatible *R* gene-triggered resistance reactions (Doke, 1985; Levine *et al.*, 1994). Collectively, ROI include H_2O_2, $^{\bullet}O_2^-$ and $OH^{\bullet}$, of which only H_2O_2 is relatively stable in solution. Biphasic increases of ROI have been reported in several cultured plant cell systems in response to treatments with bacterial or fungal *Avr* genes (Levine *et al.*, 1994; Chandra *et al.*, 1996; Baker *et al.*, 1997). In these systems, the first peak usually occurs within an hour and a second sustained peak appears four to five hours later. The first peak has been interpreted as non-specific because it is induced by both virulent and avirulent bacteria, whereas the second sustained burst was considered to be race-specific. A key feature of ROI is their dose-dependent antagonistic action: low doses induce antioxidant enzymes like glutathione *S*-transferase and glutathione peroxidase whereas high doses lead to an abrupt triggering of *R*-gene-dependent PCD (Levine *et al.*, 1994). This suggests a role for ROI as a local trigger for PCD and as a diffusible signal for the induction of cellular protectants in neighbouring cells. It is tempting to speculate that the strict spatial restriction of death in the HR is due to this dose-dependent antagonistic action of ROI (Lamb and Dixon, 1997).

One source generating ROIs in plants is thought to be produced by an enzymatic machinery similar to the mammalian neutrophil NADPH oxidase complex (Doke, 1985). The main component of NADPH oxidase consists of membrane proteins ($gp91^{phox}$ and $p22^{phox}$), cytosolic proteins $p47^{phox}$ and $p67^{phox}$, and a GTP-binding protein, Rac, which has a key function in regulating the activity of the complex (Bokoch, 1994). In plants, sequence homologues of $gp91^{phox}$ and Rac have been reported (Keller *et al.*, 1998; Torres *et al.*, 1998; Kawasaki *et al.*, 1999). However, direct enzymatic evidence for their involvement in pathogen-triggered extracellular superoxide production is missing. A plant-specific large hydrophilic N-terminal domain in the $gp91^{phox}$ homologues, containing two EF-hand motifs that are implicated in Ca^{2+} binding, suggests a potential novel regulatory mechanism but could also mean that the sequence-related proteins in plants and animals are not functionally equivalent. Transgenic rice expressing a dominant active or negative form of the rice Rac homologue, *OsRac1*, was generated to test the role of the wild-type gene (Kawasaki *et al.*, 1999). Plants expressing the dominant active version of OsRac1 produce ROIs constitutively and exhibit discrete spontaneous leaf lesions. In contrast, the dominant negative variant reduces accumulation of ROI, inhibits spontaneous cell death in a rice lesion mimic background and partially compromises *R*-gene-triggered resistance to the rice blast fungus, *Magnaporthe grisea*. This suggests a Rac-dependent control of ROI accumulation and HR in plants, similar to the Rac-dependent control of the mammalian neutrophil NADPH oxidase complex. However, *bona fide* mutants in genes encoding components of the NADPH oxidase and reconstitution of an active complex in a cell-free system will be needed to critically test their role in *R*-gene-triggered ROI production.

The function of salicylic acid (SA) as a crucial signal molecule leading to systemic acquired resistance (SAR) and pathogenesis-related (PR) gene induction has been known for many years. Recent data suggest a complex role of SA and an involvement in the activation of the HR at primary infection sites. Transgenic *NahG Arabidopsis* plants, expressing a bacterial SA hydroxylase that metabolizes SA to catechol, lose the ability to induce SAR and *PR* gene expression, and are also compromised in local *R*-gene-dependent resistance responses (Delaney *et al.*, 1994; Gaffney *et al.*, 1993). In these studies, loss of SA accumulation by *NahG* attenuated *R*-gene-dependent HR and was associated with enlarged lesions at sites of attempted infection (Gaffney *et al.*, 1993). Thus, if the speed of pathogen growth is faster than the rate of elicitation of HR, escaped pathogens keep eliciting HR in neighbouring cells, thereby resulting in an increase of primary lesion size or a 'trailing' HR phenotype.

Therefore, SA may define the rate of cell death in *R*-gene-triggered HR reactions. Indeed, pre-exposure to SA can hasten cell death and increases the number of cells committing suicide upon challenge of suspension-cultured soybean cells with an avirulent pathogen (Shirasu *et al.*, 1997). Furthermore, SA accumulation itself is subjected to a positive feedback regulation: SA potentiates induction of phenylalanine ammonia-lyase (PAL) which is a key enzyme in SA synthesis. SA depletion by salicylate hydroxylase gene expression delays the timing of cell death by 2–3 h after challenge with incompatible bacterial pathogens in tobacco (Draper, 1997). The delayed HR correlated with a reduced and delayed oxidative burst, measured by a H_2O_2-responsive defence gene promoter. Taken together, these data suggest an early role of SA in the amplification of a sustained oxidative burst in *R* gene-specified resistance. The mechanism of this defence amplification is not understood but is likely to involve changes in the phosphorylation status of a signalling pathway(s) rather than *de novo* protein synthesis (Shirasu *et al.*, 1997). It is therefore tempting to hypothesize that SA acts in concert with ROIs to define the threshold required to initiate cell death.

Further support for this hypothesis comes from studies on ozone tolerance in *Arabidopsis* (Rao and Davis, 1999). The phytotoxicity and necrosis-inducing potential of ozone is likely due to its ability to generate ROI (Runeckles and Vaartnou, 1997), while ozone tolerance is generally correlated with higher activities of antioxidant enzymes that are capable of removing ROI (Sharma and Davis, 1997). SA has been shown to play a dose-dependent role in influencing the outcome of an ozone-triggered ROI imbalance (Rao and Davis, 1999). Ozone exposure in the presence of low levels of SA induced antioxidant responses and ozone tolerance, whereas the same treatment in the presence of high SA levels triggered an oxidative burst, *PR* gene accumulation and cell death. Thus, it seems likely that high levels of SA in conjunction with their ability to amplify the ozone-induced ROI production exceed a threshold to trigger HR like cell death.

In animal cells, nitric oxide (NO) is known as a critical redox-active signalling molecule involved in various physiological conditions including inflammation, autoimmunity, and programmed cell death (Schmidt and Walter, 1994). It was only recently shown that nitric oxide (NO) also plays an important role in the induction of defence responses and HR cell death in plants (Delledonne *et al.*, 1998; Durner *et al.*, 1998). Durner and colleagues showed that in tobacco plants NO synthase (NOS) activity was enhanced upon TMV-triggered and *R*-gene-dependent HR. Furthermore, external application of NO leads to *PAL* gene activation, accumulation of SA, and increased levels of *PR* gene mRNA. Using a soybean cultured cell system, Delledonne *et al.* (1998) showed that NO accumulation occurs in an *R*-gene-dependent manner and that it triggers cell death synergistically with ROIs. Importantly, nitroprusside-generated NO steady-state levels of ca. 2 μM do not cause recognizable cell death but if administered together with 6–10 μM H_2O_2 result in a 5–10-fold increase in cell death compared to treatments with H_2O_2 only. NOS inhibitors compromise both induction of *PAL* and *R*-gene dependent HR cell death. Several questions remain to be answered. For example, is the NO- and ROI-amplified cell death identical to that seen after ROI application alone or do the cells die differently? The gene(s) encoding plant NOS remain elusive to date. Do plants utilize a NO-generating system that is different from those in animals, for example nitric reductase? Certainly, the isolation and characterization of NO-insensitive or supersensitive mutants would aid to further our understanding of NO function(s) in the HR.

Genes required for R*-gene-triggered resistance*

Mutational analyses of *R*-gene-dependent resistance pathways have been reported in a number of plants. Pioneering work was carried out by Torp and Jørgensen who described the isolation of induced susceptible barley mutants of an *Mla 12*-containing cultivar which was otherwise resistant to the powdery mildew fungus (Torp and Jørgensen, 1986; Jørgensen, 1988). In more than 90% of the mutants, susceptibility resulted from mutations in the *R* gene *Mla12* but three mutants mapped to other loci, designated *Rar1* and *Rar2* (Freialdenhoven *et al.*, 1994). The biotrophic powdery mildew fungus selectively attacks epidermal host cells and is strictly dependent on living host cells for nutrient uptake. Barley plants containing *Mla12* recognize powdery mildew isolates carrying *avrMla12* and initiate HR in attacked epidermal cells. Both susceptible *rar1-1* and *rar1-2* mutant plants have lost the capacity to execute cell death of attacked epidermal cells (Freialdenhoven *et al.*, 1994). In wild-type plants, the epidermal HR is preceded by a biphasic hydrogen peroxide accumulation, of which only the second phase is compromised in both susceptible *rar1-1* and *rar1-2* mutants (Shirasu *et al.*, 1999a). Interestingly, mutant *Rar1* plants compromise the function of multiple

R genes against the powdery mildew fungus (Jörgensen, 1996). This identifies *Rar1* as a convergence point in *R*-gene-triggered resistance and is consistent with a signalling function downstream from pathogen perception but upstream of the oxidative burst.

Recently, *Rar1* was isolated and shown to encode a small 25 kDa intracellular protein with no sequence similarity to previously characterised proteins (Shirasu *et al.*, 1999a). The Rar1 protein contains two tandemly repeated copies of a novel 60 amino acid Zn^{2+}-binding domain, designated CHORD (cysteine- and histidine-rich domain). Strikingly, Rar1 homologues, each containing the CHORD tandem array, have been found in protozoa and metazoa. Rar1 homologues are generally encoded by single-copy genes in each species, suggesting that the basic function of the protein may have been preserved throughout the evolution of eukaryotic organisms. Silencing of the *Rar1* homologue in the nematode *Caenorhabditis elegans* (*chp*) leads to semi-sterility of female adults and lethality of embryos, indicating an essential function for the gene in nematode development. Interestingly, *chp*-silenced individuals exhibit a female germline defect, gonad hyperplasia, similar to that reported in *ced-3* and *ced-4* mutants (Gumienny *et al.*, 1998). In wild-type *C. elegans*, female germline cells undergo PCD after meiosis as part of the normal process of development. This germline PCD is executed by the core apoptotic machinery containing *ced-3*, *ced-4* and *ced-9*, and is regulated by yet unknown genes. Importantly, germline PCD is not compromised by mutations in genes regulating PCD in somatic tissue which is also dependent on the core apoptotic machinery, indicating separate control of the cell death machinery in somatic and germline cells. Whether the *Rar1* homologue *chp* represents a component in this previously identified PCD pathway for female germline development remains to be tested.

Mutational analyses in *Arabidopsis* revealed two mutants, *ndr1* and *eds1*, each compromising race-specific resistance responses (Century *et al.*, 1995; Parker *et al.*, 1996). Similar to barley *rar1* mutants, defective *ndr1* and *eds1* alleles in *Arabidopsis* each compromise the function of multiple *R* genes. Both *ndr1* and *eds1* mutants compromise resistance to pathogens from totally different phyla, i.e. prokaryotes and eukaryotes. Null mutants in *NDR1* and *EDS1* preferentially compromise the action of different subfamilies of the NB-LRR *R* gene class, though these requirements are not mutually exclusive (Aarts *et al.*, 1998). *NDR1* is required for the function of several

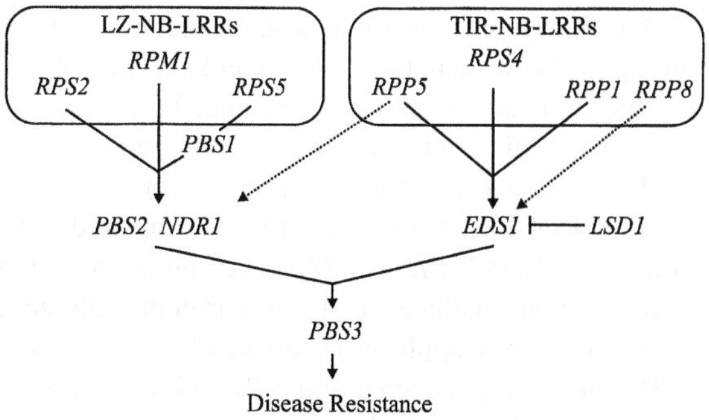

Figure 2. Skeletal *R*-gene-dependent disease resistance pathways in *Arabidopsis*. Solid lines indicate signal pathways which require full function of NDR1 or EDS1. Dashed lines represent pathways partially compromised in either *ndr1* or *eds1* mutants.

LZ-NB-LRR genes whereas *EDS1* is essential for a number of TIR-NB-LRR-type *R* genes. Thus, *NDR1* and *EDS1* define genetic convergence points of at least two different disease resistance signalling pathways which are each linked to R proteins of particular types (Figure 2).

Upon downy mildew attack, mutant *eds1* plants are compromised in triggering the HR (Parker *et al.*, 1996). However, unlike *rar1* mutations in barley, a null mutant of *NDR1* suppresses resistance but not always HR cell death, as only one out of four tested *avr/R* gene interactions (*avrRpt2*/*RPS2*) resulted in loss of both resistance and HR (Century *et al.*, 1995). Thus, if the retained cell death response in *ndr1* mutants is identical to the authentic HR seen in wild-type plants, this indicates uncoupling of resistance and cell death in at least three *avr/R* gene interactions. This may indicate branching of a cell death and resistance pathway somewhere downstream from a common *R* gene trigger.

NDR1 encodes a small putatively membrane-anchored protein with low sequence similarity to *HIN1* from tobacco (Century *et al.*, 1997). *HIN1* is rapidly induced in HR cell death responses (Gopalan *et al.*, 1996) and, like *HIN1*, transcript accumulation of *NDR1* mRNA is induced upon pathogen infection. Biochemical functions of both proteins remain unclear although *HIN1* was shown to be induced during leaf senescence, indicating potential cross-talk between these two distinct cell death pathways (Pontier *et al.*, 1999). However, an involvement of *NDR1* in senescence is unlikely since *ndr1* null mutant plants exhibit no recognizable phenotype in the absence of pathogens.

The *EDS1* protein contains sequence blocks with similarity to the catalytic site of eukaryotic lipases and the same arrangement of sequence blocks occurs also in a ferulic acid esterase from *Aspergillus niger* (Falk *et al.*, 1999). Thus, it remains to be shown whether EDS1 hydrolyses a lipid or non-lipid substrate. Like *NDR1*, *EDS1* mRNA accumulation is also induced upon challenge with an avirulent pathogen. Furthermore, SA application restores both resistance to *Peronospora parasitica* and *PR1* mRNA accumulation in *eds1* mutant plants, suggesting a function of EDS1 downstream of *R*-gene-mediated pathogen recognition but upstream of SA perception (Parker *et al.*, 1996).

The combined genetic and molecular analysis of *NDR1*, *EDS1*, and *Rar1* suggests a function for each of these genes downstream of *R*-gene-mediated pathogen recognition. What can we learn from these genes and the encoded proteins with respect to PCD in *R*-gene-triggered resistance? There is evidence that each gene functions in signalling rather than execution of the resistance response. Although *eds1* and *rar1* mutants are each compromised both in pathogen arrest and HR cell death, the *ndr1* mutant uncouples resistance from HR cell death in several *avr*/*R* interactions. Thus, not only can *R* genes trigger resistance without HR in wild-type plants (e.g. *Rx* in potato) but growth restriction of the pathogen can also be separated from HR (at least in some cases) by mutations in signalling components.

Null mutants have been isolated for each *NDR1* and *EDS1*. These plants do not exhibit recognizable phenotypes other than compromised resistance, suggesting a specialized function of the gene products in disease resistance. In addition, both *NDR1* and *EDS1* encode proteins that are not conserved in other phyla, suggesting that plants have developed some unique molecular devices to relay *R*-gene-emitted signals. In contrast, Rar1 revealed related proteins in protozoa and metazoa. In addition, only partially defective *rar1* alleles were recovered despite exhaustive screens, suggesting that *rar1* null mutants may be lethal (Torp and Jørgensen, 1986; Shirasu *et al.*, 1999a). This and the cross-phyla conservation of Rar1 proteins may define a launch pad to find out whether *R* genes have recruited a fundamental eukaryotic cellular function to relay resistance and HR.

There is a further aspect that can be learned from studies with these signalling mutants. Although NDR1- and EDS1-dependent pathways appear to be preferentially utilized in response to signals from LZ-NB-LRR- and TIR-NB-LRR- type R proteins respectively, there is also clear evidence for simultaneous use of both pathways in some cases (Aarts *et al.*, 1998). For example, *RPP8*-specified resistance to *P. parasitica* is only partially attenuated by an *eds1* null mutant (Figure 2). Likewise, the *ndr1* null mutant significantly compromises the function of *EDS1*-dependent *RPP5* resistance specificities. Thus, strength of an *R*-gene-triggered resistance response can not only be modulated by changing *R* gene dosage or amount of Avr product (see above) but also by blocking one of at least two simultaneously activated downstream signalling pathways. It is then conceivable that in some *R/Avr* interactions, a cell's decision to die or not to die is dependent upon time- and amplitude-dependent signals from parallel transduction pathways. This model allows for cases in which the flux through one of the parallel transduction pathways would be sufficient to reach the threshold for HR cell death activation. This latter scenario could explain the observation that in some *avr/R* interactions *ndr1* null mutants exhibit retained HR but compromised resistance (Century *et al.*, 1995).

Genetic analysis of three novel putative signal transduction genes (*PBS1*, *PBS2* and *PBS3*) in *Arabidopsis* supports the idea of signal integration operating at different levels upon R protein signal emission (Warren *et al.*, 1999). Of four *R* genes to *P. syringae*, mutations in *PBS1* selectively blocks the function of one, *pbs2* mutants compromise three, and the *pbs3* mutant partially blocks all four (Figure 2). Interestingly, the *pbs2* mutants appear to suppress the same set of *R* genes as *ndr1*, and the range of affected *R* gene functions in *pbs3* clearly overlaps with those blocked by *eds1*. This strongly suggests the existence of an interwoven signal relay system to translate *R* gene triggers into defence responses and cellular suicide.

Host mutations inducing disease resistance and perturbations in cell death control

A large number of mutants inducing spontaneous cell death were initially isolated in maize and classified as lesion mimics (Hoisington *et al.*, 1982). Lesion mimics may result from defects in developmental PCD pathways, pathogen-inducible HR cell death control, or physiological perturbations that cause necrosis and chlorosis in various plant tissues (Hu *et al.*, 1998). Indeed, changes in *R* loci and R proteins can lead to HR autoactivation in the absence of pathogens (Hu *et al.*, 1996; Rathjen *et al.*, 1999). Lesion mimics may or may not result in heightened pathogen resistance depending on type of lesion and life-style of pathogens. For example, one lesion mimic mutant may exhibit resistance to a biotrophic but enhanced susceptibility to a necrotrophic pathogen (see below). The *Arabidopsis acd1* mutant shows enhanced lesion formation upon pathogen infection or abiotic stresses and is susceptible to opportunistic leaf pathogens (Greenberg and Ausubel, 1993). Thus, it is conceivable that distinct types of cell death pathways are linked with disease resistance to particular pathogens. Here we review what has been learned from the type of lesion mimic mutants that are known to confer disease resistance.

Many *Arabidopsis* mutants exhibit spontaneous lesions and enhanced resistance to pathogens (Weymann *et al.*, 1995). Most of these mutants constitutively express marker genes for systemic acquired resistance and possess elevated levels of SA. In mutants *lsd6* and *lsd7*, spontaneous lesion formation, *PR* gene expression and heightened fungal resistance were all suppressed in a *NahG* background, thereby supporting a critical role for SA and SA-dependent processes in both cell death induction and pathogen resistance. Surprisingly, application of INA or SA restores spontaneous lesion formation in *lsd6/NahG* but not in *lsd7/NahG* plants. One interpretation is that *LSD6* functions in a feedback regulation of lesion formation by SA or SA-dependent processes. Interestingly, *lsd2*-, *lsd4*- and *lsd5*-dependent lesion formation is not suppressed by *nahG*, though *PR* gene expression and heightened pathogen resistance are clearly compromised (Hunt *et al.*, 1997). This suggests that SA accumulation occurs in these mutants either downstream of lesion formation or, alternatively, the pathway leading to the lesions is different from those in *lsd6* and *lsd7*. Two newly isolated dominant gain-of-function mutants, *acd6* and *ssi1*, also show lesion formation, elevated defence and enhanced resistance to pathogens (Rate *et al.*, 1999; Shah *et al.*, 1999). Similar to *lsd6* and *lsd7*, both *acd6* and *ssi1* phenotypes are suppressed in a *NahG* background. Surprisingly, a synthetic inducer of the SA pathway, benzo(1,2,3)thiadiazole-7-carbothioic acid (BTH), not only reactivates these phenotypes but also induces tumour-like abnormal growths in *acd6 NahG* plants. This indicates that the SA pathway might also have a role in coupling cell death decisions with cellular growth processes.

The *lsd1* mutant exhibits heightened resistance to normally virulent bacterial and fungal pathogens in uninoculated *lsd1* plants at the pre-lesion state (Dietrich *et al.*, 1994). Elevated resistance is associated with a spreading lesion phenotype consuming the entire leaf. In uninoculated *lsd1* plants, abiotic cues such as exposure to long days can also induce spreading lesions. Thus, *lsd1* is a conditional lesion mimic mutant. The recessive inheritance of the mutation is consistent with the hypothesis that LSD1 has a function in setting the boundaries of cell death once the programme is activated. Jabs *et al.* (1996) showed that extracellularly administered superoxide, but not H_2O_2, triggers cell death in *lsd1* mutants and also that inhibition of intrinsic superoxide formation by the NADPH oxidase inhibitor diphenyleneiodonium reduced the frequency of *lsd1* cell death initiation. This suggests that superoxide is necessary and sufficient for the initiation of lesion formation. Possibly, *lsd1* is not capable of down-regulating the amount of superoxide generated by the oxidative burst. Thus, the impaired cell death control in *lsd1* mutants may result in a hyper-responsiveness to superoxide. This has led to a model in which LSD1 acts like a rheostat in cell death initiation. In this model, LSD1 is dampening a death signal that is dependent on superoxide until a threshold for cell death initiation is reached. Superoxide itself and/or other secondary signal molecules which are dependent on the oxidative burst of attacked cells could diffuse to neighbouring cells where secondary signal molecules are produced, resulting in an amplification of the signal. Candidates for these secondary signal molecules are SA and NO which are known to act in concert with ROI (see above). Indeed, very low levels of exogenously administered SA are sufficient to trigger spreading lesion formation in *lsd1* plants (Dietrich *et al.*, 1994). LSD1 encodes a small novel protein containing three putative zinc finger domains (Dietrich *et al.*, 1997). Zinc finger domains are known to have diverse functions ranging from protein-protein communication, binding to nucleic acids, or even binding of small lipid molecules (Gaullier *et al.*, 1998). To delimit possible biochemical functions of LSD1 protein, it will be important to clarify the role(s) of the predicted zinc finger domains.

Surprisingly, the spreading lesion phenotype in *lsd1* plants is not suppressed in a *nahG* background, indicating that lesion initiation does not require SA (similar to *lsd2*, *lsd4*, and *lsd5*). This contrasts with the observation that the heightened resistance phenotype is dependent on SA (D.H. Aviv, J.E. Parker, U. Neuenschwander, J. Ryals, R.A. Dietrich and J.L. Dangl, unpublished). Thus, the *lsd1* spreading lesion phenotype can be uncoupled from disease resistance. Intriguingly, *eds1* suppresses both the resistance phenotype and the cell death phenotype of *lsd1* (D.H. Aviv, J.E. Parker, U. Neuenschwander, J. Ryals, R.A. Dietrich and J.L. Dangl, unpublished). All conditions that initiate spreading cell death in *lsd1* plants (i.e. pathogen challenge, SA treatment, shift to long-day light) do not induce lesions in the *lsd1/eds1* double mutants. Furthermore, *lsd1/eds1* plants displayed complete susceptibility when challenged with *P. syringae* carrying *avrRps4* upon recognition by its cognate TIR-NB-LRR *R* gene. In contrast, full HR type resistance was observed in the double mutants if exposed to *Avr-RPM1* in the presence of its cognate LZ-NB-LRR *R* gene. Thus, it is conceivable that EDS1 is essential in relaying potential cell death signals derived from both biotic and abiotic cues that are subsequently interpreted by LSD1 in a negative regulatory manner to establish a threshold for cell death initiation.

Similar to *lsd1* in *Arabidopsis*, *mlo* mutants in barley exhibit resistance at the pre-lesion state and confer non-race-specific resistance to the powdery mildew fungus (Buschges *et al.*, 1997). The resistance requires at least two additional genes, designated *Ror1* and *Ror2*; mutations in either *Ror* gene leads to susceptibility in a *mlo* background (Freialdenhoven *et al.*, 1996). Another phenotype of plants lacking wild-type *Mlo* is the spontaneous formation of discrete leaf lesions shortly before the onset of leaf senescence (Wolter *et al.*, 1993). The spontaneous cell death in *mlo* mutants suggests that Mlo functions as a negative regulator of leaf cell death/senescence. This may indicate a link in developmental cell death control and pathogen resistance. Indeed, *mlo*-dependent spontaneous cell death is compromised in *mlo ror1* and *mlo ror2* double mutants, indicating at least overlapping genetic pathways leading to cell death and resistance (Peterhänsel *et al.*, 1997). Is *mlo*-dependent resistance tightly linked to host cell death activation as seen in most *R*-gene-specified resistance? Contrary to *R*-gene-specified resistance to the powdery mildew fungus, the initially attacked epidermal host cell does not undergo cell death in *mlo* plants. Rather, clusters of adjacent mesophyll cells undergo cell death at later time points, indicating tissue specificity in the cell death response. Thus, it is possible that the absence of a functional *Mlo* gene lowers the threshold for cell death in mesophyll cells in pathogen-attacked or pre-senescing cells.

In *mlo* plants, cell wall appositions (CWAs) are formed adjacent to the fungal penetration site within an attacked epidermal cell (von Röpenack *et al.*, 1998). CWAs are a complex of phenolics, callose and proteins reinforced by oxidative cross-linking with ROIs (Bradley *et al.*, 1992; Thordal-Christensen *et al.*, 1997; Hückelhoven and Kogel, 1998). Once cross-linked, they become highly resistant to cell-wall-degrading enzymes and thus can function as a structural barrier against fungal penetration. Since CWA cross-linking appears to occur earlier upon infection in *mlo* than in *Mlo* plants (von Röpenack *et al.*, 1998), accelerated CWA formation is likely to be an important component in *mlo*-dependent resistance. ROIs are known to function as co-inducers of HR cell death and cell wall cross-linking (Bradley *et al.*, 1992; Levine *et al.*, 1994). Thus, accelerated ROI production in plants lacking the negative regulator *Mlo* may result in effective CWA cross-linking in epidermal cells and HR cell death in mesophyll cells.

Mlo is the founder of a novel class of plant-specific proteins with seven integral transmembrane (TM)-spanning helices (Devoto *et al.*, 1999). The protein locates in the plasma membrane with its N-terminus exposed extracellularly and the C-terminal end located in the cytoplasm (Devoto *et al.*, 1999). Transient *Mlo* expression in single epidermal cells of *mlo* leaves is sufficient to confer susceptibility, suggesting that Mlo functions in a cell-autonomous fashion (Shirasu *et al.*, 1999a). From the currently available databases, about 35 Mlo family members are predicted in the *Arabidopsis* genome (Devoto *et al.*, 1999). Interestingly, the Mlo family represents the vast majority of 7 TM proteins so far identified in higher plants. The sequence variability among family members within a species, their topology and subcellular localization are reminiscent of the most abundant class of metazoan 7 TM receptors, the G-protein-coupled receptors. These relay extracellular signals through ligand binding into amplified intracellular responses via activation of the α-subunit of heterotrimeric G proteins. All 7 TM protein families identified to date function as GPCRs, representing a total of six GPCR subfamilies (Bockaert and Pin, 1999). Thus, Mlo and its siblings represent either a seventh plant-specific GPCR subfamily or define a first example in which a 7 TM protein family functions via a novel biochemical mechanism. In either case, Mlo may reveal a novel molecular mechanism by which plants control ROI functions in response to endogenous or exogenous cues.

Interestingly, *mlo* plants exhibit enhanced susceptibility to the rice blast fungus *Magnaporthe grisea* which, unlike *Erysiphe graminis*, can colonize a broad range of grass species (Jarosch *et al.*, 1999). *Mag-*

naporthe invades both epidermal and mesophyll cells in *mlo* genotypes whereas invasive fungal hyphae are strictly limited to epidermal cells in *Mlo* plants. Enhanced susceptibility to *Magnaporthe* in *mlo* genotypes is associated with substantial host cell death of infected epidermal and mesophyll cells. The simplest explanation for the conflicting *Erysiphe* and *Magnaporthe* infection phenotypes on *mlo* plants may reside in the different life-styles of the two fungi. *Magnaporthe* is a facultative biotrophic fungus that is capable of producing toxins and may kill penetrated host cells before colonization (Lebrun *et al.*, 1990). Thus, *Magnaporthe* may have turned the primed cell death responsiveness in *mlo* leaves to its own advantage.

Similar to the broad-spectrum resistance conferred by *mlo* mutations in barley, the *edr1* mutant in *Arabidopsis* exhibits enhanced disease resistance to at least two powdery mildew fungi, *Erysiphe cichoracearum* and *E. cruciferarum* (Frye and Innes, 1998). Also, neither *mlo* nor *edr1* plants exhibit constitutive expression of *PR* genes. However, while growth arrest of the fungus in *mlo* plants occurs early, growth restriction in *edr1* plants is observed at late stages during the infection, leading to substantial reduction in fungal conidiophore and conidia formation but allowing profuse hyphal growth on the leaf surface. Like *mlo*, clusters of dead mesophyll cells are observed in *edr1* leaves in response to pathogen challenge. Because the fungus attacks exclusively epidermal tissue, the signal for mesophyll cell death may originate in the epidermal layer. Thus, the *edr1* mutation may sensitize ('hair trigger') mesophyll cells towards death, suggesting a function for the wild-type gene in controlling homeostasis in mesophyll cells. It seems plausible that the late mesophyll cell death response seen in *edr1* plants causes a shortage of nutrient supply to the biotrophic fungus, thereby affecting only late stages of fungal development such as sporulation. Interestingly, *edr1* plants retain full susceptibility to several strains of the oomycete *P. parasitica*. In contrast to powdery mildews, this fungus forms haustoria in mesophyll cells. Thus, different modes of pathogenesis of the two fungi may explain selective activation of the hair trigger cell death. Alternatively, *P. parasitica* may be able to suppress the hair trigger for cell death in *edr1* plants.

The recessive maize lethal leaf spot1 (*lls1*) mutant exhibits developmentally controlled, randomly scattered necrotic leaf lesions and enhanced resistance to fungal pathogens (Simmons *et al.*, 1998). These necrotic lesions eventually consume entire leaves and kill the plant. Lesion formation is also promoted by light and cell injury. The *LLS1* gene encodes a protein with two conserved motifs resembling those found in aromatic ring-hydroxylating (ARH) dioxygenases. Thus, it is conceivable that LLS1 functions by cleaving the ring structure of an unknown phenolic compound(s). In the absence of LLS1, this phenolic substrate may accumulate and become toxic to the cell, especially in oxidative conditions. Alternatively, LLS1 may degrade a phenolic signal molecule that promotes PCD upon environmental cues including pathogens. One possible candidate for the deduced dioxygenase function of LLS1 is SA or a related compound. The presence of two potential iron-sulfur clusters in LLS1 has led to the speculation that its activity may be redox state-dependent. Thus, LLS1 may also serve as a rheostat in SA-dependent cell death control, consistent with the indeterminate nature of the lesions in the loss of function mutant (Gray *et al.*, 1997). *LLS1* is highly expressed in leaf epidermal cells but expression in the mesophyll is low (Simmons *et al.*, 1998). This correlates with the observation that heightened disease resistance to fungal pathogens was effective at the epidermis but not in the mesophyll of *lls1* plants. Thus, it is possible that antimicrobial activity in the mutant is confined to cells where the *LLS1* gene is normally expressed at high levels.

It was recently shown that silencing of the protoporphyrinogen oxidase (PPO) in *Arabidopsis* leads to a lesion mimic phenotype, *PR* gene expression, elevated SA levels and heightened disease resistance to *Peronospora parasitica* (Molina *et al.*, 1999). PPO catalyses an enzymatic step in tetrapyrrole biosynthesis. Biosynthesis of tetrapyrroles is strictly regulated in plants to avoid accumulation of intermediates that can be photoactively oxidized. These photoactive oxidations of tetrapyrrole coincide with an accumulation of ROI. It seems plausible that a photodynamic ROI cascade, triggered by tetrapyrrole intermediate accumulation, mimics the oxidative burst seen in *R*-gene-driven resistance responses. Interestingly, treatment of *Arabidopsis* wild-type plants with sub-lethal levels of diphenyl ether herbicides, known to inhibit PPO activity, reproduced the defence responses seen in the silenced PPO plants and conferred heightened *P. parasitica* resistance (Molina *et al.*, 1999). The idea of a photodynamic ROI cascade triggering cell death is further supported by the molecular characterization of the light-dependent lesion mimic mutant *Les22* in maize (Hu *et al.*, 1998). The *Les22* mutant disrupts the gene encoding another enzyme of tetrapyrrole biosynthesis, uroporphyrinogen decarboxylase (UROD). The mutation leads to the accumulation of the tetrapyrrole intermediate uroporphyrinogen in leaf tissue that may, upon high-light regimes, induce a photodynamic ROI cascade triggering cell death.

Conclusions

Common molecular building blocks identified in R proteins from different plants conferring resistance to different classes of pathogens suggest that pathogen perception and signal emission are based on a common molecular mechanism. A fundamental question that needs to be resolved is whether R proteins recognize pathogen determinants directly or changes in host components imposed by *Avr* gene products. Irrespective of this, R proteins have an intrinsic potential for signal emission into cell death pathways. However, activation of HR cell death is not only dependent on amplitude and duration of the R protein signal but also on downstream transduction pathways. Although a simple signalling pathway was anticipated, an interwoven network of downstream signalling has been unravelled, involving parallel pathways, branching, and convergence points before a cell commits itself to suicide.

Signal amplification emerges as a further critical feature to drive cells into the commitment phase of death. SA, ROI, and NO are key molecules in this process but the molecular targets of their concerted action remains to be determined. Molecular and genetic analysis of several lesion mimic mutants underscores the critical role of ROIs in HR and has provided the first insights into how cell death thresholds are set via interplays of positive and negative regulatory components. The eminent role of the oxidative burst makes it tempting to speculate that redox-sensitive proteins may serve as molecular switches to render cell death irrevocable. Future challenges include the development of genetic tools to identify components of the commitment and execution phase in HR cell death, a full integration of SA/ROI/NO into the emerging genetic pathways, and the integration of the HR pathway into those controlling other forms of physiological PCD in plants.

Acknowledgements

We thank Nick Collins for critical reading of this manuscript, and our colleagues Jane Parker and Jeff Dangl for providing unpublished data. This work is supported by grants from the GATSBY Charitable Foundation and the BBSRC to P. S-L.

References

Aarts, N., Metz, M., Holub, E., Staskawicz, B.J., Daniels, M.J. and Parker, J.E. 1998. Different requirements for *EDS1* and *NDR1* by disease resistance genes define at least two *R* gene-mediated signaling pathways in *Arabidopsis*. Proc. Natl. Acad. Sci. USA 95: 10306–10311.

Adams, S.E., Jones, R.A.C. and Coutts, R.H.A. 1986. Expression of potato virus X resistance gene *Rx* in potato leaf protoplasts. J. Gen. Virol. 67: 2341–2345.

Aravind, L., Dixit, V.M. and Koonin, E.V. 1999. The domains of death: evolution of the apoptosis machinary. Trends Biochem. Sci. 24: 47–53.

Baker, B., Zambryski, P., Staskawicz, B. and Dinesh-Kumar, S.P. 1997. Signaling in plant-microbe interactions. Science 276: 726–733.

Bendahmane, A., Kanyuka, K. and Baulcombe, D.C. 1999. The *Rx* gene from potato controls separate virus resistance and cell death responses. Plant Cell 11: 781–791.

Bertin, J., Nir, W.J., Fischer, C.M., Tayber, O.V., Errada, P.R., Grant, J.R., Keilty, J.J., Gosselin, M.L., Robison, K.E., Wong, G.H.W., Glucksmann, M.A. and DiStefano, P.S. 1999. Human CARD4 protein is a novel CED-4/Apaf-1 cell death family member that activates NF-κB. J. Biol. Chem. 274: 12955–12958.

Bockaert, J. and Pin, J.P. 1999. Molecular tinkering of G protein-coupled receptors: an evolutionary success. EMBO J. 18: 1723–1729.

Bokoch, G.M. 1994. Regulation of the human neutrophil NADPH oxidase by the Rac GTP- binding proteins. Curr. Opin. Cell Biol. 6: 212–218.

Bradley, D.J., Kjellbom, P. and Lamb, C.J. 1992. Elicitor-induced and wound-induced oxidative cross-linking of a proline-rich plant-cell wall protein: a novel, rapid defense response. Cell 70: 21–30.

Buschges, R., Hollricher, K., Panstruga, R., Simons, G., Wolter, M., Frijters, A., van Daelen, R., van der Lee, T., Diergaarde, P., Groenendijk, J., Topsch, S., Vos, P., Salamini, F. and Schulze-Lefert, P. 1997. The barley *Mlo* gene: a novel control element of plant pathogen resistance. Cell 88: 695–705.

Century, K.S., Holub, E.B. and Staskawicz, B.J. 1995. *NDR1*, a locus of *Arabidopsis thaliana* that is required for disease resistance to both a bacterial and a fungal pathogen. Proc. Natl. Acad. Sci. USA 92: 6597–6601.

Century, K.S., Shapiro, A.D., Repetti, P.P., Dahlbeck, D., Holub, E. and Staskawicz ,B.J. 1997. *NDR1*, a pathogen-induced component required for *Arabidopsis* disease resistance. Science 278: 1963–1965.

Chandra, S., Martin, G.B. and Low, P.S. 1996. The Pto kinase mediates a signaling pathway leading to the oxidative burst in tomato. Proc. Natl. Acad. Sci. USA 93: 13393–13397.

Chinnaiyan, A.M., Chaudhary, D., Orourke, K., Koonin, E.V. and Dixit, V.M. 1997. Role of CED-4 in the activation of CED-3. Nature 388: 728–729.

Delaney, T.P., Uknes, S., Vernooij, B., Friedrich, L., Weymann, K., Negrotto, D., Gaffney, T., Gutrella, M., Kessmann, H., Ward, E. and Ryals, J. 1994. A central role of salicylic acid in plant-disease resistance. Science 266: 1247–1250.

Delledonne, M., Xia, Y.J., Dixon, R.A. and Lamb, C. 1998. Nitric oxide functions as a signal in plant disease resistance. Nature 394: 585–588.

Devoto, A., Piffanelli, P., Nilsson, I., Wallin, E., Panstruga, R., von Heijne, G. and Schulze-Lefert P. 1999. Topology, subcellular localization and sequence diversity of the Mlo family in plants. J. Biol. Chem. 274: 34993–35004.

Dietrich, R.A., Delaney, T.P., Uknes, S.J., Ward, E.R., Ryals, J.A. and Dangl, J.L. 1994. *Arabidopsis* mutants simulating disease resistance response. Cell 77: 565–577.

Dietrich, R.A., Richberg, M.H., Schmidt, R., Dean, C. and Dangl, J.L. 1997. A novel zinc finger protein is encoded by the *Arabidopsis LSD1* gene and functions as a negative regulator of plant cell death. Cell 88: 685–694.

Doke, N. 1985. NADPH-dependent O_2^- generation in membrane fraction isolated from wounded potato tubers inoculated with *Phytophthora infestans*. Physiol. Plant Path. 27: 311–322.

Draper, J. 1997. Salicylate, superoxide synthesis and cell suicide in plant defence. Trends Plant Sci. 2: 162–165.

Durner, J., Wendehenne, D. and Klessig, D.F. 1998. Defense gene induction in tobacco by nitric oxide, cyclic GMP, and cyclic ADP-ribose. Proc. Natl. Acad. Sci. USA 95: 10328–10333.

Ellis, J. and Jones, D. 1998. Structure and function of proteins controlling strain-specific pathogen resistance in plants. Curr. Opin. Plant Biol. 1: 288–293.

Ellis, J.G., Lawrence, G.J., Luck, J.E. and Dodds, P.N. 1999. Identification of regions in alleles of the flax rust resistance gene L that determine differences in gene-for-gene specificity. Plant Cell 11: 495–506.

Falk, A., Feys, B.J., Frost, L.N., Jones, J.D.G., Daniels, M.J. and Parker, J.E. 1999. *EDS1*, an essential component of *R* gene-mediated disease resistance in *Arabidopsis* has homology to eukaryotic lipases. Proc. Natl. Acad. Sci. USA 96: 3292–3297.

Flor, H.H. 1971. Current status of the gene-for-gene concept. Annu. Rev. Phytopath. 9: 275–296.

Freialdenhoven, A., Scherag, B., Hollricher, K., Collinge, D.B., Thordal-Christensen, H. and Schulze-Lefert, P. 1994. *Nar-1* and *Nar-2*, two loci required for Mla12-specified race-specific resistance to powdery mildew in barley. Plant Cell 6: 983–994.

Freialdenhoven, A., Peterhänsel, C., Kurth, J., Kreuzaler, F. and Schulze-Lefert, P. 1996. Identification of genes required for the function of non-race- specific *mlo* resistance to powdery mildew in barley. Plant Cell 8: 5–14.

Frye, C.A. and Innes, R.W. 1998. An *Arabidopsis* mutant with enhanced resistance to powdery mildew. Plant Cell 10: 947–956.

Gaffney, T., Friedrich, L., Vernooij, B., Negrotto, D., Nye, G., Uknes, S., Ward, E., Kessmann, H. and Ryals, J. 1993. Requirement of salicylic acid for the induction of systemic acquired resistance. Science 261: 754–756.

Gaullier, J.M., Simonsen, A., Darrigo, A., Bremnes, B., Stenmark, H. and Aasland, R. 1998. FYVE fingers bind Ptdins(3)P. Nature 394: 432–433.

Gopalan, S., Wei, W. and He, S.Y. 1996. *hrp* gene-dependent induction of *hin1*: a plant gene activated rapidly by both harpins and the *avrPto* gene-mediated signal. Plant J. 10: 591–600.

Görg, R., Hollricher, K. and Schulze-Lefert, P. 1993. Functional analysis and RFLP-mediated mapping of the *Mlg* resistance locus in barley. Plant J. 3: 857–866.

Gray, J., Close, P.S., Briggs, S.P. and Johal, G.S. 1997. A novel suppressor of cell death in plants encoded by the *Lls1* gene of maize. Cell 89: 25–31.

Greenberg, J.T. and Ausubel, F.M. 1993. *Arabidopsis* mutants compromised for the control of cellular damage during pathogenesis and aging. Plant J. 4: 327–341.

Gumienny, T.L., Lambie, E., Hartwieg, E., Horvitz, H.R. and Hengartner, M.O. 1998. Genetic control of programmed cell death in the *Caenorhabditis elegans* hermaphrodite germline. Development 126: 1011–1022.

Hammond-Kosack, K.E. and Jones, J.D.G. 1994. Incomplete dominance of tomato *Cf* genes for resistance to *Cladosporium fulvum*. Mol. Plant-Microbe Interact. 7: 58–70.

Hammond-Kosack, K.E., Staskawicz, B.J., Jones, J.D.G. and Baulcombe, D.C. 1995. Functional expression of a fungal avirulence gene from a modified potato virus X genome. Mol. Plant-Microbe Interact. 8: 181–185.

Hoisington, D.A., Neuffer, M.G. and Walbot, V. 1982. Disease lesion mimics in maize. 1. Effect of genetic background, temperature, developmental age, and wounding on necrotic spot formation with *Les1*. Dev. Biol. 93: 381–388.

Hu, G., Richter, T.E., Hulbert, S.H. and Pryor, T. 1996. Disease lesion mimicry caused by mutations in the rust resistance gene *rp1*. Plant Cell 8: 1367–1376.

Hu, G.S., Yalpani, N., Briggs, S.P. and Johal, G.S. 1998. A porphyrin pathway impairment is responsible for the phenotype of a dominant disease lesion mimic mutant of maize. Plant Cell 10: 1095–1105.

Hückelhoven, R. and Kogel, K.-H. 1998. Tissue-specific superoxide generation at interaction sites in resistant and susceptible near-isogenic barley lines attacked by the powdery mildew fungus (*Erysiphe graminis* f.sp. *hordei*). Mol. Plant-Microbe Interact. 11: 292–300.

Hunt, M.D., Delaney, T.P., Dietrich, R.A., Weymann, K.B., Dangl, J.L. and Ryals, J.A. 1997. Salicylate-independent lesion formation in *Arabidopsis lsd* mutants. Mol. Plant-Microbe Interact. 10: 531–536.

Jabs, T., Dietrich, R.A. and Dangl, J.L. 1996. Initiation of runaway cell death in an *Arabidopsis* mutant by extracellular superoxide. Science 273: 1853–1856.

Jarosch, B., Kogel, K.H. and Schaffrath, U. 1999. The ambivalence of the barley *Mlo* locus: mutations conferring resistance against powdery mildew (*Blumeria graminis* f. sp. *hordei*) enhance susceptibility to the rice blast fungus *Magnaporthe grisea*. Mol. Plant-Microbe Interact. 12: 508–514.

Jones, D.A., Thomas, C.M., Hammond-Kosack, K.E., Balint-Kurti, P.J. and Jones, J.D.G. 1994. Isolation of the tomato *Cf-9* gene for resistance to *Cladosporium fulvum* by transposon tagging. Science 266: 789–793.

Jørgensen, J.H. 1988. Genetic-analysis of barley mutants with modifications of powdery mildew resistance gene Ml-a12. Genome 30: 129–132.

Jørgensen, J.H. 1996. Effect of three suppressors on the expression of powdery mildew resistance genes in barley. Genome 39: 492–498.

Kawasaki, T., Henmi, K., Ono, E., Hatakeyama, S., Iwano, M., Satoh, H. and Shimamoto, K. 1999. The small GTP-binding protein Rac is a regulator of cell death in plants. Proc. Natl. Acad. Sci. USA 96: 10922–10926.

Keller, T., Damude, H.G., Werner, D., Doerner, P., Dixon, R.A. and Lamb, C. 1998. A plant homolog of the neutrophil NADPH oxidase gp91phox subunit gene encodes a plasma membrane protein with Ca^{2+} binding motifs. Plant Cell 10: 255–266.

Klement, Z. 1963. Rapid detection of the pathogenicity of phytopathogenic pseudomonads. Nature 199: 299–300.

Klement, Z. 1982. Hypersensitivity. In: Phytopathogenic Prokaryotes, Academic Press, New York, pp. 149–177.

Kobe, B. and Deisenhofer, J. 1995. A structural basis of the interactions between leucine-rich repeats and protein ligands. Nature 366: 183–186.

Köhm, B.A., Goulden, M.G., Gilbert, J.E., Kavanagh, T.A. and Baulcombe, D.C. 1993. A potato virus X resistance gene mediates an induced, nonspecific resistance in protoplasts. Plant Cell 5: 913–920.

Lamb, C. and Dixon, R.A. 1997. The oxidative burst in plant disease resistance. Annu. Rev. Plant Physiol Plant Mol. Biol. 48: 251–275.

Lebrun, M.H., Dutfoy, F., Gaudemer, F., Kunesch, G. and Gaudemer, A. 1990. Detection and quantification of the fungal phytotoxin tenuazonic acid produced by *Pyricularia oryzae*. Phytochemistry 29: 3777–3783.

Leckie, F., Mattei, B., Capodicasa, C., Hemmings, A., Nuss, L., Aracri, B., DeLorenzo, G. and Cervone, F. 1999. The specificity of polygalacturonase-inhibiting protein (PGIP): a single amino acid substitution in the solvent-exposed β-strand/β-turn region of the leucine-rich repeats (LRRs) confers a new recognition capability. EMBO J. 18: 2352–2363.

Levine, A., Tenhaken, R., Dixon, R. and Lamb, C. 1994. H_2O_2 from the oxidative burst orchestrates the plant hypersensitive disease resistance response. Cell 79: 583–593.

Martin, G.B., Brommonschenkel, S.H., Chungwongse, J., Frary, A., Ganal, M.W., Spivey, R., Wu, T., Earle, E.D. and Tanksley, S.D. 1993. Map-based cloning of a protein kinase gene conferring disease resistance in tomato. Science 262: 1432–1436.

Molina, A., Volrath, S., Guyer, D., Maleck, K., Ryals, J. and Ward, E. 1999. Inhibition of protoporphyrinogen oxidase expression in *Arabidopsis* causes a lesion-mimic phenotype that induces systemic acquired resistance. Plant J. 17: 667–678.

Olie, R.A., Durrieu, F., Cornillon, S., Loughran, G., Gross, J., Earnshaw, W.C. and Golstein, P. 1998. Apparent caspase independence of programmed cell death in *Dictyostelium*. Curr. Biol. 8: 955–958.

Parker, J.E., Holub, E.B., Frost, L.N., Falk, A., Gunn, N.D. and Daniels, M.J. 1996. Characterization of *eds1*, a mutation in *Arabidopsis* suppressing resistance to *Peronospora parasitica* specified by several different *RPP* genes. Plant Cell 8: 2033–2046.

Parniske, M., Hammond-Kosack, K.E., Golstein, C., Thomas, C.M., Jones, D.A., Harrison, K., Wulff, B.B.H. and Jones, J.D.G. 1997. Novel disease resistance specificities result from sequence exchange between tandemly repeated genes at the *Cf-4/9* locus of tomato. Cell 91: 821–832.

Peterhänsel, C., Freialdenhoven, A., Kurth, J., Kolsch, R. and Schulze-Lefert, P. 1997. Interaction analyses of genes required for resistance responses to powdery mildew in barley reveal distinct pathways leading to leaf cell death. Plant Cell 9: 1397–1409.

Pontier, D., Gan, S., Amasino, R.M., Robby, D. and Lam, E. 1999. Markers for hypersensitive response and senescence show distinct patterns of expression. Plant Mol. Biol. 39: 1243–1255.

Rao, M.V. and Davis, K.R. 1999. Ozone-induced cell death occurs via two distinct mechanisms in *Arabidopsis*: the role of salicylic acid. Plant J. 17: 603–614.

Rate, D.N., Cuenca, J.V., Bowman, G.R., Guttman, D.S. and Greenberg, J.T. 1999. The gain-of-function *Arabidopsis acd6* mutant reveals novel regulation and function of the salicylic acid signaling pathway in controlling cell death, defenses, and cell growth. Plant Cell 11: 1695–1708.

Rathjen, J.P., Chang, J.H., Staskawicz, B.J. and Michelmore, R.W. 1999. Constitutively active *Pto* induces a Prf-dependent hypersensitive response in the absence of *avrPto*. EMBO J. 18: 3232–3240.

Runeckles, V.C. and Vaartnou, M. 1997. EPR evidence for superoxide anion formation in leaves during exposure to low levels of ozone. Plant Cell Envir. 20: 306–314.

Salmeron, J.M., Barker, S.J., Carland, F.M., Mehta, A.Y. and Staskawicz, B.J. 1994. Tomato mutants altered in bacterial disease resistance provide evidence for a new locus controlling pathogen recognition. Plant Cell 6: 511–520.

Salmeron, J.M., Oldroyd, G.E.D., Rommens, C.M.T., Scofield, S.R., Kim, H.S., Lavelle, D.T., Dahlbeck, D. and Staskawicz, B.J. 1996. Tomato *Prf* is a member of the leucine-rich repeat class of plant disease resistance genes and lies embedded within the *Pto* kinase gene cluster. Cell 86: 123–133.

Schmidt, H.H.H.W. and Walter, U. 1994. No at work. Cell 78: 919–925.

Scofield, S.R., Tobias, C.M., Rathjen, J.P., Chang, J.H., Lavelle, D.T., Michelmore, R.W. and Staskawicz, B.J. 1996. Molecular basis of gene-for-gene specificity in bacterial speck disease of tomato. Science 274: 2063–2065.

Shah, J., Kachroo, P. and Klessig, D.F. 1999. The *Arabidopsis ssi1* mutation restores pathogenesis-related gene expression in *npr1* plants and renders defensin gene expression salicylic acid dependent. Plant Cell 11: 191–206.

Sharma, Y.K. and Davis, K.R. 1997. The effects of ozone on antioxidant responses in plants. Free Radical Biol. Med. 23: 480–488.

Shirasu, K., Nakajima, H., Rajasekhar, V.K., Dixon, R.A. and Lamb, C. 1997. Salicylic acid potentiates an agonist-dependent gain control that amplifies pathogen signals in the activation of defense mechanisms. Plant Cell 9: 261–270.

Shirasu, K., Lahaye, L., Tan, M.-W., Zhou, F., Azevedo, C. and Schulze-Lefert, P. 1999a. A novel class of eukaryotic zinc-binding protein is required for disease resistance signaling in barley and development in *C. elegans*. Cell 99: 355–366.

Shirasu, K., Nielsen, K., Piffanelli, P., Oliver, R. and Schulze-Lefert, P. 1999b. Cell-autonomous complementation of mlo resistance using a biolistic transient expression system. Plant J. 17: 293–299.

Simmons, C., Hantke, S., Grant, S., Johal, G.S. and Briggs, S.P. 1998. The maize lethal leaf spot 1 mutant has elevated resistance to fungal infection at the leaf epidermis. Mol. Plant-Microbe Interact. 11: 1110–1118.

Song, W.Y., Wang, G.L., Chen, L.L., Kim, H.S., Pi, L.Y., Holsten, T., Gardner, J., Wang, B., Zhai, W.X., Zhu, L.H., Fauquet, C. and Ronald, P. 1995. A receptor kinase-like protein encoded by the rice disease resistance gene, *Xa21*. Science 270: 1804–1806.

Stakman, E.C. 1915. Relation between *Puccinia graminis* and plants highly resistance to its attack. J. Agric. Res. 4: 193–199.

Tang, X.Y., Frederick, R.D., Zhou, J.M., Halterman, D.A., Jia, Y.L. and Martin, G.B. 1996. Initiation of plant disease resistance by physical interaction of AvrPto and Pto kinase. Science 274: 2060–2063.

Thomas, C.M., Jones, D.A., Parniske, M., Harrison, K., Balint-Kurti, P.J., Hatzixanthis, K. and Jones, J.D.G. 1997. Characterization of the tomato *Cf-4* gene for resistance to *Cladosporium fulvum* identifies sequences that determine recognitional specificity in *Cf-4* and *Cf-9*. Plant Cell 9: 2209–2224.

Thordal-Christensen, H., Zhang, Z.G., Wei, Y.D. and Collinge, D.B. 1997. Subcellular localization of H_2O_2 in plants. H_2O_2 accumulation in papillae and hypersensitive response during the barley-powdery mildew interaction. Plant J. 11: 1187–1194.

Thornberry, N.A. and Lazebnik, Y. 1998. Caspases: enemies within. Science 281: 1312–1316.

Torp, J. and Jörgensen, J.H. 1986. Modification of barley powdery mildew resistance gene Ml-a12 by induced mutation. Can. J. Gen. Cytol. 28: 725–731.

Torres, M.A., Onouchi, H., Hamada, S., Machida, C., Hammond-Kosack, K.E. and Jones, J.D.G. 1998. Six *Arabidopsis thaliana* homologues of the human respiratory burst oxidase (gp91phox). Plant J. 14: 365–370.

van den Ackerveken, G.F.J.M., van Kan, J.A.L. and de Wit, P.J.G.M. 1992. Molecular analysis of the avirulence gene *Avr9* of the fungal tomato pathogen *Cladosporium fulvum* fully supports the gene-for-gene hypothesis. Plant J. 2: 359–366.

van der Biezen, E.A. and Jones, J.D.G. 1998. The NB-ARC domain: a novel signalling motif shared by plant resistance gene products and regulators of cell death in animals. Curr. Biol. 8: R226–R227.

von Röpenack, E., Parr, A. and Schulze-Lefert, P. 1998. Structural analyses and dynamics of soluble and cell wall-bound phenolics in a broad spectrum resistance to the powdery mildew fungus in barley. J. Biol. Chem. 273: 9013–9022.

Warren, R.F., Merrit, P.M., Holub, E. and Innes, R.W. 1999. Identification of three putative signal transduction genes involved in *R* gene-specific resistance in *Arabidopsis*. Genetics 152: 401–412.

Weymann, K., Hunt, M., Uknes, S., Neuenschwander, U., Lawton, K., Steiner, H.Y. and Ryals, J. 1995. Suppression and restoration of lesion formation in *Arabidopsis lsd* mutants. Plant Cell 7: 2013–2022.

Whitham, S., Dinesh-Kumar, S.P., Choi, D., Hehl, R., Corr, C. and Baker, B. 1994. The product of the tobacco mosaic virus resistance gene *N*: similarity to Toll and the interleukin-1 receptor. Cell 78: 1011–1115.

Wolter, M., Hollricher, K., Salamini, F. and Schulze-Lefert, P. 1993. The *mlo* resistance alleles to powdery mildew infection in barley trigger a developmentally controlled defence mimic phenotype. Mol. Gen. Genet. 239: 122–128.

Yu, I.C., Parker, J. and Bent, A.F. 1998. Gene-for-gene disease resistance without the hypersensitive response in *Arabidopsis dnd1* mutant. Proc. Natl. Acad. Sci. USA 95: 7819–7824.

Plant Molecular Biology **44:** 387–397, 2000.
E. Lam, H. Fukuda and J. Greenberg (Eds.), Programmed Cell Death in Higher Plants.

Endonucleases

Munetaka Sugiyama[1,*], Jun Ito[2], Shigemi Aoyagi[3,4] and Hiroo Fukuda[2]
[1]*Botanical Gardens, Graduate School of Science, The University of Tokyo, Hakusan 3-7-1, Bunkyo-ku, Tokyo 112-0001, Japan (*author for correspondence; e-mail: sugiyama@ns.bg.s.u-tokyo.ac.jp);* [2]*Department of Biological Sciences, Graduate School of Science, The University of Tokyo, Hongo 7-3-1, Bunkyo-ku, Tokyo 113-0033, Japan;* [3]*Biological Institute, Graduate School of Science, Tohoku University, Aramaki Aza Aoba, Aoba-ku, Sendai 980-8578, Japan;* [4]*present address: Pharmacology Research, R&D, Kissei Pharmaceutical Co., Ltd., Kashiwabara 4365-1, Hotaka, Minamiazumi, Nagano 399-8304, Japan.*

Key words: Ca^{2+}-dependent endonuclease, DNA hydrolysis, programmed cell death, Zn^{2+}-dependent endonuclease

Abstract

Programmed cell death (PCD) involves hydrolysis of genomic DNA, which must be catalyzed by endonuclease(s) capable of digesting dsDNA. Plants have two major classes of endonucleases active towards dsDNA, Zn^{2+}-dependent endonuclease and Ca^{2+}-dependent endonuclease. Both classes are found among endonucleases nominated for machineries of PCD in plants. Survey of plant endonucleases in relation to PCD leads to a possibility that a different class of endonuclease reflects a different phase of PCD-associated DNA hydrolysis.

Introduction

At the final stage of programmed cell death (PCD), genomic DNA suffers hydrolysis to be lost, which strikes the last blow to a cell. During this hydrolysis process, DNA becomes fragmented and the ends of DNA increase. The increased ends can be usually detected by the TUNEL method, in which labeled dUMP is added to each of 3′-hydroxyl termini of DNA by using terminal deoxynucleotidyl transferase. Thus, PCD-associated DNA hydrolysis must involve endonucleolytic enzyme(s) that can cleave dsDNA to produce 3′-hydroxyl termini. This article reviews such plant endonucleases likely responsible for DNA hydrolysis during PCD and proposes a possible relationship between the type of endonuclease and the phase of DNA hydrolysis.

Classification of plant endonucleases

Many enzymes capable of cleaving native dsDNA in an endonucleolytic manner have been purified from plant sources and characterized. Biochemical properties of these enzymes are summarized in Table 1. (Note that this table does not include unpurified endonucleases or endonucleases implicated in DNA repair.) With the requirement for divalent cations and a pH optimum as major criteria, they can be categorized into the following two classes.

Zn^{2+}-dependent endonuclease

The first class is Zn^{2+}-dependent endonuclease, which is obviously equivalent to a group of plant nuclease I defined by Wilson (1975, 1982). Enzymes of this class are characterized primarily by requirement for Zn^{2+} and by a pH optimum in the acidic range. It should be noted, however, that Zn^{2+} dependence varies apparently under different experimental conditions, as pointed out by Wilson (1975). The Zn^{2+} ion is required for stabilization of the enzymes and for reactivation of EDTA-inactivated enzymes but does not always stimulate their activities when simply added to the standard reaction mixture.

Zn^{2+}-dependent endonucleases prefer RNA and ssDNA to dsDNA as a substrate. Although some Zn^{2+}-dependent endonucleases were described to be active specifically towards single-stranded nucleic

Table 1. Biochemical properties of plant endonucleases.

Class	Name of enzyme[a]	Plant species	Source material	Mol. mass (kDa)	Glycosylation	Nucleotidase	Substrate preference	Product end groups	pH optimum[b]	Cation dependence	References
1	scallion nuclease	*Allium chinense*	bulbs	38	ND[c]	3′-nucleotidase	RNA, ssDNA>dsDNA	5′-P/3′-OH	6.0	Zn^{2+}	Uchida *et al.*, 1993
1	BEN1	*Hordeum vulgare*	secretion from aleurone layers	35	+	3′-nucleotidase	RNA, DNA	5′-P/3′-OH	6.0	Zn^{2+}	Brown and Ho, 1986, 1987; Aoyagi *et al.*, 1998
1	acid nuclease	*Medicago sativa*	germinated seeds	37	ND	3′-nucleotidase	ssDNA>RNA>dsDNA	5′-OH/3′-P	5.5	Zn^{2+}	Yuspanis *et al.*, 1996; Christou *et al.*, 1998
1	pollen nuclease	*Petunia hybrida*	secretion from pollen	34	ND	ND	RNA>ssDNA>dsDNA	ND	5.0	Zn^{2+}	Westhuizen *et al.*, 1987
1	wheat seedling nuclease	*Triticum eastivum*	seedlings	44	ND	3′-nucleotidase	ssDNA>RNA>dsDNA	5′-P/3′-OH	5.0	Zn^{2+}	Hanson and Fairley, 1969; Kroeker *et al.*, 1975
1	mung bean nuclease	*Vigna radiata*	sprouts	39	+	3′-nucleotidase	RNA, ssDNA>dsDNA	5′-P/3′-OH	5.0	Zn^{2+}	Sung and Laskowski, 1962; Kowalski *et al.*, 1976
1	ZEN1	*Zinnia elegans*	developing TEs in culture	43	+	ND	RNA, ssDNA>dsDNA	ND	5.5	Zn^{2+}	Thelen and Northcote, 1989; Aoyagi *et al.*, 1998
1 (?)	*Avena* nuclease	*Avena sativa*	seedling leaves	33	ND	ND	RNA, ssDNA, dsDNA	5′-P/3′-OH	5.5–6.0	?	Wyen *et al.*, 1971
2	malt DNase A, B	*Hordeum vulgare*	solid malt diastase	32	ND	ND	DNA>RNA=0	5′-P/3′-OH	Mg^{2+} 6.0 Mn^{2+} 7.5	Ca^{2+}, Mg^{2+}, Mn^{2+}	Liao *et al.*, 1977
2	nuclease L	*Lens culinaris*	seedlings	39	ND	3′-nucleotidase	ssDNA>dsDNA>RNA	5′-P/3′-OH	8.0	Ca^{2+}	Kefalas and Yuspanis, 1995
2	neutral nuclease	*Medicago sativa*	germinated seeds	41	ND	3′-nucleotidase	ssDNA, RNA>dsDNA	5′-P/3′-OH	7.0	Ca^{2+}	Yuspanis *et al.*, 1996; Christou *et al.*,1998
2	NUCIII	*Nicotiana tabacum*	shoots with HR lesions	35	+	ND	ssDNA, dsDNA	ND	ND	Ca^{2+}	Mittler and Lam, 1995
2	pea seed nuclease	*Pisum sativum*	germinated seeds	30+24	+	3′-nucleotidase	ssDNA>dsDNA	ND	6.5–8.0	Ca^{2+}, Mg^{2+}	Wani and Hadi, 1979; Naseem and Hadi, 1987
2 (?)	NUC35	*Nicotiana tabacum*	shoots with HR lesions	35	+	ND	ssDNA, dsDNA	ND	ND	Ca^{2+}(?)	Mittler and Lam, 1997
2 (?)	tobacco nuclease 1 (TC NUC)	*Nicotiana tabacum*	secretion from cultured cells	35	+	3′-nucleotidase	ssDNA, RNA>dsDNA	5′-P/3′-OH	5.7	Ca^{2+}(?)	Oleson *et al.*, 1974, 1982; Mittler and Lam, 1997
2 (?)	wheat chloroplast nuclease	*Triticum vulgare*	chloroplasts from seedling leaves	29	ND	ND	ssDNA>RNA>dsDNA	5′-P/3′-OH	7.8	?	Kuligowska *et al.*, 1988

[a]If no specific names are given for the purified enzymes, tentative names are presented.
[b] Determined for DNase activity (towards ssDNA in most cases).
[c] Not determined.

```
3'-NT/Nu    1:MAMSQLSSRSTDVSLHPPAHSAYTVRGGPFTSHDVRQDLLALNCLCVASVSLASLLFILS 60

BEN1        1:                                        MGLL-L-LLQVLLVAAAARA 18
ZEN1        1:                                        .A.IR.SIISC.GFFMINNY 20
3'-NT/Nu   61:SPHREANLSHYVLYSDSSAPVFLAHTHTHTHSTVHATADMARARF.Q..L.T.TLLSTA. 120

BEN1       19:PGAQAWGKEGHYMTCKIADGFLTS---EALTGVKALLPSW----ANGELAEVCSWADSQR 71
ZEN1       21:NAV...S....V...Q..QEL.SP---D.AHA.QM...DY----VK.N.SAL.V.P.QI. 73
S1          1:     ..NL..ETVAY..QS.VA.---STESFCQNI.GDD----STSY..N.AT...TYK 48
P1          1:     ..AL..ATVAYV.QHYVSP---..ASWAQGI.G.S----SSSY..SIA....EY. 48
3'-NT/Nu  121:LPVS..WSK..MSVAL..KRHMGASLV.KAELAAKV.SFSGPYPKSPDMVQTAP...DIK 180
                   *   **      *                  *                 * *

BEN1       72:F--RYRWSRSLHFADTPGD-CKFSYARDCHDTKGNKNVCVVGAINNYTAAL---QDSSS- 124
ZEN1       74:HWY....TSP...I...D.A.S.D.T.....SN.MVDM..A...K.F.SQ.SHY.HGT.D 133
S1         49:YTDAGEF.KPY..I.AQDN-PPQ.CGV.Y-.RDCGSAG.SIS..Q...NI.---LE.PN- 102
P1         49:LTSAGK..A....I.AEDN-PPTNCNV.Y-ERDCGSSG.SIS..A...QRV---S...L- 102
3'-NT/Nu  181:TIGLKTL.-TW.YIT..YY-TDEDFTL.V-SPVQTV..AS.--.PMLQT.I---EKPTA- 231
                         *               *               *

BEN1      125:-PFN-PTESLMFLAHFVGDVHQPMHCGHV-------DDLGGNTIKLRWYRRKSN--LHHV 173
ZEN1      134:RRY.-M..A.L.VS..M..I.....V.FT-------T.E.....D...F.H...--.... 183
S1        103:-GSEA-LNA.K.VV.II..T...L.DENL--------EA...G.DVTYDGETT.--...I 150
P1        103:-SSENHA.A.R..V..I..MT..L.DEAY--------AV...K.NVTFDGYHD.--..SD 151
3'-NT/Nu  232:-NSDVIVQ..AL.L..M..I...L.NVNLFSNQYPES.....KQLVVIDSKGTKML..AY 290
                       *    *  **  ** *              ***              **

BEN1      174:WDSDVITQTMKDFFDKDQDAMIESIQRNITDD-WSSEEKQWETCRSKTTTCAEKY-AQES 231
ZEN1      184:..REI.LTAASELY...MESLQKA..A.F.HGL..DDVNS.KD.-DDISN.VN..-.K.. 241
S1        151:..TNMPEEAAGG--YSLSV.K-----TYADLL-TERIKTGTYSSKKDSW.DGIDI-KDPV 201
P1        152:..TYMPQKLIGG--HALS..E-----SWAKTL-VQNI.SGNY.AQAIGWIKGDNI-SEPI 202
3'-NT/Nu  291:...MAEGKSGE.VPRPLSE.DYDDLNNFADYL--EATYASTL.DKE.NLVDTTEISKETF 348
              **

BEN1      232:AVLACD-AYEGVE--------------QDDTLGDEYYFKALPVVQKRLAQGGLRLAAILN 276
ZEN1      242:IA...KWG.....--------------AGE..S.D.FDSRM.I.M..I....V..SM... 287
S1        202:STSMIW-.ADANTYVCSTVLDDGLAYINSTD.SG...D.SQ..FEELI.KA.Y....W.D 260
P1        203:TTATRW-.SDANALVCTVVMPHGAAAL.TGD.YPT..DSVIDTIELQI.K..Y...NWI. 261
3'-NT/Nu  349:DLALKY-..P.AD--------------NGA..SN..KTN.KKISERQVLLA.Y...KM.. 393
                                             *   *               * **

BEN1      277:RIFSGNGRLQSI                                                 288
ZEN1      288:.V.GSSSS.EDALVPT                                             303
S1        261:L.A.QPS                                                      267
P1        262:E.HGSEIAK                                                    270
3'-NT/Nu  394:TTLKSVSMDTILQGLKSIQSEVDTENKAEVHNHYDQKGISAAVTAIVAVALFIAGIIIAT 453

3'-NT/Nu  454:LVVLALKCYLPKRDRFGSYEHVAL                                     477
```

Figure 1. Alignment of amino acid sequences of Zn^{2+}-dependent endonucleases, nuclease S1 from *Aspergillus oryzae* (Iwamatsu *et al.*, 1991), nuclease P1 from *Penicillium citrinum* (Maekawa *et al.*, 1991), and 3′-nucleotidase/nuclease (3′-NT/Nu) of *Leishmania donovani* (Debrabant *et al.*, 1995), with deduced sequences of BEN1 and ZEN1. Hyphens represent gaps introduced to optimize sequence alignment. Amino acid residues identical to the corresponding ones in BEN1 are indicated by dots. Conserved residues among all five sequences are marked with asterisks. Boxes show amino acid residues that have been reported to be involved in binding of nuclease P1 to Zn^{2+} ion (Volbeda *et al.*, 1991). From Aoyagi *et al.* (1998).

acids, they are not specific in a strict sense. Under the appropriate conditions, any Zn^{2+}-dependent endonuclease can nick and linearize supercoiled dsDNA. In addition, several enzymes were clearly shown to split dsDNA into small fragments at a considerable efficiency (Brown and Ho, 1987; Thelen and Northcote, 1989; Uchida *et al.*, 1993). The degradation product possesses a phosphomonoester group at the 5′ end and a hydroxyl group at the 3′ end. As an exception, acid nuclease of alfalfa was reported to produce nucleotides bearing 3′-phosphate (Yuspanis *et al.*, 1996). All Zn^{2+}-dependent endonucleases tested are accompanied by 3′-nucleotidase activity that catalyzes the hydrolysis of 3′-phosphoester linkage of 3′-nucleotides. Zn^{2+}-dependent endonucleases share physical properties as well as enzymatic properties. They are monomeric glycoproteins with similar molecular sizes in the range of 33–44 kDa. A representative example for this class is mung bean nuclease, the best-characterized plant endonuclease. And the greater part of plant endonucleases purified to date belong to this class (Table 1). Besides the enzymes listed in Table 1, many unpurified DNase activities probably due to Zn^{2+}-dependent endonucleases have been reported from various tissues of diverse plant species (Blank and McKeon, 1989; Grafi *et al.*, 1991; Yen and Green, 1991). Thus, Zn^{2+}-dependent endonuclease may be a widespread class of enzymes in the plant kingdom. Zn^{2+}-dependent endonucleases have also been isolated from fungi and protists, which include the famous endonuclease, nuclease S1 from *Aspergillus oryzae*. Then, enzymes of this class are sometimes referred to as S1-type endonucleases. Interestingly, Zn^{2+}-dependent endonucleases have never been discovered from metazoans to our knowledge.

Recently, cDNAs for the barley 35 kDa nuclease (BEN1) and zinnia 43 kDa nuclease (ZEN1) were isolated and sequenced (Aoyagi *et al.*, 1998). The deduced amino acid sequences of BEN1 and ZEN1 showed significant similarity with each other and with those of Zn^{2+}-dependent endonucleases from fungi and protists (Figure 1). Analysis of the crystal structure of nuclease P1 from *Penicillium citrinum* revealed that its binding to Zn^{2+} involves Trp-1, His-6, Asp-45, His-60, His-116, Asp-120, His-126, His-149, and Asp-153 residues (Volbeda *et al.*, 1991). All of these nine residues are conserved among Zn^{2+}-dependent endonucleases presented in Figure 1. These data suggest that enzymes of this class are relatives evolved from a common ancestral enzyme.

By searching a protein data base with the amino acid sequence of BEN1 or ZEN1 as a query sequence, five more proteins (amino acid sequences deduced from nucleotide sequences of correspondent cDNAs) were found to exhibit high sequence similarity with BEN1 and ZEN1: one from daylily, two from arabidopsis, and two from zinnia. These two putative endonucleases of zinnia are not identical to ZEN1. This indicates that multiple Zn^{2+}-dependent endonucleases exist in a single plant species.

Ca^{2+}-dependent endonuclease

The second class is Ca^{2+}-dependent endonuclease. Ca^{2+}-dependent endonucleases are distinguished from Zn^{2+}-dependent endonucleases literally by the requirement for Ca^{2+} instead of Zn^{2+}. In some cases, Zn^{2+} was shown to be highly inhibitory against DNA hydrolysis by Ca^{2+}-dependent endonucleases (e.g., Mittler and Lam, 1995). Most Ca^{2+}-dependent endonucleases have pH optima in the neutral region and substrate preferences of ssDNA far over RNA, which can also distinguish enzymes of the present class from typical Zn^{2+}-dependent endonucleases.

Ca^{2+}-dependent endonucleases produce nucleotides bearing phosphomonoester groups at the 5′ ends as a result of DNA hydrolysis. They are always accompanied by 3′-nucleotidase activity. These characteristics are common to Zn^{2+}-dependent and Ca^{2+}-dependent endonucleases of plant origin.

There are relatively few papers demonstrating clearly the existence of Ca^{2+}-dependent endonucleases in plants as compared with Zn^{2+}-dependent endonucleases. According to Nakamura *et al.* (1987), however, Ca^{2+}-dependent activities of DNA hydrolysis at pH 8.5 were found from diverse taxons ranging from green algae to angiospermae. This implies that Ca^{2+}-dependent endonucleases occur generally in plants.

Enzymologically, plant Ca^{2+}-dependent endonucleases are somewhat similar with several mammalian Ca^{2+}/Mg^{2+}-dependent endonucleases, including DNase I and DNase γ. Liao (1977) compared malt DNases A and B with pancreatic DNase I in their pH-activity profiles as influenced by Mg^{2+} and Mn^{2+} and their amino acid compositions, a result which indicates striking similarities among these three DNases.

In spite of the above-mentioned properties shared by Ca^{2+}-dependent endonucleases, it is still questionable whether they belong to truly a single class. Ca^{2+}-dependent endonuclease is rather a tentative class, which might be subdivided given more detailed information for each individual enzyme, particularly its primary structure.

Endonucleases and various PCDs in plants

Are various PCDs in plants driven by the same basic mechanism or by several different mechanisms? This question remains to be answered. In consideration of a possibility that plant PCDs are a mixture of outwardly similar but fundamentally different phenomena, PCD-associated endonucleases should not be lumped together. In this section, endonucleases that may participate in DNA hydrolysis during PCD are described on a case-by-case basis.

Endosperm degeneration

Endosperm degenerates during seed maturation and germination to support the growth of embryo through mobilization of its reserves. This endosperm degeneration is a kind of PCD for endosperm cells.

The relationship between endosperm degeneration and endonuclease activation has been studied with a few cereal species. In cereals, the endosperm degeneration process consists of three distinct stages as follows (Fincher, 1989; Lopes and Larkins, 1993). At the first stage, starchy endosperm cells degenerate to lose physiological activities during seed maturation. Consequently, the starchy endosperm cells of mature grains are non-living. However, they still maintain cellular structures and contain the remnants of nuclei, ribosomes, endoplasmic reticulum, and RNA. These remnants are cleaned up with the aid of degradative enzymes secreted from the aleurone layers which are dependent upon gibberellic acid at the second stage that commences after seed imbibition. Thereafter, the endosperm enters the third stage in which cells of the aleurone layer undergo PCD.

With respect to the first stage of endosperm degeneration, changes in the activities of endonucleases were investigated with maize kernels (Young *et al.*, 1997). DNase activity gel staining revealed three nucleases with masses of 33.5, 36.0, and 38.5 kDa present in the endosperm. Comparison of nuclease profiles between the wild type and the *shrunken2* (*sh2*) mutant, in which internucleosomal fragmentation of DNA in the endosperm is accelerated, suggested that the 33.5 kDa nuclease is most closely associated with DNA fragmentation. Although cation requirement of these three nucleases was not assessed in detail, they may belong to the class of Ca^{2+}-dependent endonuclease since they are active in the presence of Ca^{2+} at pH 6.8 and not below pH 5.0.

An endonuclease probably responsible for DNA degradation in the endosperm at the second stage was purified from the secretion of the emryo-less half seeds of barley (Brown and Ho, 1986). This enzyme was shown to be a 35 kDa monomeric glycoprotein with Zn^{2+}-dependent endonuclease and 3′-nucleotidase activities (Brown and Ho, 1986, 1987), and later it was designated BEN1 (Aoyagi *et al.*, 1998). The amino acid sequence of BEN1 deduced from the nucleotide sequence of its cDNA has an additional hydrophobic region at the N-terminus when compared with the partial amino acid sequence determined for the mature BEN1 (Aoyagi *et al.*, 1998). This region is likely

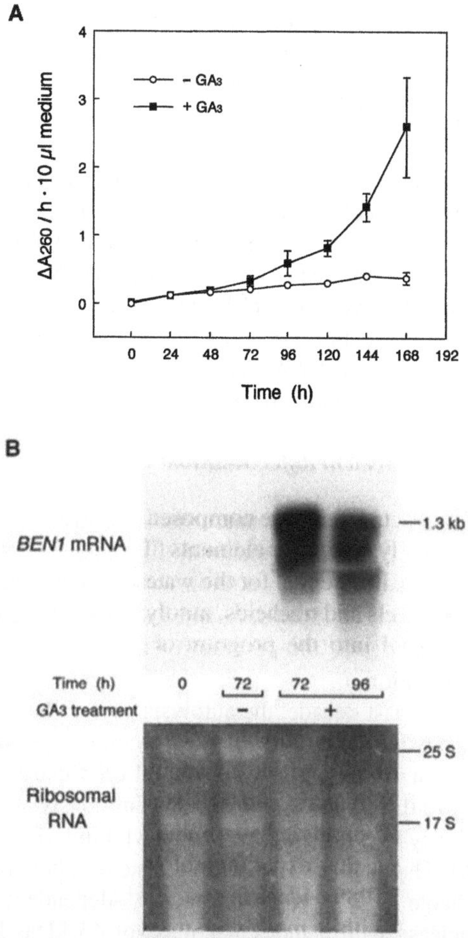

Figure 2. Changes in the activity of DNase and the amount of mRNA encoding BEN1 during incubation of embryo-less half-seeds of barley in the presence or absence of gibberellic acid (GA_3). A. DNase activity towards ssDNA in the medium. B. RNA gel blot analysis showing the accumulation of mRNA for BEN1 in the cultured half-seeds. From Aoyagi *et al.* (1998) with unpublished data.

to act as a signal sequence for secretion, which was recently supported by experiments with transgenic tobacco cells expressing BEN1 (J. Ito, M. Sugiyama and H. Fukuda, unpublished).

Both the BEN1 protein and the total DNase activity attributable to BEN1 increased in the secretion of the isolated aleurone layers or the embryo-less half-seeds upon addition of gibberellic acid to the culture medium (Brown and Ho, 1986; Brown *et al.*, 1988). The accumulation of mRNA encoding BEN1 in the embryo-less half-seeds was also induced by the administration of gibberellic acid (Aoyagi *et al.*, 1998; Figure 2). These results suggest that during germination of barley seeds, BEN1 is synthesized through gene expression *de novo* in the aleurone layer in response to gibberellic acid, which is supplied from the embryo normally, and then secreted into the starchy endosperm to degrade its DNA.

PCD in the aleurone layer at the third stage was demonstrated to be associated with internucleosomal fragmentation of DNA under the control of gibberellic acid and its antagonist, abscisic acid, with germinating barley seeds (Wang *et al.*, 1996, 1998). Although Kalinski *et al.* (1986) proposed the involvement of some linker-specific and EDTA-insensitive nuclease(s) in this DNA fragmentation, such nucleases have not yet been identified.

Tracheary element differentiation

Vessels and tracheids are composed of dead and empty cells, namely, tracheary elements (TEs). As the emptiness of TEs is essential for the water-conducting function of vessels and tracheids, autolysis, a kind of PCD, is integrated into the program of tracheary element differentiation.

In the past decade, the autolysis process has been studied extensively with an *in vitro* culture system of zinnia, in which isolated mesophyll cells can be induced to differentiate into TEs synchronously at high frequency by supplying auxin and cytokinin (Fukuda, 1996). Using this experimental system, Thelen and Northcote (1989) identified a Zn^{2+}-dependent endonuclease with a molecular mass of 43 kDa, later designated ZEN1 (Aoyagi *et al.*, 1998), whose activity was tightly correlated with autolysis during TE differentiation. That is, the activity of ZEN1 showed a transient increase at the time of autolysis specifically in the culture inducing the differentiation (Thelen and Northcote, 1989). The accumulation of RNA encoding ZEN1 was also specific to the pre-autolysis stage of TE differentiation (Aoyagi *et al.*, 1998; Figure 3). Moreover, ZEN1 was the sole endonuclease detectable in the DNase activity gel staining at the autolysis stage (Thelen and Northcote, 1989). All these results strongly suggest that ZEN1 is responsible for DNA degradation during autolysis in the course of TE differentiation.

Comparison of the deduced amino acid sequence of ZEN1 with the partial sequence of the mature ZEN1 revealed the presence of hydrophobic region at the N-terminus upstream of the cleavage site as is the case for BEN1 of barley (Aoyagi *et al.*, 1998). Unlike BEN1, however, ZEN1 seems to be transported to the vacuole because most of the activity of ZEN1 expressed in transgenic tobacco cells was recovered in the purified vacuoles (J. Ito, M. Sugiyama and H. Fukuda, unpublished). This is consistent with the hypothesis that the collapse of the tonoplast results in the release of hydrolytic enzymes accumulated in the vacuole and subsequently complete degradation of intracellular components including genomic DNA during TE differentiation (Fukuda, 1996; Groover *et al.*, 1997).

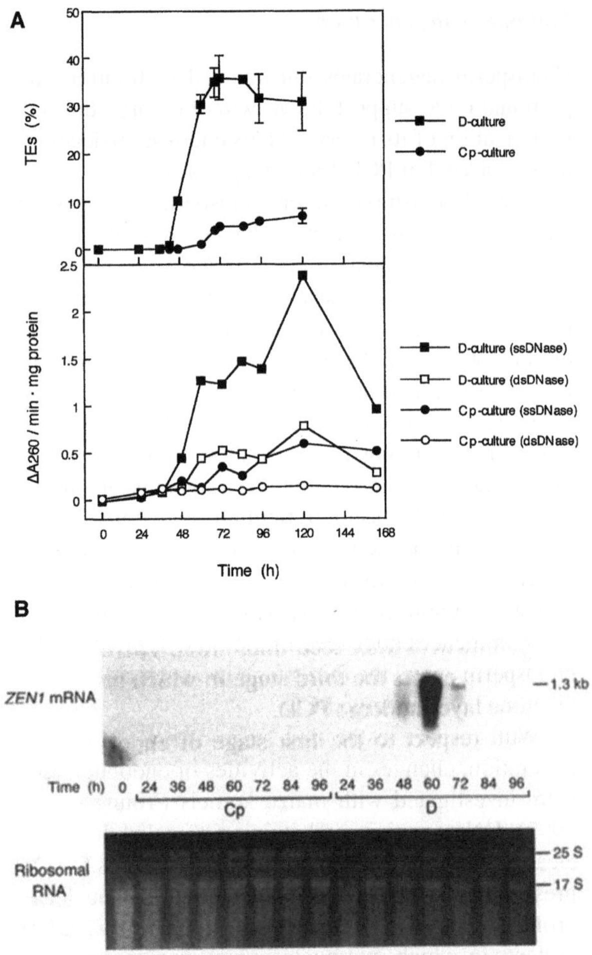

Figure 3. Changes in the proportion of TEs, the activity of DNase, and the amount of mRNA encoding ZEN1 during culture of isolated mesophyll cells of zinnia in Cp or D medium. The Cp medium contained 0.001 mg/l benzyladenine and 0.1 mg/l naphthaleneacetic acid as phytohormones whereas the D medium contained 0.2 mg/l benzyladenine and 0.1 mg/l naphthaleneacetic acid. A. The proportion of TEs in total cells and the activity of DNase towards ssDNA or dsDNA. B. RNA gel blot analysis showing the accumulation of mRNA for ZEN1. From Aoyagi *et al.* (1998) with unpublished data.

Leaf senescence

Leaf senescence causes death of leaf cells to liberate the reserves from the dying cells. Recently, this senescence-associated death was shown to be a type of PCD accompanied with internucleosomal fragmentation of DNA (Yen and Yang, 1998). Endonucleases whose activity was changed in concert with the progression of the senescence were identified in wheat and barley. In the case of wheat, single-strand-preferring endonucleases with apparent molecular masses ranging from 32 to 38 kDa increased in activity during the dark-induced senescence of primary leaves and decreased during the regreening of the senescent leaves induced by their exposure to the light (Blank and McKeon, 1989). These endonucleases are considered to belong to the class of Zn^{2+}-dependent endonuclease in the light of their sensitivity to EDTA and enhancement of their renaturation by the presence of Zn^{2+} ion. In contrast, endonucleases activated with the senescence of barley leaves seem to be Ca^{2+}-dependent since they were slightly stimulated by Ca^{2+} and inhibited by Zn^{2+} (Wood *et al.*, 1998).

A number of genes of which expression is up-regulated during senescence have been isolated from various plant species. Out of them, DSA6 from daylily and Bnuc1 from barley encode putative Zn^{2+}-dependent endonucleases with signal peptides at their N-termini (Panavas *et al.*, 1999; Muramoto *et al.*, 1999). The amino acid sequence of Bnuc1 is similar to but not identical with that of BEN1.

Hypersensitive response

Hypersensitive response (HR) of plants to the infection of avirulent pathogens involves PCD, which brings about the formation of the zone of dead cells around the site of infection. Mittler and Lam (1995) identified three nucleases different in apparent molecular mass, NUCI (100.5 kDa), NUCII (38 kDa), and NUCIII (36 kDa), that were induced in a coordinated manner with PCD during HR of tobacco. All of these nucleases were shown to be inhibited by EDTA, EGTA, and Zn^{2+}. The EDTA- or EGTA-inhibited activities of NUCI and NUCIII were restored by Ca^{2+}, indicating that these belong to a class of Ca^{2+}-dependent endonuclease. NUCIII, which accounted for the majority of DNase activity detected, was found in purified nuclei. From these data, it was inferred that fragmentation of nuclear DNA associated with PCD during HR could be attributed to the HR-specific increase in the activities of Ca^{2+}-dependent endonucleases, particularly NUCIII.

In later papers (Mittler *et al.*, 1997; Mittler and Lam, 1997), however, Mittler and co-workers suggested a more complicated scenario for the DNA degradation process during HR-associated PCD. With respect to the fashion of DNA degradation, their findings are summarized as follows.

1. HR leads to leakage of electrolytes and cleavage of nuclear DNA to 50 kb and then to smaller fragments in tobacco leaves. The DNA cleavage does not precede the electrolyte leakage. Necrotic cell death caused by a freeze-thaw treatment also induces rapid appearance of 50 kb fragments.
2. The increase in the DNase activity induced upon HR of tobacco leaves does not always correlate with DNA fragmentation. It can be induced by various treatments of tobacco leaves such as infiltration with 1 M sucrose that result in neither electrolyte leakage nor DNA fragmentation.
3. Most of the increased activity of DNase during HR of tobacco leaves is present in the apoplastic fraction. The apoplastic fraction contains at least two endonucleases: an inducible one and a constitutive one (designated NUC35).

These findings have uncoupled different aspects implicated in DNA degradation during HR. The stimulation of nuclear endonucleases such as NUCIII is induced upon HR but may not play an essential role in the extensive fragmentation of nuclear DNA at the late stage of HR. This DNA fragmentation seems to be catalyzed mainly by apoplastic endonucleases, which can reach the nucleus after the integrity of the plasma membrane is compromised. The increase in activity of apoplastic endonucleases induced by various biotic or abiotic stresses may accelerate DNA fragmentation.

The apoplastic endonucleases found in tobacco leaves undergoing HR were reported to be Ca^{2+}-dependent because of their sensitivity to EGTA (Mittler and Lam, 1997). However, there remains a possibility that these endonucleases are Zn^{2+}-dependent, in the light of our knowledge that EGTA can chelate both Zn^{2+} and Ca^{2+} ions with similar efficiency. Of the apoplastic endonucleases, NUC35 is closely related with TC NUC, an endonuclease that is released from tobacco cells into the culture medium, since antibody raised against purified NUC35 recognized TC NUC as well as NUC35 (Mittler and Lam, 1997). This TC NUC must be identical to tobacco nuclease I, which was previously purified from the medium used for a culture of tobacco cells (Oleson *et al.*, 1974,

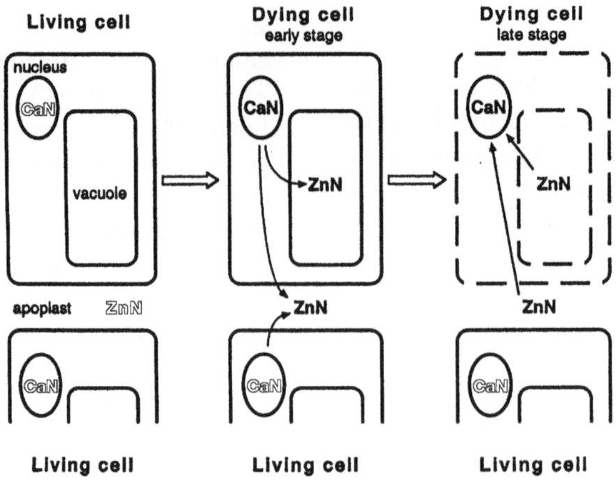

Figure 4. Hypothetical scheme for the process of DNA hydrolysis during plant PCD involving different classes of endonucleases. CaN: Ca^{2+}-dependent endonnulcease. ZnN: Zn^{2+}-dependent endonuclease. Open characters indicate the basal level of the activity of the corresponding enzyme.

1982). The tobacco nuclease I was shown to have a pH optimum in the acidic region like typical Zn^{2+}-dependent endonucleases instead of Ca^{2+}-dependent endonucleases. Thus, the close relationship between NUC35 and TC NUC (tobacco nuclease I) raises a question about cation dependence of the apoplastic endonucleases including NUC35.

Concluding remarks

Plants have at least two distinct classes of endonucleases capable of digesting DNA, namely Zn^{2+}-dependent endonuclease and Ca^{2+}-dependent endonuclease. Of plant endonucleases that have been nominated for enzymes catalyzing DNA degradation during PCD, some belong to the former class and others belong to the latter class. Does the difference in the class of PCD-associated endonuclease imply the difference in the type of PCD? The answer to this question would be 'No', considering the cases of starchy endosperm degeneration and leaf senescence which appear to involve endonucleases of both classes.

An alternative explanation is that endonucleases of the different classes play different roles in DNA degradation during PCD. If this is true, how do they share in the work of DNA degradation? The pH optima and subcellular localization of endonucleases can give hints as to this question. All Zn^{2+}-dependent endonucleases exhibit maximal activity in the acidic pH region ranging from pH 5.0 to 6.5. Consistent with this property, Zn^{2+}-dependent endonucleases are found in the acidic compartments, i.e., extracellular fluid and vacuole. Accordingly, Zn^{2+}-dependent endonucleases are supposed to attack nuclear DNA only after the collapse of the plasma membrane or tonoplast. By contrast with Zn^{2+}-dependent endonucleases, typical Ca^{2+}-dependent endonucleases have pH optima in the neutral region. Although information about subcellular localization of Ca^{2+}-dependent endonucleases is very poor, it can be speculated that they may occur in nuclei, referring to the observation that NUCIII from tobacco was detected in purified nuclei. Thus, a hypothetical scheme can be drawn for PCD-associated DNA degradation as shown in Figure 4. At first, the activities of Ca^{2+}-dependent endonucleases increase in the nuclei of cells undergoing PCD and cause rather limited fragmentation of nuclear DNA. On the other hand, PCD induces the accumulation of Zn^{2+}-dependent endonucleases in the apoplast or vacuoles. Finally, membrane systems are broken down and nuclear DNA is exposed to extensive degradation by apoplastic or vacuolar Zn^{2+}-dependent endonucleases.

The above-described scheme coincides with the process of degeneration of starchy endosperm in cereals. Studies with maize kernels indicated that the activities of Ca^{2+}-dependent endonucleases are elevated in association with internucleosomal fragmentation of DNA at the first stage of starchy endosperm degeneration (Young *et al.*, 1997). At the second stage during germination, DNA in the starchy endosperm is subjected to DNA hydrolysis by Zn^{2+}-dependent endonuclease secreted from the aleurone layer, which has been demonstrated through experiments with barley seeds (Brown and Ho, 1986, 1987).

In addition, two-step DNA fragmentation suggested for HR of tobacco leaves (Mittler and Lam, 1997) can be accounted for straightly by the presented scheme, assuming that apoplastic endonucleases found in pathogen-infected leaves are not Ca^{2+}-dependent but Zn^{2+}-dependent. In the case of TE differentiation, a slightly modified version of the scheme may be suitable. Since DNA fragmentation does not precede the collapse of tonoplast in developing TEs (Groover *et al.*, 1997), the program of DNA degradation linked with TE differentiation appears to skip the first stage involving Ca^{2+}-dependent endonucleases.

No obvious similarities can be seen between DNA-degrading machineries for plant PCD and those for apoptosis of mammalian cells at present. In the case of apoptosis, a peculiar type of endonuclease, namely, caspase-activated DNase (CAD), which preexists as an inactive complex with its specific inhibitor (ICAD) in living cells and is activated upon the digestion of ICAD by caspase 3, is considered to initiate DNA fragmentation (Enari *et al.*, 1998; Sakahira *et al.*, 1998). To date, neither CAD-like activity nor genes encoding CAD-like proteins have been reported from plants. DNase γ, another candidate for the enzyme responsible for DNA fragmentation during apoptosis, requires Ca^{2+} and Mg^{2+} for reaction (Shiokawa *et al.*, 1994). This cation dependence suggests only a possible relationship between DNase γ and Ca^{2+}-dependent endonucleases associated with plant PCDs. Zn^{2+}-dependent endonuclease, which probably takes a major part in DNA degradation at the late stage of plant PCDs, is absent in mammalian cells.

Taking a larger view of the DNA degradation system, however, mammalian apoptosis can be regarded to employ a procedure consisting of two phases like PCD in plants. The first phase proceeds autonomously both in mammalian cells and in plant cells. In this phase, nuclear DNA of dying cells suffers relatively limited fragmentation by CAD or DNase γ in mammalian apoptosis and by Ca^{2+}-dependent endonucleases in plant PCDs. In the second phase of apoptotic DNA degradation, dead cells are engulfed into macrophages and nuclear DNA is hydrolyzed by the lysosomal function, which may correspond to the hydrolysis of DNA by apoplastic or vacuolar Zn^{2+}-dependent endonucleases in plant PCDs.

As the readers can see in this article, information available for the relationship between plant endonucleases and PCDs is quite fragmentary. In some experimental systems, only the activities of Zn^{2+}-dependent endonucleases have been assessed, and only Ca^{2+}-dependent endonucleases in others. More comprehensive studies on endonucleases in various PCD models would examine the generality of the two-phase hypothesis discussed above. In this context, future studies on PCD-associated endonucleases should pay attention to their classes, their localization, and the timing of the rise in their activities.

The molecular mechanism that links the induction of endonuclease activities to the specific stage of PCD is one of the key devices of plant PCD to be elucidated. Based on the studies of BEN1 and ZEN1, it can be presumed that the increase in activities of Zn^{2+}-dependent endonucleases results from gene expression *de novo*. Thus, promoter analysis of the genes for Zn^{2+}-dependent endonucleases would give important hints at the molecular level about the mechanism responsible for PCD-associated induction of these enzymes. With respect to Ca^{2+}-dependent endonucleases, it is totally unknown whether the induction of their activities is attributed to gene expression *de novo* or activation of pre-existing enzymes. In order to solve this problem, it will be necessary to isolate molecular probes for plant Ca^{2+}-dependent endonucleases such as cDNAs and antibodies.

References

Aoyagi, S., Sugiyama, M. and Fukuda, H. 1998. *BEN1* and *ZEN1* cDNAs encoding S1-type DNases that are associated with programmed cell death in plants. FEBS Lett. 429: 134–138.

Blank, A. and McKeon, T.A. 1989. Single-strand-preferring nuclease activity in wheat leaves is increased in senescence and is negatively photoregulated. Proc. Natl. Acad. Sci. USA 86: 3169–3173.

Brown, P.H. and Ho, T.-H.D. 1986. Barley aleurone layers secrete a nuclease in response to gibberellic acid: purification and partial characterization of the associated ribonuclease, deoxyribonuclease, and 3′-nucleotidase activities. Plant Physiol. 82: 801–806.

Brown, P.H. and Ho, T.-H.D. 1987. Biochemical properties and hormonal regulation of barley nuclease. Eur. J. Biochem. 168: 357–364.

Brown, P.H., Mecham, R.P. and Ho, T.-H.D. 1988. Hormonal regulation of barley nuclease: investigation using a monoclonal antibody. Plant Cell Envir. 11: 747–753.

Christou, A., Mantrangou, C. and Yupsanis, T. 1998. Similarities and differences in the properties of alfalfa endonucleases. J. Plant Physiol. 153: 16–24.

Debrabant, A., Gottlieb, M. and Dwyer, D.M. 1995. Isolation and charcterization of the gene encoding the surface membrane 3′-nucleotidase/nuclease of *Leishmania donovani*. Mol. Biochem. Parasitol. 71: 51–63.

Enari, M., Sakahira, H., Yokoyama, H., Okawa, K., Iwamatsu, A. and Nagata, S. 1998. A caspase-activated DNase that degrades

DNA during apoptosis, and its inhibitor ICAD. Nature 391: 43–50.

Fincher, G.B. 1989. Molecular and cellular biology associated with endosperm mobilization in germinating cereal grains. Annu. Rev. Plant Physiol. Plant Mol. Biol. 40: 305–346.

Fukuda, H. 1996. Xylogenesis: initiation, progression, and cell death. Annu. Rev. Plant Physiol. Plant Mol. Biol. 47: 299–325.

Grafi, G., Meller, E., Sher, N. and Sela, I. 1991. Characterization of S1/mung-bean-type nuclease activity in plant cell suspensions. Plant Sci. 74: 107–114.

Groover, A., DeWitt, N., Heidel, A. and Jones, A. 1997. Programmed cell death of plant tracheary elements differentiating in vitro. Protoplasma 196: 197–211.

Hanson, D.M. and Fairley, J.M. 1969. Enzymes of nucleic acid metabolism from wheat seedlings. I. Purification and general properties of associated deoxyribonuclease, ribonuclease, and 3′-nucleotidase activities. J. Biol. Chem. 244: 2440–2449.

Iwamatsu, A., Aoyama, H., Dibó, G., Tsunasawa, S. and Sakiyama, F. 1991. Amino acid sequence of nuclease S1 from *Aspergillus oryzae*. J. Biochem. 110: 151–158.

Kalinski, A., Chandra, G.R. and Muthukrishnan, S. 1986. Study of barley endonucleases and α-amylase genes. J. Biol. Chem. 261: 11393–11397.

Kefalas, P.S. and Yupsanis, T. 1995. Properties and specificity of a calcium dependent endonuclease from germinated lentil (*Lens culinaris*). J. Plant Physiol. 146: 1–9.

Kowalski, D., Kroeker, W.D. and Laskowski, M. Sr. 1976. Mung bean nuclease I. Physical, chemical, and catalytic properties. Biochemistry 15: 4457–4463.

Kroeker, W.D., Hanson, D.M. and Fairley, J.L. 1975. Activity of wheat seedling nuclease toward single-stranded nucleic acids. J. Biol. Chem. 250: 3767–3772.

Kuligowska, E., Klarkowska, D. and Szarkowski, J.W. 1988. Purification and properties of endonuclease from wheat chloroplasts, specific for single-stranded DNA. Phytochemistry 27: 1275–1279.

Liao, T.-H. 1977. Isolation and characterization of multiple forms of malt deoxyribonuclease. Biochemistry 16: 1469–1474.

Lopes, M.A. and Larkins, B.A. 1993. Endosperm origin, development, and function. Plant Cell 5: 1383–1399.

Maekawa, K., Tsunasawa, S., Dibó, G. and Sakiyama, F. 1991. Primary structure of nuclease P1 from *Penicillium citrinum*. Eur. J. Biochem. 200: 651–661.

Mittler, R. and Lam, E. 1995. Identification, characterization, and purification of a tobacco endonuclease activity induced upon hypersensitive response cell death. Plant Cell 7: 1951–1962.

Mittler, R. and Lam, E. 1997. Characterization of nuclease activities and DNA fragmentation induced upon hypersensitive response cell death and mechanical stress. Plant Mol. Biol. 34: 209–221.

Mittler, R., Simon, L. and Lam, E. 1997. Pathogen-induced programmed cell death in tobacco. J. Cell Sci. 110: 1333–1344.

Muramoto, Y., Watanabe, A., Nakamura, T. and Takabe, T. 1999. Enhanced expression of a nuclease gene in leaves of barley plants under salt stress. Gene 234: 315–321.

Nakamura, S., Ogawa, K. and Kuroiwa, T. 1987. Survey of Ca^{2+}-dependent nuclease in green plants. Plant Cell Physiol. 28: 545–548.

Naseem, I. and Hadi, S.M. 1987. Single-strand-specific nuclease of pea seeds: glycoprotein nature and associated nucleotidase activity. Arch. Biochem. Biophys. 255: 437–445.

Oleson, A.E., Janski, A.M. and Clark, E.T. 1974. An extracellular nuclease from suspension cultures of tobacco. Biochim. Biophys. Acta 366: 89–100.

Oleson, A.E., Janski, A.M., Fahrlander, P.D. and Wiesner, T.A. 1982. Nuclease I from suspension-cultured *Nicotiana tabacum*: purification and properties of the extracellular enzyme. Arch. Biochem. Biophys. 216: 223–233.

Panavas, T., Pikula, A., Reid, P.D., Rubinstein, B. and Walker, E.L. 1999. Identification of senescence-associated genes from daylily petals. Plant Mol. Biol. 40: 237–248.

Sakahira, H., Enari, M. and Nagata, S. 1998. Cleavage of CAD inhibitor in CAD activation and DNA degradation during apoptosis. Nature 391: 96–99.

Shiokawa, D., Ohyama, H., Yamada, T., Takahashi, K. and Tanuma, S. 1994. Identification of an endonuclease responsible for apoptosis in rat thymocytes. Eur. J. Biochem. 226: 23–30.

Sung, S.-C. and Laskowski, M. Sr. 1962. A nuclease from mung bean sprouts. J. Biol. Chem. 237: 506–511.

Thelen, M.P. and Northcote, D.H. 1989. Identification and purification of a nuclease from *Zinnia elegans* L.: a potential molecular marker for xylogenesis. Planta 179: 181–195.

Uchida, H., Wu, Y.-D., Takadera, M., Miyashita, S. and Nomura, A. 1993. Purification and some properties of plant endonuclease from scallion bulbs. Biosci. Biotechnol. Biochem. 57: 2139–2143.

Volbeda, A., Lahm, A., Sakiyama, F. and Suck, D. 1991. Crystal structure of *Penicillium citrinum* P1 nuclease at 2.8 Å resolution. EMBO J. 10: 1607–1618.

Wang, M., Oppedijk, B.J., Lu, X., Duijn, B.V. and Schilperoort, R.A. 1996. Apoptosis in barley aleurone during germination and its inhibition by abscisic acid. Plant Mol. Biol. 32: 1125–1134.

Wang, M., Oppedijk, B.J., Caspers, M.P.M., Lamers, G.E.M., Boot, M.J., Geerlings, D. N.G., Bakhuizen, B., Meijer, A.H. and Duijn, B.V. 1998. Spatial and temporal regulation of DNA fragmentation in the aleurone of germinating barley. J. Exp. Bot. 49: 1293–1301.

Wani, A.A. and Hadi, S.M. 1979. Partial purification and properties of endonucleases from germinating pea seeds specific for single-stranded DNA. Arch. Biochem. Biophys. 196: 138–146.

Westhuizen, A.J., Gliemeroth, A.K., Wenzel, W. and Hess, D. 1987. Isolation and partial characterization of an extracellular nuclease from pollen of *Petunia hybrida*. J. Plant Physiol. 131: 373–384.

Wilson, C.M. 1975. Plant nucleases. Annu. Rev Plant Physiol. 26: 187–208.

Wilson, C.M. 1982. Plant nucleases: biochemistry and development of multiple molecular forms. Isozymes Curr. Topics Biol. Med. Res. 6: 33–54.

Wood, M., Power, J.B., Davey, M.R., Lowe, K.C. and Mulligan, B.J. 1998. Factors affecting single strand-preferring nuclease activity during leaf aging and dark-induced senescence in barley (*Hordeum vulgare* L.). Plant Sci. 131: 149–159.

Wyen, N.V., Erdei, S. and Farkas, G.L. 1971. Isolation from *Avena* leaf tissues of a nuclease with the same type of specificity towards RNA and DNA: accumulation of the enzyme during leaf senescence. Biochim. Biophys. Acta 232: 472–483.

Yen, C.-H. and Yang, C.-H. 1998. Evidence for programmed cell death during leaf senescence in plants. Plant Cell Physiol. 39: 922–927.

Yen, Y. and Green, P.J. 1991. Identification and properties of the major ribonucleases of *Arabidopsis thaliana*. Plant Physiol. 97: 1487–1493.

Young, T.E., Gallie, D.R. and DeMason, D.A. 1997. Ethylene-mediated programmed cell death during maize endosperm development of wild-type and *shrunken2* genotypes. Plant Physiol. 115: 737–751.

Yupsanis, T., Eleftheriou, P. and Kelepiri, Z. 1996. Separation and purification of both acid and neutral nucleases from germinated alfalfa seeds. J. Plant Physiol. 149: 641–649.

Plant Molecular Biology **44:** 399–415, 2000.
E. Lam, H. Fukuda and J. Greenberg (Eds.), Programmed Cell Death in Higher Plants.

Plant proteolytic enzymes: possible roles during programmed cell death

Eric P. Beers*, Bonnie J. Woffenden and Chengsong Zhao
Department of Horticulture, Virginia Polytechnic Institute and State University, Blacksburg, VA 24061, USA
*(*author for correspondence; fax 540-231-3083; e-mail: ebeers@vt.edu)*

Key words: endopeptidases, peptidases, plants, programmed cell death, proteases, proteinases

Abstract

Proteolytic enzymes are known to be associated with developmentally programmed cell death during organ senescence and tracheary element differentiation. Recent evidence also links proteinases with some types of pathogen- and stress-induced cell suicide. The precise roles of proteinases in these and other plant programmed cell death processes are not understood, however. To provide a framework for consideration of the importance of proteinases during plant cell suicide, characteristics of the best-known proteinases from plants including subtilisin-type and papain-type enzymes, phytepsins, metalloproteinases and the 26S proteasome are summarized. Examples of serine, cysteine, aspartic, metallo- and threonine proteinases linked to animal programmed cell death are cited and the potential for plant proteinases to act as mediators of signal transduction and as effectors of programmed cell death is discussed.

Introduction

Proteolytic enzymes are known to be principal players in animal cell suicide programs. Less is known about the proteolytic enzymes associated with plant programmed cell death (PCD). This review provides background information about the roles of serine, cysteine, aspartic, metallo- and threonine proteinases as regulators of animal PCD and considers the existing evidence linking these enzymes to plant PCD in developing flowers, senescing organs and differentiating tracheary elements and in response to stress. The discovery that cysteine proteases related to the *ICE/CED3* family are important regulators of some types of cell death (Yuan *et al.*, 1993) led to a rapid increase in efforts to understand the roles of proteases during PCD. Although the exact nature of the involvement of cell death proteases remains obscure, it is understood that cysteine proteases with specificity for Asp residues (the caspases) and other proteolytic systems (the 26S proteasome, granzyme B, calpain, cathepsin D and matrix metalloproteinases) are important to PCD processes (Greenberg, 1996; Thornberry *et al.*, 1997; Werb, 1997; Grimm and Osborne, 1999; Roberts *et al.*, 1999). The processes leading to PCD can be divided into three connected parts: inducing signals, signal transduction pathways and a final common effector or execution pathway (Roberts *et al.*, 1999). Proteases may generate inducing signals specific to PCD by processing or releasing bioactive molecules or by activating cell surface receptors. It is clear that proteases, specifically the caspases, have major roles in the signal transduction process (Muzio, 1998). With regard to the execution pathway, it appears that two functionally distinct groups of proteins are the targets of cell death proteases: those involved in the organization and maintenance of cell structure, and enzymes that function in homeostatic pathways (Thornberry *et al.*, 1997). In contrast to what is known from animal model systems, there are currently no reports that directly establish a principal role for a specific protease in plant cell suicide. There are, however, numerous accounts that link mechanistically distinct proteases with plant cell suicide pathways. Because high background levels of proteases and a lack of specific inhibitors often prevent definitive identification of the enzymes being investigated, this review will focus primarily on reports describing proteases (at the protein or mRNA/cDNA levels) that can be assigned to specific families within the five sub-subclasses listed below.

There are several terms commonly used to describe proteolytic enzymes. Peptidase, the general term for both exopeptidase and endopeptidase, is the term recommended by the Nomenclature Committee of the International Union of Biochemistry and Molecular Biology (NCIUBMB) for any enzyme that hydrolyzes peptide bonds. Exopeptidases act near the terminus of polypeptide chains and endopeptidases act internally in polypeptide chains. The terms 'protease' and 'proteinase' are frequently used in the literature and are synonymous with peptidase and endopeptidase, respectively. For recent reviews of plant proteolysis, readers are referred to articles by Vierstra (1996) and Callis (1997). The most thoroughly characterized cell death proteases are endopeptidases, which are divided into the following sub-subclasses based on catalytic mechanism: serine endopeptidases (EC 3.4.21), cysteine endopeptidases (EC 3.4.22), aspartic endopeptidases (EC 3.4.23), metalloendopeptidases (EC 3.4.24), and threonine endopeptidases (EC 3.4.25). The peptidase family designations used in this review are those of Rawlings and Barrett (1999).

Serine endopeptidases (EC 3.4.21)

Evidence from animal models supporting a role for serine endopeptidases during PCD is limited to granzyme B, an S1 family, trypsin-type enzyme (Greenberg, 1996). A trypsin-like activity was recently implicated during plant PCD. By assaying the culture medium from differentiating tracheary elements, Groover and Jones (1999) detected the presence of a secreted 40 kDa peptidase. The 40 kDa enzyme was maximally active at pH 5.0 and was inhibited by soybean trypsin inhibitor. The appearance of this peptidase in the culture medium was coordinated with secondary cell wall synthesis. Interestingly, PCD of tracheary elements was prevented by the inclusion of trypsin inhibitor in the culture medium. This observation prompted Groover and Jones (1999) to conclude that extracellular proteolysis is essential to PCD during tracheary element differentiation, analogous perhaps to PCD of the epidermal keratinocyte, also reported to be dependent on a secreted serine peptidase (Marthinuss *et al.*, 1995). The chloroplastic, ATP-dependent ClpP protease (family S14) has been evaluated with respect to its role during senescence, and it does not appear to have a specialized role as a regulator of senescence (Shanklin *et al.*, 1995; Crafts-Brandner *et al.*, 1996; Weaver *et al.*, 1999). Plant subtilisin- and kexin-type (family S8) serine endopeptidases have been the focus of several investigations during the past 5 years. Animal and yeast kexin-type serine endopeptidases are proprotein convertases involved in the generation of bioactive peptides by proteolytic processing of inactive precursors (Seidah *et al.*, 1998). As plant kexin-type enzymes have been implicated in plant defense responses and are associated with plant PCD, they are considered here in some detail.

Plant subtilisins are sometimes referred to as cucumisin-like peptidases, in recognition of the first subtilisin isolated from a plant (Yamagata *et al.*, 1994). Totally conserved residues are those of the catalytic triad, Asp, His and Ser. A schematic diagram of the structure of the tomato protein P69B (Tornero *et al.*, 1997) is shown in Figure 1 as a representative example of plant subtilisin-type peptidases. Subtilisins are synthesized as preproenzymes (zymogens). *P69B* is predicted to encode a 745 amino acid residue protein with a 22-residue signal sequence and a 92-residue prodomain. After cleavage of the signal peptide, subtilisins are ultimately activated by cleavage of the propeptide producing the mature active enzyme (Taylor *et al.*, 1997). For the plant enzymes, it is not yet known whether the prodomain is cleaved via an intra- or inter-molecular mechanism. Subtilisins from other sources are capable of intra-molecular prodomain cleavage (Shinde *et al.*, 1995), and the prodomain can remain associated with the mature enzyme as a competitive inhibitor prior to final degradation of the prodomain and release of the active mature protein (Jordan *et al.*, 1995). In some cases, C-terminal extensions may be autolytically processed (e.g., Yamagata *et al.*, 1994). Plant subtilisins are also characterized by large (ca. 220–250 residue) insertions within the catalytic domain between the stabilizing Asn and reactive Ser residues. These insertions are of unknown function and account for a greater than two-fold increase in catalytic domain size of plant subtilisins relative to that of bacterial enzymes (Siezen and Leunissen, 1997).

Plant cell wall-localized enzymes with kexin-like cleavage specificity have been partially characterized. The S1 subsite of the proprotein convertase kexin is highly selective for Arg. For example, the best synthetic substrates for kexin are based on cleavage sites in pro-α-factor: Pro-Met-Tyr-Lys-Arg/Glu-Ala-Glu-Ala (Fuller, 1998). (Peptide bond cleavage is indicated

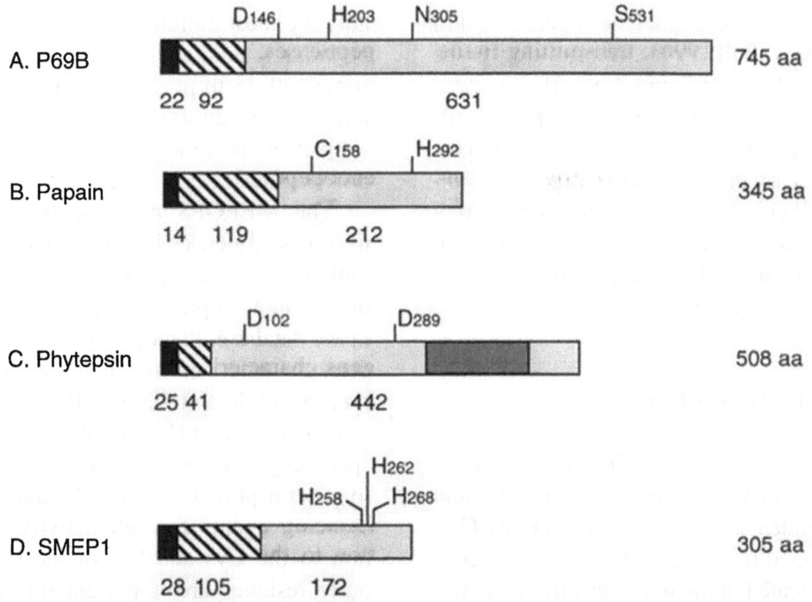

Figure 1. Diagram of three-domain structure of endopeptidases. A. Serine endopeptidase, subtilisin type P69B (Tornero *et al.*, 1997). B. Cysteine endopeptidase, papain. C. Aspartic endopeptidase, barley phytepsin (Glathe *et al.*, 1998). D. Metalloendopeptidase, soybean SMEP1 (Pak *et al.*, 1997). Identities and positions of catalytic amino acid residues are shown above each diagram. The non-catalytic stabilizing Asn residue is also shown for P69B. Black bar, a possible signal sequence; striped bar, N-terminal pro-sequence; light gray bar, mature protein; dark gray bar, plant-specific domain.

by '/'.) Cleavage of systemin *in vitro* at its Lys-Arg/Asp site by a tomato membrane-bound peptidase has been reported, and a systemin-binding protein, also localized to the plasma membrane, is recognized by an antiserum against a *Drosophila* kexin-type peptidase (Schaller and Ryan, 1994). In addition, the KP6 antifungal toxin, when synthesized and secreted by transgenic tobacco plants, was processed at its Lys-Arg/Gly site, consistent with the colocalization of KP6 and a kexin-type serine endopeptidase in the cell wall (Kinal *et al.*, 1995). The results with systemin and KP6 combined with the observation that the plant subtilisin P69 catalyzes cleavage of a cell wall leucine-rich repeat protein (Tornero *et al.*, 1996) indicate that plant kexin-type proteins may process bioactive cell wall peptides to initate signal transduction pathways that originate in the extracellular matrix.

Plant subtilisins have been cloned from melon (*Cucumis melo*) (Yamagata *et al.*, 1994), tomato (Tornero *et al.*, 1997; Meichtry *et al.*, 1999), *Alnus* (Ribeiro *et al.*, 1995), lilly (Kobayashi *et al.*, 1994; Taylor *et al.*, 1997) and *Arabidopsis* (Ribeiro *et al.*, 1995; Neuteboom *et al.*, 1999; Zhao *et al.* 2000). Subtilisins exist as large multigene families in *Arabidopsis* and tomato (Meichtry *et al.*, 1999). Distinct expression patterns for each of the five tomato subfamilies were observed at the organ level (Meichtry *et al.*, 1999), and characterization of a genomic cluster of 4 genes revealed pathogen- and development-specific regulation of plant subtilisins (Jorda *et al.*, 1999). Together these data suggest requirements for subtilisins in a number of proteolytic events throughout the life of a plant.

There is no direct evidence supporting a role for the family S8 subtilisins in plant PCD. However, cell death-associated expression of plant subtilisins has been noted. The lily subtilisin, LIM9, accumulates in anthers and specifically in the tapetum (Taylor *et al.*, 1997). The tapetum is just one of five cell types (the circular cell cluster, the tapetum, epidermal cells adjacent to the stomium, the endothecium, and the connective) within anthers that undergo degeneration leading to dehiscence (Bonner and Dickinson, 1989). The expression of subtilisin *LeSBT1* in roots and flowers of tomato (Meichtry *et al.*, 1999) may indicate a role for this plant subtilisin during PCD in these organs where several cell types are engaged in cell suicide. For example, the emergence of lateral roots has been correlated with apoptosis in adjacent cortical cells of the primary root (Kosslak *et al.*, 1997) and subtilisin mRNA accumulates during lateral root emergence (Neuteboom *et al.*, 1999). In addition to anther cell death, examples of PCD in specific cell types of flowers and developing seeds include degeneration of haploid megaspores (Bell, 1996), transmitting tissue cells (Wang *et al.*, 1996), synergids (recently reviewed by Beers, 1997) and nucellar and endosperm cells (Young and Gallie, 2000). Finally, characterizations of PCD during tracheary element differentiation in *Zinnia* mesophyll cell cultures have reported activity of a 60 kDa (the approximate expected molecular mass for mature plant subtilisin) serine endopeptidase (Ye and Varner, 1996 Beers and Freeman, 1997).

Cysteine endopeptidases (EC 3.4.22)

The two families of cysteine endopeptidases known to be important in animal PCD, caspase (family C14) and the calcium-dependent peptidase calpain (family C2), have not been cloned from plants. Caspase-like activity has been implicated in the hypersensitive response (del Pozo and Lam, 1998) and caspases are reviewed elsewhere in this issue (Lam and del Pozo, 2000). Calcium-dependent peptidase activities have been detected in root (Safadi *et al.*, 1997) and stromal (Bushnell *et al.*, 1993) extracts, and calpeptin, an inhibitor of calpain, prevents victorin-induced proteolysis of the large subunit of Rubisco (Navarre and Wolpert, 1999). Two additional families of cysteine endopeptidases, ubiquitin (Ub) C-terminal hydrolases (family C12) and Ub peptidases (family C19), are components of the Ub-proteasome-dependent pathway of proteolysis (see Threonine peptidases). Although members from both families cleave after the C-terminal Gly of Ub, C19 enzymes are able to cleave Ub from larger leaving groups, i.e. C19 enzymes catalyze the deubiquitination of proteins in a regulatory process that has been compared to reversible protein phosphorylation (Wilkinson, 1997). As is true for protein kinases/phosphatases, Ub peptidases have been linked to several vital cellular processes (D'Andrea and Pellman, 1998). To date, however, functional significance in plant or animal PCD has not been demonstrated for these enzymes. Recent work localizing barley nucellain, a legumain homologue (family C13), to the walls of degenerating cells in the nucellus has raised the question of whether this apoplastic peptidase contributes to PCD (Linnestad *et al.*, 1998). Prior to this report, legumains were thought of as enzymes involved in storage protein degradation and proprotein processing (Shimada *et al.*, 1994) in dicots, where they are known as vacuolar processing enzymes (VPEs). Papain-like peptidases (family C1) are unquestionably the most thoroughly investigated family of plant peptidases, and, as these enzymes appear to regulate some cell death programs in plants and are probably required for autolysis during PCD, they will be the subject of the remainder of this section on cysteine endopeptidases.

The MEROPS database (http://www.merops.co.uk/) lists sequences for more than 50 plant papain homologues from 26 species, making this the largest and most widely represented plant endopeptidase family in the database. Plant papain-type enzymes are zymogens characterized by small-size (40–50 kDa for the prepropeptide and 22–35 kDa for the mature, active enzyme), acidic pH optimum, broad *in vitro* substrate specificity, sensitivity to cysteine peptidase inhibitors (e.g. leupeptin, E-64, TPCK) and a requirement for reducing agents for full activity *in vitro*. In addition to the Cys and His of the catalytic dyad, two other residues are important for catalysis, Gln and either Asn or Asp. The order of these residues in the sequence is Gln, Cys, His, Asn/Asp. Although these peptidases are considered to have broad substrate specificity, there is preference by papain for substrates containing a bulky non-polar side-chain (e.g. Phe) at the P2 position and a partial preference by the S1 subsite for Lys and Arg (Menard and Storer, 1998, and references therein).

Knowing the subcellular localization of peptidases may provide clues to the nature of the co-localized potential substrates and suggest specific functions performed by the peptidase during PCD. Papain-type enzymes possess short signal peptides that presumably direct co-translational delivery to the lumen of the endoplasmic reticulum (ER) (Mitsuhashi and Minamikawa, 1989). Experiments with purified native proteins, recombinant proteins and antibodies against well characterized plant papain-type enzymes have defined different secretory pathway localizations for plant papain-type enzymes including the lumen of the ER (Okomoto *et al.*, 1999), a putative lytic compartment derived from the ER, the ricinosome, of castor bean (Schmid *et al.*, 1998), the vacuole (Holwerda *et al.*, 1990, 1992) and the apoplast (Holwerda *et al.*, 1992; Jones and Mullet, 1995; Koehler and Ho, 1988). Plant papain-type peptidases appear to be processed from propeptides to active peptidases in multi-step fashion in post-Golgi, lytic compartment(s) (Holwerda *et al.*, 1990; Okomoto *et al.*, 1999), such as the vacuole (Okomoto *et al.*, 1999) or the cell wall. These enzymes are not expected to be active in the lumen of the ER where the pH and reducing conditions are not favorable.

Based on the amino acid sequences of their prodomains, two distinct subfamilies of papain-type enzymes can be recognized (Karrer *et al.*, 1993). The first, similar to animal cathepsins H and L, is characterized by a ca. 110-residue propeptide containing an $\mathbf{E}X_3\mathbf{R}X_3\mathbf{F}X_2\mathbf{N}X_3\mathbf{I}X_3\mathbf{N}$ (ERFNIN) motif. The second, similar to cathepsin B, has a shorter propeptide (ca. 60 residues) and lacks the ERFNIN signature. The prodomains of these two subfamilies show no sequence homology, and yet, crystal structures of enzymes in these subfamilies are similar (Groves *et al.*, 1998, and references therein), revealing that the prodomain extends along the substrate-binding site, occupying S and S′ subsites. The orientation of the prodomain, however, is opposite to that of the peptidyl substrate. This antiparallel orientation has the effect of protecting the prodomain from cleavage while shielding the active site from access by the substrate. This mechanism of enzyme inhibition is specific with regard to prodomain-parent enzyme interactions and represents an exciting area of investigation concerning the design of specific inhibitors (Groves *et al.*, 1998). Three other functions of papain-type prodomains are known: (1) proper protein folding (Tao *et al.*, 1994), (2) stabilization of the protein in neutral or alkaline environments (Mach *et al.*, 1994), and (3) targeting of the enzyme. The latter case was demonstrated for a plant enzyme, aleurain (Holwerda *et al.*, 1992), where the prodomain contained an N-terminal vacuolar targeting domain, NPIR, also conserved in a gibberellin-induced rice seed enzyme, oryzain γ (Watanabe *et al.*, 1991) and tomato (Drake *et al.*, 1996) and maize (Griffiths *et al.*, 1997) senescent leaf papain-type peptidases.

Some plant papain homologues also possess C-terminal prodomains, such as actinidin (Paul *et al.*, 1995), rice oryzains α and β (Watanabe *et al.*, 1991) and pea Tpp (Granell *et al.*, 1992). For actinidin, the C-terminal extension was necessary for proper autolytic cleavage of the proprotein (Paul *et al.*, 1995). Several plant papain-type enzymes possess putative ER retention signals at their C-termini, i.e. KDEL, RDEL or TDEL residues (Valpuesta *et al.*, 1995; Nadeau *et al.*, 1996; Becker *et al.*, 1997; Guerrero *et al.*, 1998). The KDEL sequence can be post-translationally removed (Becker *et al.*, 1997; Okomoto *et al.*, 1999), perhaps allowing accumulated inactive proprotein to proceed to the vacuole or other acid compartment for activation (Okomoto *et al.*, 1999). Removal of KDEL C-termini, however, is not always necessary for secretion of plant proteins to occur (Jones and Herman, 1993).

Primary sequence motifs and posttranslational processing of papain-type enzymes are clearly important determinants of localization and activation. Nevertheless, promoter analysis has shown that a very diverse set of factors is known to regulate plant papain-type enzymes at the transcriptional level as well. Promoters for two intronless barley genes, *EPB1* and *EPB2* (Mikkonen *et al.*, 1996), and wheat gene *A121* (Cejudo *et al.*, 1992) are sufficient to confer GA-responsive expression of GUS in a barley transient expression assay. For all three promoters, ABA could reverse the effect of GA. Promoter-reporter fusions based on other genes for papain-type peptidases have demonstrated developmental regulation in germinating seeds (Yamauchi *et al.*, 1996), fruit (Lin *et al.*, 1993) and senescing leaves (Gan and Amasino, 1995). In cereal aleurone, a MYB transcription factor activates expression of a cathepsin B-like peptidase, as well as other hydrolases (Gubler *et al.*, 1999).

Numerous examples of increases in papain-type peptidase activity in developing and germinating seeds and growing seedlings have been reported (recently reviewed by Granell, 1998). Both cathepsin L/H-type (Watanabe *et al.*, 1991; Becker *et al.*, 1994; Domoto *et al.*, 1995; Kato and Minamikawa, 1996; Tranbarger and Misra, 1996; Shintani *et al.*, 1997) and cathepsin B-type (Cejudo *et al.*, 1992) papain homologues have been cloned from embryos and germinating seeds. The degradation of seed storage proteins by papain-type enzymes has been recognized for several plant species (Shutov and Vaintraub, 1987; de Barros and Larkins, 1990; Koehler and Ho, 1990; Kato and Minamikawa, 1996). However, proteinase A from vetch, originally implicated in early stages of storage protein cleavage, is neither colocalized with storage proteins in protein bodies nor are its expression and accumulation coordinated with storage protein degradation (Becker *et al.*, 1997). This absence of temporal and spatial coordination of proteinase A with its proposed substrate indicates that *in vitro* studies with peptidases and potential substrates should be interpreted with caution, especially in the absence of subcellular localization data.

Recent efforts to identify leaf, flower and fruit programmed senescence genes have resulted in the cloning of several full-length cDNAs predicting papain homologues expressed in senescing organs from daylily (Valpuesta *et al.*, 1995; Guerrero *et al.*, 1998),

Arabidopsis (Lohman *et al.*, 1994), maize (Griffiths *et al.*, 1997), carnation (Jones *et al.*, 1995b), tomato (Drake *et al.*, 1996), eggplant (Xu and Chye, 1999) and kiwi (Lin *et al.*, 1993). Based on the presence of the ERFNIN motif, these papain-type enzymes are similar to mammalian cathepsins H and L rather than to cathepsin B (Karrer *et al.*, 1993). In addition to germinating seeds and senescing organs, a diverse set of organs, tissues and cell types undergoing PCD exhibit increased accumulation of transcripts encoding papain-type peptidases. These include several cell types in tobacco anthers (Koltunow *et al.*, 1990), senescing ovaries of pea (Granell *et al*, 1992; Cercos *et al.*, 1999), senescing pea nodules (Kardailsky and Brewin, 1996) and differentiating tracheary elements (Ye and Varner, 1996; Xu and Chye, 1999). Increases in cysteine peptidase activity and gene expression associated with environmental stress conditions and plant-pathogen interactions have also been noted (Koizumi *et al.*, 1993; Williams *et al.*, 1994; Jones and Mullet, 1995; Moriyasu and Ohsumi, 1996; D'Silva *et al.*, 1998; Solomon *et al.*, 1999).

That some papain-type proteins are targeted to the large central vacuole or other intracellular lytic compartments is consistent with peptidases that function both in routine intracellular protein turnover and in specialized roles during PCD. The rationale for such dual functionality for single-gene products lies in the proposed ability of vacuoles or their related provacuoles and secondary vacuoles to serve as (1) lytic compartments analogous to the animal lysosomal system (Matile, 1978; Marty *et al.*, 1980; Swanson *et al.*, 1998) when intact, and (2) reservoirs of hydrolases that, upon tonoplast rupture, are released into the cytosol. Release of hydrolases into the cytosol may serve a signaling function or provide cell death/autolysis effectors. Plasma membrane or tonoplast rupture is a recognized marker for some cell death programs, such as synergid death, suspensor death, pith autolysis, tracheary element differentiation, leaf and petal senescence (reviewed by Beers, 1997) and programmed death of barley aleurone cells (Bethke *et al.*, 1999). Moreover, all vacuoles are not created equal, and the presence of small, cysteine peptidase-containing vacuoles of unknown function is developmentally regulated (Paris *et al.*, 1996; Swanson *et al.*, 1998) and may also be controlled by carbon status (Moriyasu and Ohsumi, 1996). The role played by extracellular papain-type enzymes in the events surrounding PCD is even less certain. They may catalyze the production of signaling molecules as proposed for kexin-type (Schaller and Ryan, 1994) or trypsin-like (Groover and Jones, 1999) serine peptidases, or they may simply degrade peptides that leak from dying cells (microheterophagy).

Recent biochemical evidence supports requirements for cysteine endopeptidases in the regulation of stress-induced PCD and during autolysis of intracellular components. Solomon *et al.* (1999) reported the activation of cysteine peptidases in soybean cells in response to oxidative stress sufficient to trigger PCD. The partially purified PCD-associated cysteine peptidase was sensitive to cystatin, a protein inhibitor of papain-type enzymes. Ectopic expression of cystatin effectively blocked H_2O_2-induced PCD. Synthetic inhibitors of papain-type enzymes, however, did not effectively block H_2O_2-induced PCD (Levine *et al.*, 1996). Synthetic inhibitors of cysteine peptidases inhibited intracellular protein degradation during sucrose starvation-induced autolysis (Moriyasu and Ohsumi, 1996). A peptide-aldehyde inhibitor of papain-type enzymes prevented the complete removal of intracellular contents of differentiated tracheary elements (Woffenden *et al.*, 1998), suggesting that papain-type enzymes are necessary for autolysis during plant PCD.

Aspartic endopeptidases (EC 3.4.23)

Other than a small number of retrotransposon endopeptidases (families A11 and A12), phytepsins (plant pepsin-type enzymes, family A1) are the only plant aspartic endopeptidases listed in the MEROPS database, and the only aspartic peptidase family implicated in plant PCD. Most of the available information about phytepsin comes from characterizations of enzymes from barley, rice, *Cynara cardunculus*, *Arabidopsis* and *Brassica*. A detailed review of plant aspartic peptidases is available (Kervinen, 1995). Aspartic endopeptidases differ from cysteine and serine peptidases in that the nucleophile that attacks the scissile peptide bond of the substrate is an activated water molecule, rather than a nucleophilic side-chain of an amino acid. They also appear to be encoded by small gene families (D'Hondt *et al.*, 1997), rather than the large multigene families observed for papain- and subtilisin-type enzymes. Two catalytic Asp residues (one in each lobe of the bilobed mature protein) act as ligands for the activated water molecule. Cleavage of model substrates (insulin B chain, glucagon and melittin) by barley phytepsin occurs between two residues possessing large hydrophobic side-chains (Leu, Ile, Val, Phe) or next to one hydrophobic residue (Kervinen *et al.*, 1993). Phytepsins also cleave Asp-Tyr (Kervinen *et al.*, 1993) and Asp-Asp (D'Hondt *et al.*, 1993) bonds. Pepstatin is the most commonly used inhibitor of phytepsin. Substrate analogues that inhibit cathpsin D are also effective inhibitors of phytepsin (Sarkkinen *et al.*, 1992). Aspartic peptidases are sometimes referred to as acid proteinases reflecting the acid pH (3.5–4.0) optimum for this family of peptidases.

Phytepsins are synthesized as inactive prepropeptides that exhibit a high degree of similarity to animal cathepsin D (Runeberg-Roos *et al.*, 1991) and microbial aspartic peptidases, with the exception of approximately 100 plant-specific residues near the C-terminus. This C-terminal extension is similar to mammalian soposins (Kervinen *et al.*, 1999). As with papain-type enzymes, the N-terminal prodomain may regulate proteolytic activity by binding to the active site. Processing of barley phytepsin (Figure 1) has been described in detail (Glathe *et al.*, 1998). A signal sequence of 25 residues is removed from the 54 kDa preproenzyme after phytepsin enters the rough ER. The enzyme undergoes proteolytic processing yielding the mature two-chain forms (31 + 15 kDa and 27 + 9 kDa) observed in barley grain. Both autocatalytic and heterocatalytic processing may be required to produce the mature enzyme *in vivo*.

Barley phytepsin is widely distributed, and expression has been detected in seeds, seedlings, flowers, stems, leaves and roots (Tormakangas *et al.*, 1994). Tomato phytepsin expression is increased in response to wounding and methyl jasmonate and systemin application (Schaller and Ryan, 1996). Phytepsin has been localized to vacuoles in leaves, roots and stigmatic papillae cells and to protein bodies in aleurone cells (Runeberg-Roos *et al.*, 1994; Marttila *et al.*, 1995; Bethke *et al.*, 1996; Glathe *et al.*, 1998; Ramalho-Santos *et al.*, 1997) where it may process seed storage proteins (D'Hondt *et al.*, 1993) or lectins (Runeberg-Roos *et al.*, 1994). A pepstatin-sensitive peptidase has been localized to cell walls where it was reported to degrade pathogenesis-related proteins (Rodrigo *et al.*, 1991). Aspartic endopeptidase extracted from the dried flowers of *Cynara carduncculus* is used as a milk-clotting enzyme for the production of a traditional sheep's milk cheese produced in Portugal and Spain (Cordeiro *et al.*, 1995).

The mammalian aspartic endopeptidase cathepsin D has been linked with PCD, suggesting a new role for an enzyme previously thought to be restricted to bulk proteolysis within the lysosomal compartment. For example, cathepsin D-deficient mice experience intestinal atrophy and destruction of lymphoid cells while maintaining bulk lysosomal proteolysis, suggesting that limited proteolysis of biologically active proteins by cathepsin D is important for the regulation of cell turnover (Saftig *et al.*, 1995). High levels of antisense cathepsin D or pepstatin protected cells from interferon-γ- and Fas/APO-1-induced apoptosis, while ectopic expression of cathepsin D induced cell death (Deiss *et al.*, 1996). Evidence linking aspartic endopeptidase with plant PCD has been presented. Aspartic peptidase mRNA levels increase during senescence of leaves (Buchanan-Wollaston, 1997) and petals (Panavas *et al.*, 1999). In barley, phytepsin is expressed in tracheary elements of the metaxylem and in developing sieve elements (Runeberg-Roos and Saarma, 1998). Also in barley, a novel gene encoding an aspartic peptidase-like protein is expressed in nucellar cells after pollination, concomitant with nucellar degeneration (Chen and Foolad, 1997). Although the predicted protein, named nucellin, shares only 20% identity with phytepsins, it exhibits several regions of near or complete identity with phytepsins including both the N- and C-terminal catalytic site motifs for aspartic peptidases.

Metalloendopeptidases (EC 3.4.24)

Compared to the serine, cysteine, aspartic and threonine endopeptidases, relatively little is known about plant metallopeptidases. As is the case for aspartic endopeptidases, metallopeptidases are hydrolases that use a water molecule for the nucleophilic attack on a peptide bond. A divalent cation, usually zinc but sometimes cobalt or manganese, activates the water molecule. The metal ion is held in place by amino acid ligands (His, Glu, Asp or Lys), typically three in number. In addition to these metal ligands, at least one other residue is required for catalysis. That residue is Glu in many of the metalloendopeptidases. Extensive characterizations have been conducted for two chloroplastic metallopeptidases, the chloroplast homologue of the bacterial, ATP-dependent metallopeptidase FtsH (family M14) (Lindahl *et al.*, 1996; Ostersetzer and Adam, 1997) and a chloroplast-processing enzyme (CPE) (Richter and Lamppa, 1998).

A soybean metalloendopeptidase (SMEP1) has been characterized in detail (Graham *et al.*, 1991; McGeehan *et al.*, 1992; Pak *et al.*, 1997) since its original detection by Ragster and Chrispeels (1979). SMEP1, originally identified as an azocoll-degrading, extracellular, EDTA-sensitive enzyme, was purified from leaves of soybean and shown to be a 19 kDa protein localized almost exclusively to the apoplast of leaves and most abundant during the late stages of leaf expansion (Graham *et al.*, 1991). The *in vitro* optimum pH for activity was 8 to 9. The purified protein was sequenced and found to share approximately 40% identity with animal matrix metallopeptidases (MMPs), including the $\mathbf{HEX_2HX_2GX_2HS}$ zinc-binding catalytic motif (McGeehan *et al.*, 1992). Consequently, SMEP1, and a recently cloned MMP from *Arabidopsis* (Liu *et al.*, 1998), have been listed with the M10A family of MMPs. Family M10A includes a number of animal endopeptidases, such as collagenases, gelatinases and stromalysins, involved in the degradation of extracellular matrix (ECM) proteins. A diagram of the SMEP1 zymogen is shown in Figure 1. As is true for the other peptidases shown in Figure 1, the prodomain is likely to serve as a regulatory domain (Stocker *et al.*, 1995, and references therein). When tested against fluorogenic and chromogenic synthetic peptide substrates developed for the human fibroblast MMP, collagenase (Stack and Gray, 1989), SMEP1 cleaved both substrates at the expected Gly/Leu scissile bond. In addition, a naturally occurring mammalian tissue inhibitor of metallopeptidases, TIMP-1, was shown to be a potent inhibitor of SMEP1 (McGeehan *et al.*, 1992).

Although the roles of M10 peptidases in plants are not yet known, work with animal MMPs suggests that plant enzymes such as SMEP1 and AtMMP may be important regulators of growth and development and may even participate in plant PCD. Degradation of ECM proteins is required during morphogenesis, differentiation and wound healing and also is a component of pathological conditions such as metastasis and tumor invasion (for a recent review see Werb, 1997). Extracellular metallopeptidases also modulate some types of PCD, and the balance between MMPs and TIMPs is likely to play an important role in intracellular ECM interactions during PCD. For example, apoptosis associated with postlactational involution of the mammary gland is a process that depends on extracellular remodeling (Lefebvre *et al.*, 1992). In mice, the MMP stromalysin-3 was not expressed in mammary glands during gestation or lactation. After weaning, however, stromalysin-3 (Lefebvre *et al.*, 1992) and stromalysin-1 (Lund *et al.*, 1996) were specifically expressed in fibroblasts adjacent to degenerating mammary ducts, consistent with a role for these ECM peptidases in basement membrane remodeling common to apoptosis. Matrix metallopeptidase-induced apoptosis in mammary glands was suppressed by co-expression of a TIMP-1 transgene (Alexander *et al.*, 1996). Whether plant MMPs have similar important roles in plant growth and development remains to be determined.

Threonine endopeptidases (EC 3.4.25)

Proteasome genes and proteasome functions are being characterized by several labs and consequently several excellent reviews are available, including an extensive review by Coux *et al.* (1996) and a short review by DeMartino and Slaughter (1999). The proteasome was also recently reviewed within the context of protein degradation in plants by Callis (1997) and Vierstra (1996). Rather than duplicate information in these recent reviews, we will provide a brief introduction to the 20S and 26S proteasome followed by a review of the proteasome as a regulator of PCD in animals and as a regulator of growth, development and PCD in plants.

The proteasome belongs to the T1 family of N-terminal nucleophile (Ntn) hydrolases. The 20S proteasome consists of four stacked heptameric rings arranged such that two adjacent interior rings of β subunits are each flanked by a ring of exterior α subunits ($\alpha_7\beta_7\beta_7\alpha_7$). Together the α and β subunits form a 700 kDa cylinder through which unfolded polypeptides are channeled and cleaved internally to produce peptides of 4 to 10 residues in length (Dick *et al.*, 1991; Akopian *et al.*, 1997). In eukaryotes the 20S proteasome subunits are not identical, but they are homologous. For example, in *Arabidopsis*, a superfamily of 23 genes encodes the 14 proteasome subunits (Genschik *et al.*, 1992; Shirley and Goodman, 1993; Fu *et al.*, 1998). Association of the 20S proteasome with another multisubunit protein, the 19S cap, creates the ATP-dependent, 2100 kDa, 26S proteasome. Partial characterizations of the 19S cap have shown that at least one of its functions is to recognize ubiquitinated proteins and present them to the 20S catalytic core for degradation.

For a long time after its discovery, the proteasome's catalytic mechanism was unclear. This was because the inhibitor sensitivities of the assembled proteasome and the amino acid sequences for its subunits were unlike those of known peptidases. It was Lowe *et al.* (1995) who determined that binding of the peptide-aldehyde-inhibitor Ac-LLnL-al occurred at the N-terminal threonine of each β subunit, suggesting that the hydroxyl group of this threonine is the reactive nucleophile catalyzing peptide bond cleavage. The mature, active β subunit has a single threonine residue active site at the N-terminus, which becomes exposed and capable of catalysis through autolytic processing of the propeptide. The catalytically inactive α subunits possess an N-terminal extension in place of the prosequence and are not processed. Three main types of proteolytic specificities are known for the eukaryotic proteasome: after basic residues (trypsin-like), after large hydrophobic residues (chymotrypsin-like), and after acidic residues (peptidylglutamyl peptide-hyrolyzing, PGPH) (Wilk and Orlowski, 1980). Two additional cleavage specificities have been reported for mammalian proteasomes: cleavage after branched-chain amino acids (BrAAP activity) and cleavage between small neutral amino acids (SNAAP activity) (Cordozo *et al.*, 1992; Orlowski *et al.*, 1993). The most commonly used synthetic substrate for the proteasome is the fluorogenic peptide Suc-Leu-Leu-Val-Tyr-Amc.

The 26S proteasome is located in both the nucleus and cytoplasm and is responsible for degrading proteins covalently ligated to Ub molecules. Ub is a 76 amino acid protein that becomes covalently ligated to cellular proteins via an isopeptide bond between the C-terminal Gly of Ub and the ϵ-amino group of Lys residues of specifically recognized target proteins (Varshavsky, 1997). Polyubiquitination of proteins is sufficient to target them for degradation by the 26S proteasome. In addition to its critical role in the degradation of ubiquitinated proteins in mammals, the proteasome also provides peptides for MHC class I antigen display (Rock *et al.*, 1994) and is also capable of degrading non-ubiquitinated proteins, such as ornithine decarboxylase (Murakami *et al.*, 1992). The 26S complex is thus considered to be the main non-lysosomal proteolytic pathway of eukaryotic cells.

The role of the proteasome in PCD in animals has been reviewed by Orlowski (1999). One of the best-studied examples of animal PCD that involves changes in the proteasome and its activities is that of intersegmental muscle (ISM) degeneration during metamorphosis of the hawkmoth, *Manduca sexta*. Initial observations included increases in polyubiquitin gene expression (Schwartz *et al.*, 1990), elevated levels of ubiquitinated proteins, and the induction of enzymes involved in Ub attachment to cellular proteins (Haas *et al.*, 1995) coincident with commitment of ISMs to death. Also using the hawkmoth model, Jones *et al.* (1995a) observed the appearance of four newly synthesized proteasome subunits, an eightfold increase in proteasome levels, and increases in two of the peptidase activities of the proteasome (caseinolytic, PGPH). Dawson *et al.* (1995) reported a similar increase in proteasome abundance in ISM cells undergoing PCD, as well as a greater than two-fold increase in the mRNA levels of *MS73*, an ATPase regulatory subunit of the 19S cap of the 26S proteasome. These changes were correlated with dramatic increases in three of the peptidase activities (chymotrypsin-like, trypsin-like, and PGPH) of the 26S proteasome (Dawson *et al.*, 1995). Together, these studies strongly support a role for the proteasome in the degeneration of hawkmoth ISMs.

The hawkmoth studies raise the question of whether plant PCD, and particularly the disassembly of cells during leaf senescence, also depends on the Ub-proteasome pathway. Expression of genes encoding Ub-proteasome pathway components is upregulated in some plant PCD models. Increases in mRNA for Ub-conjugating enzymes (E2s) and Ub (Genschik *et al.*, 1994) and in GUS expression driven by a Ub promoter (Garbarino and Belknap, 1994) were observed in senescing leaves. Other reports, however, do not support a role for the proteasome during senescence. No changes in the levels of several Ub pathway components, including the 26S proteasome component RPN10/MBP1 (van Nocker *et al.*, 1996), were detected during senescence of daylily petals (Stephenson and Rubinstein, 1998). Recently, Bahrami and Gray (1999) cloned an α-type proteasome subunit from tobacco and described its spatial and temporal expression patterns. High levels of expression in young, expanding tissues and decreasing levels during senescence of leaves and flowers led these authors to conclude that the proteasome does not degrade and recycle protein during senescence. The accumulation of Ub and/or ubiquitinated proteins (Li *et al.*, 1995) paralleled that of the papain-type peptidase *TA56* (Koltunow *et al.*, 1990) in degenerating anthers. Levels of Ub and/or ubiquitinated proteins also increased during tracheary element differentiation in wounded *Coleus* stems (Stephenson *et al.*, 1996), and members of one *Arabidopsis* E2 family display vascular tissue-specific expression (Thoma *et al.*, 1996), while members of other *Arabidopsis* E2 families have more generalized expression patterns (Vierstra, 1996). No clear picture concerning the role of the proteasome in plant PCD has emerged from these experiments with senescing organs, degenerating anthers, and differentiating tracheary elements.

The discovery and development of proteasome inhibitors, lactacystin (Omura *et al.*, 1991) and several peptide aldehydes, has led to a proliferation of investigations of the proteasome as a regulator of PCD in animal models. Use of proteasome inhibitors has led to some confusion, however, as treatment may either promote or prevent PCD, depending on the system in question. For example, proteasome activity was shown to be required for PCD of primary mouse thymocytes in response to various cell death inducers (Grimm *et al.*, 1996), of NGF-deprived sympathetic neurons following withdrawal of nerve growth factor (Sadoul *et al.*, 1996), and of T cells activated by T-cell receptor cross-linking (Cui *et al.*, 1997). In contrast, proteasome activity apparently prevents PCD in response to various PCD-eliciting stimuli in certain cancer cell lines, including multiple human leukemia cell lines (Shinohara *et al.*, 1996; Drexler, 1997; Dou *et al.*, 1999; Masdehors *et al.*, 1999; Zhang *et al.*, 1999), human prostatic carcinoma cells (Herrmann *et al.*, 1998), and glioma cells (Kitagawa *et al.*, 1999).

These apparently contradictory data may be explained, in part, by a model summarized by Grimm and Osborne (1999) which suggests that proteasome inhibition may have different effects on proliferating versus non-proliferating cells due to cell cycle regulation by the Ub-proteasome pathway. In non-cycling (quiescent or terminally differentiated) cells, the proteasome may be responsible for directly or indirectly activating the caspase cascade to induce cell death. In this case, inhibition of the proteasome would prevent cell death. In cycling cells, on the other hand, proteasome inhibition results in the accumulation of cell cycle regulators normally degraded via the proteasome. Subsequent dysregulation of the cell cycle may drive cells into PCD, potentially involving participation of the caspase cascade. This model is consistent with much of the available data. It has also been suggested that prevention versus induction of PCD by proteasome inhibitors depends on the concentration of inhibitor (Rock *et al.*, 1994; Lin *et al.*, 1998; Meriin *et al.*, 1998). Obviously, the role of the proteasome in PCD is complex and experiments that do not rely on inhibitors are needed before more detailed models can be proposed and tested.

Inhibitor-based studies of the involvement of the proteasome in plant PCD lag behind those in animal systems. Proteasome activity has been implicated in the development of tracheary elements in cell suspension cultures of *Zinnia* (Woffenden *et al.*, 1998). In the *Zinnia* tracheary element study, application of the proteasome inhibitor lactacystin at culture initiation completely prevented tracheary element differentiation, while inhibition of proteasome activity following commitment to differentiation did not prevent tracheary element formation but resulted in a delay in development. The tracheary element cell death program was neither blocked nor induced prematurely in these experiments. That specific proteasome inhibition late during differentiation did not preclude autolytic clearing of tracheary elements is consistent with the conclusion that the proteasome does not participate in bulk autolysis of differentiating tracheary elements.

Reverse genetic approaches to understanding the functions of the Ub-proteasome pathway in plants are limited to a few examples. The *Arabidopsis* deletion mutant *tt3* which does not express α subunit *PAF1* (Fu *et al.*, 1998) was similar to wild-type plants with respect to seedling germination, growth and fertility (Shirley and Goodman, 1993). Over-expression of a mutant form of Ub unable to form polyubiquitin chains resulted in aberrant vascular tissue (Bachmair *et al.*, 1990) and spontaneous local lesion formation in response to mild stress (Becker *et al.*, 1993). Using homologous recombination to create a gene disruption, Fu *et al.* (1999) demonstrated a requirement for *RPN10/MBP1* during development in the moss *Physcomitrella*. Strains harboring the *RPN10/MBP1* knockout arrested developmentally and were unable to initiate gametophorogenesis to complete their life cycle. These studies support a role for the proteasome in some aspects of plant growth and development possibly including PCD.

Conclusions

This review provides basic information about five mechanistically distinct endopeptidases with an emphasis on plant enzymes. Specific examples illustrating the possible involvement of peptidases in plant PCD have been cited and selected references are presented in Table 1. With the exception of the 26S proteasome, evidence for a cytosolic/nuclear proteolytic system able to fulfill the roles ascribed to the caspases is conspicuously absent from plants. It is possible that undiscovered proteolytic systems exist in plants, perhaps including caspase homologues or analogues. It has also been suggested that specialized vacuoles or other lytic compartments may release into

Table 1. Plant programmed cell death models and selected peptidase references.

PCD model	Peptidase	Reference
Oxidative stress	cysteine	Solomon *et al.*, 1999
Hypersensitive response	cysteine (caspase)	del Pozo and Lam, 1998
Leaf and flower senescence	cysteine	Lohman *et al.*, 1994; Jones *et al.*, 1995b; Drake *et al.*, 1996; Griffiths *et al.*, 1997; Guerrero *et al.*, 1998
	aspartic	Ramalho-Santos *et al.*, 1997; Panavas *et al.*, 1999
Tracheary elements	aspartic	Runeberg-Roos and Saarma, 1998
	cysteine	Minami and Fukuda, 1995; Ye and Varner, 1996; Beers and Freeman, 1997; Xu and Chye, 1999
	serine	Ye and Varner, 1996; Beers and Freeman, 1997; Groover and Jones, 1999
Anther	cysteine	Kultonow *et al.*, 1990; Taylor *et al.*, 1997; Xu and Chye, 1999
Nucellus	cysteine	Linnestad *et al.*, 1998; Xu and Chye, 1999
	aspartic	Chen and Foolad, 1997; Törmäkangas *et al.*, 1994
Aleurone	cysteine	Koehler and Ho, 1988; Holwerda *et al.*, 1990; Holwerda *et al.*, 1992; Swanson *et al.*, 1998

the cytosol peptidases important to the regulation of aleurone PCD (Bethke *et al.*, 1999). Whether such a mechanism does operate in the aleurone or other plant cells and could provide peptidases to fill the niche occupied by caspases remains to be determined. That subtilisins, papain homologues, phytepsins, and plant MMPs all have been localized to the apoplast is consistent with potential roles for these enzymes in cell wall remodeling and signaling important to plant growth, development and PCD. Direct evidence supporting regulatory roles for plant cell wall peptidases has not been presented, however.

An extensive discussion of endogenous protein inhibitors of plant peptidases was not presented here. Yet based on the apparent ability of the cysteine peptidase inhibitor cystatin to inhibit PCD in soybean (Solomon *et al.*, 1999), and the potential for exploiting peptidase prodomains as highly specific inhibitors (Groves *et al.*, 1998), it appears that this area of investigation may soon provide important insights into both natural and engineered regulation of plant PCD. In addition to cystatin, endogenous inhibitors of serine and aspartic peptidases have been characterized (Maganja *et al.*, 1992; Ishikawa *et al.*, 1994; Kreft *et al.*, 1997). Although these inhibitors have been implicated as defense molecules against plant predators, and have been used to enhance pest resistance (Koiwa *et al.* 1998; Urwin *et al.*, 1998), their potential as regulators of endogenous plant peptidases has not been fully explored.

Animal PCD models have set a powerful precedent for the expectation that plant peptidases are important regulators of plant PCD and yet direct genetic evidence in support of this assumption is currently lacking. Now that several plant PCD models are being exploited, and associated peptidase activities have been partially characterized, it is expected that reverse genetics experiments aimed at defining roles for specific peptidase genes will soon confirm the importance of peptidases in plant PCD.

Acknowledgements

We thank Dr. Brenda Winkel for critical reading of the manuscript and Vanessa Funk for reference management. Work in the laboratory of E.P.B. is supported by NSF and NRICGP.

References

Akopian, T.N., Kisselev, A.F. and Goldberg, A.L. 1997. Processive degradation of proteins and other catalytic properties of the proteasome from *Thermoplasma acidophilum*. J. Biol. Chem. 272: 1791–1798.

Alexander, C.M., Howard, E.W., Bissell, M.J. and Werb, Z. 1996. Rescue of mammary epithelial cell apoptosis and entactin degradation by a tissue inhibitor of metalloproteinases-1 transgene. J. Cell Biol. 135: 1669–1677.

Bachmair, A., Becker, F., Masterson, R.V. and Schell, J. 1990. Perturbation of the ubiquitin system causes leaf curling, vascular tissue alteration and necrotic lesions in a higher plant. EMBO J. 9: 4543–4549.

Bahrami, A.R. and Gray, J.E. 1999. Expression of a proteasome α-type subunit gene during tobacco development and senescence. Plant Mol. Biol. 39: 325–333.

Becker, C., Fischer, J., Nong, V.H. and Münitz, K. 1994. PCR cloning and expression analysis of cDNAs encoding cysteine proteinases from germinating seeds of *Vicia sativa* L. Plant Mol. Biol. 26: 1207–1212.

Becker, C., Sneyuk, V.I., Shutov, A.D., Nong, V.H., Fischer, J., Horstmann, C. and Muntz, K. 1997. Proteinase A, a storage-globulin-degrading endopeptide of vetch (*Vicia sativa* L.) seeds, is not involved in early steps of storage-protein mobilization. Eur. J. Biochem. 248: 304–312.

Becker, F., Buschfeld, E., Schell, J. and Bachmair, A. 1993. Altered response to viral infection by tobacco plants perturbed in ubiquitin system. Plant J. 3: 875–881.

Beers, E.P. 1997. Programmed cell death during plant growth and development. Cell Death Differ. 4: 649–661.

Beers, E.P. and Freeman, T.B. 1997. Proteinase activity during tracheary element differentiation in *Zinnia* mesophyll cultures. Plant Physiol. 113: 873–880.

Bell, P.R. 1996. Megaspore abortion: a consequence of selective apoptosis? Int. J. Plant Sci. 157: 1–7.

Bethke, P.C., Hillmer, S. and Jones, R.L. 1996. Isolation of intact protein storage vacuoles from barley aleurone. Plant Physiol. 110: 521–529.

Bethke, P.C., Lonsdale, J.E., Fath, A. and Jones, R.L. 1999. Hormonally regulated programmed cell death in barley aleurone cells. Plant Cell 11: 1033–1045.

Bonner, L.J. and Dickinson, H.G. 1989. Anther dehiscence in *Lycopersicon esculentum* Mill. New Phytol. 113: 97–115.

Buchanan-Wollaston, V. 1997. The molecular biology of leaf senescence. J Exp. Bot. 48: 181–199.

Bushnell, T.P., Bushnell, D. and Jagendorf, A.T. 1993. A purified zinc protease of pea chloroplasts, EP1, degrades the large subunit of ribulose-1,5-bisphosphate carboxylase/oxygenase. Plant Physiol. 103: 585–591.

Callis, J. 1997. Regulation of protein degradation in plants. Genet. Eng. 19: 121–148.

Cejudo, F.J., Ghose, T.K., Stabel, P. and Baulcombe, D.C. 1992. Analysis of the gibberellin-responsive promoter of a cathepsin B-like gene from wheat. Plant Mol. Biol. 20: 849–856.

Cercos, M., Santamaria, S. and Carbonell, J. 1999. Cloning and characterization of TPE4A, a thiol-protease gene induced during ovary senescence and seed germination. Plant Physiol. 119: 1341–1348.

Chen, F. and Foolad, M.R. 1997. Molecular organization of a gene in barley which encodes a protein similar to aspartic protease and its specific expression in nucellar cells during degeneration. Plant Mol. Biol. 35: 821–831.

Cordeiro, M.C., Xue, Z.-T., Pietrzak, M., Paris, M.S. and Brodelius, P.E. 1995. Plant aspartic proteinases from *Cynara cardunculus* ssp. *flavescens* cv. Cardoon; nucleotide sequence of a cDNA encoding cyprosin and its organ-specific expression. In: T. Takahashi (Ed.) Aspartic Proteinases, Plenum, New York, pp. 367–372.

Cordozo, C., Vinitsky, A., Hidalgo, M.C., Michaud, C. and Orlowski, M. 1992. A 3,4-dichloroisocoumarin-resistant component of the multicatalytic proteinase complex. Biochemistry 31: 7373–7380.

Coux, O., Tanaka, K. and Goldberg, A.L. 1996. Structure and functions of the 20S and 26S proteasomes. Annu. Rev. Biochem. 65: 801–847.

Crafts-Brandner, S.J., Klein, R.R., Klein, P., Hölzer, R. and Feller, U. 1996. Coordination of protein and mRNA abundances of stromal enzymes and mRNA abundances of the Clp protease subunits during senescence of *Phaseolus vulgaris* (L.) leaves. Planta 200: 312–318.

Cui, H., Matsui, K., Omura, S., Schauer, S.L., Matulka, R.A., Sonenshein, G.E. and Ju, S.-T. 1997. Proteasome regulation of activation-induced T cell death. Proc. Natl. Acad. Sci. USA 94: 7515–7520.

D'Andrea, A. and Pellman, D. 1998. Deubiquitinating enzymes: a new class of biological regulators. Crit. Rev. Biochem. Mol. Biol. 33: 337–352.

D'Hondt, K., Bosch, D., Van Damme, J., Goethals, M., Vanderkerckhove, J. and Krebbers, E. 1993. An aspartic proteinase present in seeds cleaves *Arabidopsis* 2 S albumin precursors *in vitro*. J. Biol. Chem. 268: 20884–20891.

D'Hondt, K.D., Stack, S., Gutteridge, S., Vandekerckhove, J., Krebbers, E. and Gal, S. 1997. Aspartic proteinase genes in the Brassicaceae *Arabidopsis thaliana* and *Brassica napus*. Plant Mol. Biol. 33: 187–192.

D'Silva, I., Poirier, G. and Heath, M.C. 1998. Activation of cysteine proteases in cowpea plants during the hypersensitive response: a form of programmed cell death. Exp. Cell Res. 245: 389–399.

Dawson, S.P., Arnold, J.E., Mayer, N.J., Reynolds, S.E., Billett, M.A., Gordon, C., Colleaux, Kloetzel, P.M., Tanaka, K., Mayer, R.J. 1995. Developmental changes of the 26S proteasome in abdominal intersegmental muscles of *Manduca sexta* during programmed cell death. J. Biol. Chem. 270: 1850–1858.

de Barros, E.G. and Larkins, B.A. 1990. Purification and characterization of zein-degrading proteases from endosperm of germinating maize seeds. Plant Physiol. 94: 297–303.

Deiss, L.P., Galinka, H., Berissi, H., Cohen, O. and Kimchi, A. 1996. Cathepsin D protease mediates programmed cell death induced by interferon-γ, Fas/APO-1 and TNF-α. EMBO J. 15: 3861–3870.

del Pozo, O. and Lam, E. 1998. Caspases and programmed cell death in the hypersensitive response of plants to pathogens. Curr. Biol. 8: 1129–1132.

DeMartino, G.N. and Slaughter, C.A. 1999. The proteasome, a novel protease regulated by multiple mechanisms. J. Biol. Chem. 274: 22123–22126.

Dick, L.R., Moomaw, C.R., DeMartino, G.N. and Slaughter, C.A. 1991. Degradation of oxidized insulin B chain by the multiproteinase complex macropain (proteasome). Biochemistry 30: 2725–2734.

Domoto, C., Watanabe, H., Abe, M., Abe, K. and Arai, S. 1995. Isolation and characterization of two distinct cDNA clones encoding corn seed cysteine proteinases. Biochim. Biophys. Acta 1263: 241–244.

Dou, Q.P., McGuire, T.F., Peng, Y. and An, B. 1999. Proteasome inhibition leads to significant reduction of bcr-abl expression and subsequent induction of apoptosis in K562 human chronic myelogenous leukemia cells. J. Pharm. Exp. Ther. 289: 781–790.

Drake, R., John, I., Farrell, A., Cooper, W., Schuch, W. and Grierson, D. 1996. Isolation and analysis of cDNAs encoding tomato cysteine proteases expressed during leaf senescence. Plant Mol. Biol. 30: 755–767.

Drexler, H.C.A. 1997. Activation of the cell death program by inhibition of proteasome function. Proc. Natl. Acad. Sci. USA 94: 855–860.

Fu, H., Doelling, J.H., Arendt, C.S., Hochstrasser, M. and Vierstra, R.D. 1998. Molecular organization of the 20S proteasome gene family from *Arabidopsis thaliana*. Genetics 149: 677–692.

Fu, H., Girod, P.-A., Doelling, J.H., van Nocker, S., Hochstrasser, M., Finley, D. and Vierstra, R.D. 1999. Structural and functional analyses of the 26S proteasome subunits from plants. Plant Mol. Biol. Rep. 26: 137–146.

Fuller, R.S. 1998. Kexin. In: A.J. Barrett, N.D. Rawlings, and J.F. Woessher (Eds.) Handbook of Proteolytic Enzymes, Academic Press, New York, (CD-ROM), chapter 115.

Gan, S. and Amasino, R.M. 1995. Inhibition of leaf senescence by autoregulated production of cytokinin. Science 270: 1986–1988.

Garbarino, J.E. and Belknap, W.R. 1994. Isolation of a ubiquitin-ribosomal protein gene (ubi3) from potato and expression of its promoter in transgenic plants. Plant Mol. Biol. 24: 119–127.

Genschik, P., Durr, A. and Fleck, J. 1994. Differential expression of several E2-type ubiquitin carrier protein genes at different developmental stages in *Arabidopsis thaliana* and *Nicotiana sylvestris*. Mol. Gen. Genet. 244: 548–556.

Genschik, P., Philipps, G., Gigot, C. and Fleck, J. 1992. Cloning and sequence analysis of a cDNA clone from *Arabidopsis thaliana* homologous to a proteasome α subunit from *Drosophila*. FEBS Lett. 309: 311–315.

Glathe, S., Kervinen, J., Nimtz, M., Li, G.H., Tobin, G.J., Copeland, T.D., Ashford, D.A., Wlodawer, A. and Costa, J. 1998. Transport and activation of the vacuolar aspartic proteinase phytepsin in barley (*Hordeum vulgare* L.). J. Biol. Chem. 273: 31230–31236.

Graham, J.S., Xiong, J. and Gillikin, J.W. 1991. Purification and developmental analysis of a metalloendoproteinase from the leaves of *Glycine max*. Plant Physiol. 97: 786–792.

Granell, A. 1998. Plant cysteine proteinases in germination and senescence. In: A.J. Barrett, N.D. Rawlings and J.F. Woessner (Eds.) Handbook of Proteolytic Enzymes, Academic Press, New York (CD-ROM), Chapter 199.

Granell, A., Harris, N., Pisabarro, A.G. and Carbonell, J. 1992. Temporal and spatial expression of a thiolprotease gene during pea ovary senescence, and its regulation by gibberellin. Plant J. 2: 907–915.

Greenberg, A.H. 1996. Activation of apoptosis pathways by granzyme B. Cell Death Differ. 3: 269–274.

Griffiths, C.M., Hosken, S.E., Oliver, D., Chojecki, J. and Thomas, H. 1997. Sequencing, expression pattern and RFLP mapping of a senescence-enhanced cDNA from *Zea mays* with high homology to oryzain γ and aleurain. Plant Mol. Biol. 34: 815–821.

Grimm, L.M., Goldberg, A.L., Poirier, G.G., Schwartz, L.M. and Osborne, B.A. 1996. Proteasomes play an essential role in thymocyte apoptosis. EMBO J. 15: 3835–3844.

Grimm, L.M. and Osborne, B.A. 1999. Apoptosis and the proteasome. Res. Probl. Cell Differ. 23: 209–228.

Groover, A. and Jones, A.M. 1999. Tracheary element differentiation uses a novel mechanism coordinating programmed cell death and secondary cell wall synthesis. Plant Physiol. 119: 375–384.

Groves, M.R., Coulomber, R., Jenkins, J. and Cygler, M. 1998. Structural basis for specificity of papain-like cysteine protease proregions toward their cognate enzymes. Prot. Struct. Funct. Genet. 32: 504–514.

Gubler, F., Raventos, D., Keys, M., Watts, R., Mundy, J. and Jacobsen, J.V. 1999. Target genes and regulatory domains in the GAMYB transcriptional activator in cereal aleurone. Plant J. 17: 1–9.

Guerrero, C., de la Calle, M., Reid, M.S. and Valpuesta, V. 1998. Analysis of the expression of two thiolprotease genes from daylily (*Hemerocallis* spp.) during flower senescence. Plant Mol. Biol. 36: 565–571.

Haas, A.L., Baboshina, O., Williams, B. and Schwartz, L.M. 1995. Coordinated induction of the ubiquitin conjugation pathway accompanies the developmentally programmed death of insect skeletal muscle. J. Biol. Chem. 270: 9407–9412.

Herrmann, J.L., Briones, J., F., Brisbay, S., Logothetis, C.J. and McDonnell, T.J. 1998. Prostate carcinoma cell death resulting from inhibition of proteasome activity is independent of functional Bcl-2 and p53. Oncogene 17: 2889–2899.

Holwerda, B.C., Galvin, N.J., Baranski, T.J. and Rogers, J.C. 1990. *In vitro* processing of aleurain. A barley vacuolar thiol protease. Plant Cell 2: 1091–1106.

Holwerda, B.C., Padgett, H.S. and Rogers, J.C. 1992. Proaleurain vacuolar targeting is mediated by short contiguous peptide interactions. Plant Cell 4: 307–318.

Ishikawa, A., Ohta, S., Matsuoka, K., Hattori, T. and Nakamura, K. 1994. A family of potato genes that encode Kunitz-type proteinase inhibitors: structural comparisons and differential expression. Plant Cell Physiol. 35: 303.

Jones, A.M. and Herman, E.M. 1993. KDEL-containing auxin-binding protein is secreted to the plasma membrane and the cell wall. Plant Physiol. 101: 595–606.

Jones, J.T. and Mullet, J.E. 1995. A salt- and dehydration-inducible pea gene, *cyp15a*, encodes a cell-wall protein with sequence similarity to cysteine proteases. Plant Mol. Biol. 28: 1055–1065.

Jones, M.E., Haire, M.F., Kloetzel, P.M., Mykles, D.L. and Schwartz, L.M. 1995a. Changes in the structure and function of the muticatalytic proteinase (proteasome) during programmed cell death in the intersegmental muscles of the hawkmoth, *Manduca sexta*. Dev. Biol. 169: 436–447.

Jones, M.L., Larsen, P.B. and Woodson, W.R. 1995b. Ethylene-regulated expression of a carnation cysteine proteinase during flower petal senescence. Plant Mol. Biol. 28: 505–512.

Jorda, L., Coego, A., Conejero, V. and Vera, P. 1999. A genomic cluster containing four differentially regulated subtilisin-like processing protease genes is in tomato plants. J Biol. Chem. 274: 2360–2365.

Jordan, F., Hu, Z. and Haghjoo, K. 1995. Interaction of subtilisin with its pro-sequence: versatility of an N-terminal extension. In: U. Shinde and I. Masayor (Eds.) Intramolecular Chaperones and Protein Folding, Springer-Verlag, New York, pp. 113–144.

Kardailsky, I.V. and Brewin, N.J. 1996. Expression of cysteine protease genes in pea nodule development and senescence. Mol. Plant-Microbe Interact. 9: 689–695.

Karrer, K.M., Peiffer, S.L. and DiToms, M.E. 1993. Two distinct gene subfamilies within the family of cysteine protease genes. Proc. Natl. Acad. Sci. USA 90: 3063–3067.

Kato, H. and Minamikawa, T. 1996. Identification and characterization of a rice cysteine endopeptidase that digests glutelin. Eur. J. Biochem. 239: 310–316.

Kervinen, J. 1995. Structure and possible function of aspartic proteinases in barley and other plants. In: K. Takahashi (Ed.) Advances in Experimental Medicine and Biology, Plenum, New York, pp. 241–254.

Kervinen, J., Sarkkinen, P., Kalkkinen, N., Mikola, L. and Saamara, M. 1993. Hydrolytic specificity of the barley grain aspartic proteinase. Phytochemistry 32: 799–803.

Kervinen, J., Tobin, G.J., Costa, J., Waugh, D.S., Wlodawer, A. and Zdanov, A. 1999. Crystal structure of plant aspartic proteinase prophytepsin: inactivation and vacuolar targeting. EMBO J. 18: 3947–3955.

Kinal, H., Park, C., Berry, J.O., Koltin, Y. and Bruenn, J.A. 1995. Processing and secretion of a virally encoded antifungal toxin in transgenic tobacco plants: evidence for a Kex2p pathway in plants. Plant Cell 7: 677–688.

Kitagawa, H., Tani, E., Ikemoto, H., Azaki, I., Nakano, A. and Omura, S. 1999. Proteasome inhibitors induce mitochondria-independent apoptosis in human glioma cells. FEBS Lett. 443: 181–186.

Kobayashi, T., Kobayashi, E., Sato, S., Hotta, Y., Miyajima, N., Tanaka, A. and Tabata, S. 1994. Characterization of cDNAs induced in meiotic prophase in lily microsporocytes. DNA Res. 1: 15–26.

Koehler, S. and Ho, T.D. 1988. Purification and characterization of gibberellic acid-induced cysteine endoproteases in barley aleurone layers. Plant Physiol. 87: 95–103.

Koehler, S.M. and Ho, T.D. 1990. A major gibberellic acid-induced barley aleurone cysteine proteinase which digests hordein. Plant Physiol. 94: 251–258.

Koiwa, H., Shade, R.E., Zhu-Salzman, K., Subramanian, L., Murdock, L.L., Nielsen, S.S., Bressan, R.A. and Hasegawa, P.M. 1998. Phage display selection can differentiate insecticidal activity of soybean cystatins. Plant J. 14: 371–379.

Koizumi, M., Yamaguchi-Shinozaki, K., Tsuji, H. and Shinozaki, K. 1993. Structure and expression of two genes that encode distinct drought-inducible cysteine proteinases in *Arabidopsis thaliana*. Gene 129: 175–182.

Koltunow, A.M., Truettner, J., Cox, K.H., Wallroth, M. and Goldberg, R.B. 1990. Different temporal and spatial gene expression patterns occur during anther development. Plant Cell 2: 1201–1224.

Kosslak, R.M., Chamberlin, M.A., Palmer, R.G. and Bowen, B.A. 1997. Programmed cell death in the root cortex of soybean root necrosis mutants. Plant J. 11: 729–745.

Kreft, S., Ravnikar, M., Mesko, P., Pungercar, J., Umek, A., Kregar, I. and Strukelj, B. 1997. Jasmonic acid inducible aspartic proteinase inhibitors from potato. Phytochemistry 44: 1001–1006.

Lam, E. and del Pozo, O. 2000. Caspase-like protease involvement in the control of plant cell death. Plant Mol. Biol., this issue.

Lefebvre, O., Wolf, C., Limacher, J.-M., Pascal, H., Wendling, C., LeMeur, M., Basset, P. and Rio, M.-C. 1992. The breast cancer-associated stromelysin-3 gene is expressed during mouse mammary gland apoptosis. J. Cell Biol. 119: 997–1002.

Levine, A., Pennell, R.I., Alvarez, M.E., Palmer, R. and Lamb, C. 1996. Calcium-mediated apoptosis in a plant hypersensitive disease resistance response. Curr. Biol. 6: 427–437.

Li, Y.Q., Southworth, D., Linskens, H.F., Mulcahy, D.L. and Cresti, M. 1995. Localization of ubiquitin in anthers and pistils of *Nicotiana*. Sex. Plant Reprod. 8: 123–128.

Lin, E., Burns, D.J. and Gardner, R.C. 1993. Fruit development regulation of the kiwifruit actinidin promoter is conserved in transgenic petunia plants. Plant Mol. Biol. 23: 489–499.

Lin, K.I., Baraban, J.M. and Ratan, R.R. 1998. Inhibition versus induction of apoptosis by proteasome inhibitors depends on concentration. Cell Death Differ. 5: 577–583.

Lindahl, M., Tabak, S., Cseke, L., Pichersky, E., Andersson, B. and Adam, Z. 1996. Identification, characterization and molecular cloning of a homologue of the bacterial FtsH protease in chloroplasts of higher plants. J. Biol. Chem. 271: 29329–29334.

Linnestad, C., Doan, D.N.P., Brown, R.C., Lemmon, B.E., Meyer, D.J., Jung, R. and Olsen, O. 1998. Nucellain, a barley homolog of the dicot vacuolar-processing protease, is localized in nucellar cell walls. Plant Physiol. 118: 1169–1180.

Liu, C.Y., Xu, H. and Graham, J.S. 1998. Cloning and characterization of an *Arabidopsis thaliana* cDNA homologous to the matrix metalloproteinases. Plant Physiol. 117: 1127.

Lohman, K.N., Gan, S., John, M.C. and Amasino, R.M. 1994. Molecular analysis of natural leaf senescence in *Arabidopsis thaliana*. Physiol. Plant. 92: 322–328.

Lowe, J., Stock, D., Jap, B., Zwickl, P., Baumeister, W. and Huber, R. 1995. Crystal structure of the 20S proteasome from the archaeon *T. acidophilum* at 3.4 Å resolution. Science 268: 533–539.

Lund, L.R., Romer, J., Thomasset, N., Solberg, H., Pyke, C., Bissell, M.J., Dano, K. and Werb, Z. 1996. Two distinct phases of apoptosis in mammary gland involution: proteinase-independent and -dependent pathways. Development 122: 181–193.

Mach, L., Mort, J.S. and Glossl, J. 1994. Noncovalent complexes between the lysosomal proteinase cathepsin B and its propeptide account for stable, extracellular, high molecular mass forms of the enzyme. J. Biol. Chem. 269: 13036–13040.

Maganja, D.B., Strukelj, B., Pungercar, J., Gubensek, F., Turk, V. and Kregar, I. 1992. Isolation and sequence analysis of the genomic DNA fragment encoding an aspartic proteinase inhibitor homolog from potato (*Solanum tuberosum* L.). Plant Mol. Biol. 20: 311–313.

Marthinuss, J., Andrade-Gordon, P. and Seiberg, M. 1995. A secreted serine protease can induce apoptosis in Pam212 keratinocytes. Cell Growth Diff. 6: 807–816.

Marttila, S., Jones, B.L. and Mikkonen, A. 1995. Differential localization of two acid proteinases in germinating barley (*Hordeum vulgare*) seed. Physiol. Plant 93: 317–327.

Marty, F., Branton, D. and Leigh, R.A. 1980. Plant vacuoles. In: N.E. Tolbert (Ed.) The Biochemistry of Plants, Academic Press, New York, pp. 625–658.

Masdehors, P., Omura, S., Merle-Beral, H., Mentz, F., Cosset, J.-M., Dumont, J., Magdelenat, H. and Delic, J. 1999. Increased sensitivity of CLL-derived lymphocytes to apoptotic death acitvation by the proteasome-specific inhibitor lactacystin. Br. J. Haematol. 105: 752–757.

Matile, P. 1978. Biochemistry and function of vacuoles. Annu. Rev. Plant Physiol. 29: 193–213.

McGeehan, G., Burkhart, W., Anderegg, R., Becherer, J.D., Gilikin, J.W. and Graham, J.S. 1992. Sequencing and characterization of the soybean leaf metalloproteinase. Plant Physiol. 99: 1179–1183.

Meichtry, J., Amrhein, N. and Schaller, A. 1999. Charazterization of the subtilase gene family in tomato (*Lycopersicon esculentum* Mill.). Plant Mol. Biol. 39: 749–760.

Menard, R. and Storer, A.C. 1998. Papain. In: A.J. Barrett, N.D. Rawlings and J.F. Woessner (Eds.) Handbook of Proteolytic Enzymes, Academic Press. New York (CD-ROM), Chapter 187.

Meriin, A.B., Gabai, A.L., Yaglom, J., Shifrin, V.I. and Sherman, M.Y. 1998. Proteasome inhibitors activate stress kinases and induce hsp72. Diverse effects on apoptosis. J. Biol. Chem. 273: 6373–6379.

Minami, A. and Fukuda, H. 1995. Transient and specific expression of a cysteine endoproteinase associated with autolysis during differentiation of *Zinnia* mesophyll cells into tracheary elements. Plant Cell Physiol. 36: 1599–1606.

Mikkonen, A., Porali, I., Cercos, M. and Ho, T.D. 1996. A major cysteine proteinase, EPB, in germinating barley seeds: structure of the two intronless genes and regulation of expression. Plant Mol. Biol. 31: 239–254.

Mitsuhashi, W. and Minamikawa, T. 1989. Synthesis and posttranslational activation of sulfhydryl-endopeptidase in cotyledons of germinating *Vigna mugo* seeds. Plant Physiol. 89: 274–279.

Moriyasu, Y. and Ohsumi, Y. 1996. Autophagy in tobacco suspension-cultured cells in response to sucrose starvation. Plant Physiol. 111: 1233–1241.

Murakami, Y., Matsufuji, S., Kameji, T., Hayashi, S., Igarashi, K., Tamura, T., Tanaka, K. and Ichihara, A. 1992. Ornithine decarboxylase is degraded by the 26S proteasome without ubiquitination. Nature 360: 597–599.

Muzio, M. 1998. Signalling by proteolysis: death receptors induce apoptosis. Int. J. Clin. Lab. Res. 28: 141–147.

Nadeau, J.A., Zhang, X.S., Li, J. and O'Neill, S.D. 1996. Ovule development: identification of stage-specific and tissue-specific cDNAs. Plant Cell 8: 213–239.

Navarre, D.A. and Wolpert, T.J. 1999. Victorin induction of an apoptotic/senescence-like response in oats. Plant Cell 11: 237–249.

Neuteboom, L.W., Ng, J.M.Y., Kuyper, M., Clijesdale, O.R., Hooykaas, P.J.J. and van der Zaal, B.J. 1999. Isolation and characterization of cDNA clones corresponding with mRNAs that accumulate during auxin-induced lateral root formation. Plant Mol. Biol. 39: 273–287.

Okamoto, T., Minamikawa, T., Edward, G. and Vakharia, V. 1999. Posttranslational removal of the carboxy-terminal KDEL of the cysteine protease SH-EP occurs prior to maturation of the enzyme. J. Biol. Chem. 274: 11390–11398.

Omura, S., Fujimoto, T., Otoguro, K., Matsuzaki, K., Moriguchi, R., Tanaka, H. and Sasaki, Y. 1991. Lactacystin, a novel microbial metabolite, induces neuritogenesis of neuroblastoma cells. J. Antibiot. 44: 113–116.

Orlowski, M., Cardozo, C. and Michaud, C. 1993. Evidence for the presence of five distinct proteolytic components in the pituitary multicatalytic proteinase complex. Properties of two components cleaving bonds on the carboxyl side of branched chain and small neutral amino acids. Biochemistry 32: 1563–1572.

Orlowski, R.Z. 1999. The role of the ubiquitin-proteasome pathway in apoptosis. Cell Death Differ. 6: 303–313.

Ostersetzer, O. and Adam, Z. 1997. Light-stimulated degradation of an unassembled Rieske FeS protein by a thylakoid-bound protease: the possible role of the FtsH protease. Plant Cell 9: 957–965.

Pak, J.H., Liu, C.Y., Huangpu, J. and Graham, J.S. 1997. Construction and characterization of the soybean leaf metalloproteinase cDNA. FEBS Lett. 404: 283–288.

Panavas, T., Pikla, A., Reid, P.D., Rubinstein, B. and Walker, E.L. 1999. Identification of senescence-associated genes from daylily petals. Plant Mol. Biol. 40: 237–248.

Paris, N., Stanley, C.M., Jones, R.L. and Rogers, J.C. 1996. Plant cells contain two functionally distinct vacuolar compartments. Cell 85: 563–572.

Paul, W., Amiss, J., Try, R., Praekelt, U., Scott, R. and Smith, H. 1995. Correct processing of the kiwifruit protease actinidin in transgenic tobacco requires the presence of the C-terminal propeptide. Plant Physiol. 108: 261–268.

Ragster, L.V. and Chrispeels, M.J. 1979. Azocoll digesting proteinases in soybean leaves: characteristics and changes during leaf maturation and senescence. Plant Physiol. 64: 857–862.

Ramalho-Santos, M., Pissarra, J., Verissimo, P., Pereira, S., Salema, R., Pires, E. and Faro, C. 1997. Cardosin A, an abundant aspartic proteinase, accumulates in protein storage vacuoles in the stigmatic papillae of *Cynara cardunculus* L. Planta 203: 204–212.

Rawlings, N.D. and Barrett, A.J. 1999. MEROPS: the peptidase database. Nucl. Acids Res. 27: 325–331.

Ribeiro, A., Akkermans, A.D.L., van Kammen, A., Bisseling, T. and Pawloski, K. 1995. A nodule-specific gene encoding a subtilisin-like protease is expressed in early stages of actinorhizal nodule development. Plant Cell 7: 785–794.

Richter, S. and Lamppa, G.K. 1998. A chloroplast processing enzyme functions as the general stromal processing peptidase. Proc. Natl. Acad. Sci. USA 95: 7463–7468.

Roberts, L.R., Adjei, P.N. and Gores, G.J. 1999. Cathepsins as effector proteases in hepatocyte apoptosis. Cell Biochem. Biophys. 30: 71–88.

Rock, K.L., Gramm, C., Rothstein, L., Clark, K., Dick, L., Hwang, D. and Goldberg, A.L. 1994. Inhibitors of the proteasome block the degradation of most cell proteins and the generation of peptides presented on MHC class I molecules. Cell 78: 761–771.

Rodrigo, I., Vera, P., van Loon, L.C. and Conejero, V. 1991. Degradation of tobacco pathogenesis-related proteins in plants. Plant Physiol. 95: 616–622.

Runeberg-Roos, P. and Saarma, M. 1998. Phytepsin, a barley vascular aspartic proteinase, is highly expressed during autolysis of developing tracheary elements and sieve cells. Plant J. 15: 139–145.

Runeberg-Roos, P., Törmäkangas, K. and Östman, A. 1991. Primary structure of a barley-grain aspartic proteinase: a plant aspartic proteinase resembling mammalian cathespin D. Eur. J. Biochem. 202: 1021–1027.

Runeberg-Roos, P., Kervinen, J., Kovaleva, V., Raikhel, N.V. and Gal, S. 1994. The aspartic proteinase of barley is a vacuolar enzyme that processes probarley lectin *in vitro*. Plant Physiol. 105: 321–329.

Sadoul, R., Ferdandez, P.-A., Quiquerez, A.-L., Martinou, I., Maki, M., Schroter, M., Becherer, J.D., Irmler, M., Tschopp, J. and Martinou, J.-C. 1996. Involvement of the proteasome in the programmed cell death of NGF-deprived sympathetic neurons. EMBO J. 15: 3845–3852.

Safadi, F., Mykles, D.L. and Reddy, A.S.N. 1997. Partial purification and characterization of a Ca^{2+}-dependent proteinase from *Arabidopsis* roots. Arch. Biochem. Biophys. 348: 143–151.

Saftig, P., Hetman, M., Schmahl, W., Weber, K., Heine, L., Mossmann, H., Koster, A., Hess, B., Evers, M., von Figura, K. and Peters, C. 1995. Mice deficient for the lysosomal proteinase cathespin D exhibit progressive atrophy of the intestinal mucosa and profound destruction of lymphoid cells. EMBO J. 14: 3599–3608.

Sarkkinen, P., Kalkkinen, N., Tilgmann, C., Siuro, J., Kervinen, J. and Mikola, L. 1992. Aspartic proteinase from barley grains is related to mammalian lysosomal cathespin D. Planta 186: 317–323.

Schaller, A. and Ryan, C.A. 1994. Identification of a 50-kDa systemin-binding protein in tomato plasma membranes having Kex2p-like properties. Proc. Natl. Acad. Sci. USA 91: 11802–11806.

Schaller, A. and Ryan, C.A. 1996. Molecular cloning of a tomato leaf cDNA encoding an aspartic protease, a systemic wound response protein. Plant Mol. Biol. 31: 1073–1077.

Schmid, M., Simpson, D., Kalousek, F. and Gietl, C. 1998. A cysteine endopeptidase with a C-terminal KDEL motif isolated from castor bean endosperm is a marker enzyme for the ricinosome, a putative lytic compartment. Planta 206: 466–475.

Schwartz, L.M., Myer, A., Kosz, L., Engelstein, M. and Maier, C. 1990. Activation of polyubiquitin gene expression during developmentally programmed cell death. Neuron 5: 411–419.

Seidah, N.G., Day, R., Marcinkiewica, M. and Chrétien, M. 1998. Precursor convertase: an evolutionarily ancient, cell-specific, combinatorial mechanism yielding diverse bioactive peptides and proteins. Ann. NY Acad. Sci. 839: 9–24.

Shanklin, J., DeWitt, N.D. and Flanagan, J.M. 1995. The stroma of higher plant plastids contain ClpP and ClpC, functional homologs of *Escherichia coli* ClpP and ClpAl: an archetypal two-component ATP-dependent protease. Plant Cell 7: 1713–1722.

Shimada, T., Hiraiwa, N., Nishimura, M. and Hara-Nishimura, I. 1994. Vacuolar processing enzyme of soybean that converts proproteins to the corresponding mature forms. Plant Cell Physiol. 35: 713–718.

Shinde, U., Li, Y. and Inouye, M. 1995. The role of the N-terminal propeptide in mediating folding of subtilisin E. In: U. Shinde and M. Inouye (Eds.) Intramolecular Chaperones and Protein Folding, Springer-Verlag, New York, pp. 11–34.

Shinohara, K., Tomioka, M., Nakano, H., Tone, S., Ito, H. and Kawashima, S. 1996. Apoptosis induction resulting from proteasome inhibition. Biochem. J. 317: 385–388.

Shintani, A., Kato, H. and Minamikawa, T. 1997. Hormonal regulation of expression of two cysteine endopeptidase genes in rice seedlings. Plant Cell Physiol. 38: 1242–1248.

Shirley, B.W. and Goodman, H.M. 1993. An *Arabidopsis* gene homologous to mammalian and insect genes encoding the largest proteasome subunit. Mol. Gen. Genet. 241: 586–594.

Shutov, A.D. and Vaintraub, I.A. 1987. Degradation of storage proteins in germinating seeds. Phytochemistry 26: 1557–1566.

Siezen, R.J. and Leunissen, J.A.M. 1997. Subtilases: the superfamily of subtilisin-like serine proteases. Protein Sci. 6: 501–523.

Solomon, M., Belenghi, B., Delledonne, M., Menachem, E. and Levine, A. 1999. The involvement of cysteine proteases and protease inhibitor genes in the regulation of programmed cell death in plants. Plant Cell 11: 431–443.

Stack, M.S. and Gray, R.D. 1989. Comparison of vertebrate collagenase and gelatinase using a new fluorogenic substrate peptide. J. Biol. Chem. 264: 4277–4281.

Stephenson, P. and Rubinstein, B. 1998. Characterization of proteolytic activity during senescence in daylilies. Physiol. Plant. 104: 463–473.

Stephenson, P., Collins, B.A., Reid, P.D. and Rubinstein, B. 1996. Localization of ubiquitin to differentiating vascular tissues. Am. J. Bot. 83: 140–147.

Stocker, W., Grams, F., Baumann, U., Reinemer, P., Gomis Ruth, F.-X., McKay, D.B. and Bode, W. 1995. The metzincins: topological and sequential relations between the astacins, asamalysins, serralysin and metrixins (collagenases) define a superfamily of zinc-peptidases. Protein Sci. 4: 823–840.

Swanson, S.J., Bethke, P.C. and Jones, R.L. 1998. Barley aleurone cells contain two types of vacuoles: characterization of lytic organelles by use of fluorescent probes. Plant Cell 10: 685–698.

Tao, K., Stearns, N.A., Dong, J., Wu, Q.L. and Sahagian, G.G. 1994. The proregion of cathepsin L is required for proper folding, stability, and ER exit. Arch. Biochem. Biophys. 311: 19–27.

Taylor, A.A., Horsch, A., Rzepczyk, A., Hasenkampf, C.A. and Riggs, C.D. 1997. Maturation and secretion of a serine proteinase is associated with events of late microsporogenesis. Plant J. 12: 1261–1271.

Thoma, S., Sullivan, M.L. and Vierstra, R.D. 1996. Members of two gene families encoding ubiquitin-conjugating enzymes, AtUBC1-3 and AtUBC4-6, from *Arabidopsis thaliana* are differentially expressed. Plant Mol. Biol. 31: 493–505.

Thornberry, N.A., Rosen, A. and Nicholson, D.W. 1997. Control of apoptosis by proteases. Adv. Pharm. 41: 155–177.

Tormakangas, K., Kervinen, J., Ostman, A. and Teeri, T. 1994. Tissue-specific localization of aspartic proteinase in developing and germinating barley grains. Planta 195: 116–125.

Tornero, P., Conejero, V. and Vera, P. 1997. Identification of a new pathogen-induced member of the subtilisin-like processing protease family from plants. J Biol. Chem. 272: 14412–14419.

Tornero, P., Mayda, E., Gomez, M.D., Canas, L., Conejero, V. and Vera, P. 1996. Characterization of LRP, a leucine-rich repeat (LRR) protein from tomato plants that is processed during pathogenesis. Plant J. 10: 315–330.

Tranbarger, T.J. and Misra, S. 1996. Structure and expression of a developmentally regulated cDNA encoding a cysteine protease (pseudotzain) from Douglas fir. Gene 172: 221–226.

Urwin, P.E., McPherson, M.J. and Atkinson, H.J. 1998. Enhanced transgenic plant resistance to nematodes by dual proteinase inhibitor constructs. Planta 204: 472–479.

Valpuesta, V., Lange, N.E., Guerrero, C. and Reid, M.S. 1995. Up-regulation of a cysteine protease accompanies the ethylene-insensitive senescence of daylily (*Hemerocallis*) flowers. Plant Mol. Biol. 28: 575–582.

van Nocker, S., Deveraux, Q., Rechsteiner, M. and Vierstra, R.D. 1996. Arabidopsis MBP1 gene encodes a conserved ubiquitin recognition component of the 26S proteasome. Proc. Natl. Acad. Sci. USA 93: 856–860.

Varshavsky, A. 1997. The ubiquitin system. Trends. Biochem. Sci. 22: 383–387.

Vierstra, R.D. 1996. Proteolysis in plants: mechanisms and functions. Plant Mol. Biol. 32: 275–302.

Wang, H., Wu, H.M. and Cheung, A.Y. 1996. Pollination induces mRNA poly(A) tail-shortening and cell deterioration in flower transmitting tissue. Plant J. 9: 715–727.

Watanabe, H., Abe, K., Emori, Y., Hosoyama, H. and Arai, S. 1991. Molecular cloning and gibberellin-induced expression of multiple cysteine proteinases of rice seeds (oryzains). J. Biol. Chem. 266: 16897–16902.

Weaver, L.M., Froehlich, J.E. and Amasino, R.M. 1999. Chloroplast-targeted ERD1 protein declines but its mRNA increases during senescence in *Arabidopsis*. Plant Physiol. 119: 1209–1216.

Werb, Z. 1997. ECM and cell surface proteolysis: regulating cellular ecology. Cell 91: 439–442.

Wilk, S. and Orlowski, M. 1980. Cation-sensitive neutral endopeptidase: isolation and specificity of the bovine pituitary enzyme. J. Neurochem. 35: 172–1182.

Wilkinson, K.D. 1997. Regulation of ubiquitin-dependent processes by deubiquitinating enzymes. FASEB J. 11: 1245–1256.

Williams, J., Bulman, M., Huttly, A., Phillips, A. and Neill, S. 1994. Characterization of a cDNA from *Arabidopsis thaliana* encoding a potential thiol protease whose expression is induced independently by wilting and abscisic acid. Plant Mol. Biol. 25: 259–270.

Woffenden, B.J., Freeman, T.B. and Beers, E.P. 1998. Proteasome inhibitors prevent tracheary element differentiation in *Zinnia* mesophyll cell cultures. Plant Physiol. 118: 419–430.

Xu, F. and Chye, M. 1999. Expression of cysteine proteinase during developmental events associated with programmed cell death in brinjal. Plant J. 17: 321–327.

Yamagata, H., Masuzawa, T., Nagaoka, Y., Ohnishi, T. and Iwasaki, T. 1994. Cucumisin, a serine protease from melon fruits, shares structural homology with subtilisin and is generated from a large precursor. J Biol. Chem. 269: 32725–32731.

Yamauchi, D., Terasaki, Y., Okamoto, T. and Minamikawa, T. 1996. Promoter regions of cysteine endopeptidase genes from legumes confer germination-specific expression in transgenic tobacco seeds. Plant Mol. Biol. 30: 321–329.

Ye, Z.-H. and Varner, J.E. 1996. Induction of cysteine and serine proteinases during xylogenesis in *Zinnia elegans*. Plant Mol. Biol. 30: 1233–1246.

Young, T.E. and Gallie, D.R. 2000. Programmed cell death during endosperm development. Plant Mol. Biol., this issue.

Yuan, J., Shaham, S., Ledoux, S., Ellis, H.M. and Horvitz, H.R. 1993. The *C. elegans* cell death gene ced-3 encodes a protein similar to mammalian interleukin-1β-converting enzyme. Cell 75: 641–652.

Zhang, X.M., Lin, H., Chen, C. and Chen, B.D.-M. 1999. Inhibition of ubiquitin-proteasome pathway activates a caspase-3-like protease and induces Bcl-2 clavage in human M-07e leukaemic cells. Biochem. J. 340: 127–133.

Zhao, C., Johnson, B.J., Kositsup, B. and Beers, E.P. 2000. Exploiting secondary growth in *Arabidopsis*. Construction of xylem and bark cDNA libraries and cloning of three xylem endopeptidases. Plant Physiol. 123: 1185–1196.

Plant Molecular Biology **44:** 417–428, 2000.
E. Lam, H. Fukuda and J. Greenberg (Eds.), Programmed Cell Death in Higher Plants.

Caspase-like protease involvement in the control of plant cell death

Eric Lam* and Olga del Pozo
Biotech Center, Foran Hall, Cook College, 59 Dudley Road, Rutgers University, New Brunswick, NJ 08903, USA
*(*author for correspondence; fax: (732) 932 6535; e-mail: lam@aesop.rutgers.edu)*

Key words: apoptosis, baculovirus p35, Bcl-2-like proteins, caspases, cell death, hypersensitive response, mitochondria

Abstract

Cell death as a highly regulated process has now been recognized to be an important, if not essential, pathway that is ubiquitous in all multicellular eukaryotes. In addition to playing key roles in the morphogenesis and sculpting of the organs to give rise to highly specialized forms and shapes, cell death also participates in the programmed creation of specialized cell types for essential functions such as the selection of B cells in the immune system of mammals and the formation of tracheids in the xylem of vascular plants. Studies of apoptosis, the most well-characterized form of animal programmed cell death, have culminated in the identification of a central tripartite death switch the enzymatic component of which is a conserved family of cysteine proteases called caspases. Studies in invertebrates and other animal models suggest that caspases are conserved regulators of apoptotic cell death in all metazoans. In plant systems, the identities of the main executioners that orchestrate cell death remain elusive. Recent evidence from inhibitor studies and biochemical approaches suggests that caspase-like proteases may also be involved in cell death control in higher plants. Furthermore, the mitochondrion and reactive oxygen species may well constitute a common pathway for cell death activation in both animal and plant cells. Cloning of plant caspase-like proteases and elucidation of the mechanisms through which mitochondria may regulate cell death in both systems should shed light on the evolution of cell death control in eukaryotes and may help to identify essential components that are highly conserved in eukaryotes.

Abbreviations: AOX, alternative oxidase; BLP, Bcl-2-like proteins; CLP, caspase-like protease; PARP, poly(ADP-ribose) polymerase; PCD, programmed cell death; ROS, reactive oxygen species; TUNEL, terminal deoxynucleotidyltransferase-mediated dUTP nick end-labeling; VDAC, voltage-dependent anion channel

Introduction

Proteolysis provides an important level of control for gene expression and plays a fundamental role in development, homeostasis, physiology and survival at the organismal level. Specific proteases function in multiple regulatory pathways found in all organisms (Vierstra, 1996). Proteolysis is a mechanism well suited for cellular adaptation to a constantly changing environment, such as the potential invasion of pathogens. The effective deployment of proteolysis as key signaling switches relies on the speed and irreversibility of the process. Thus, inactive regulatory proteases are kept as dormant swords waiting to be unsheathed for the decisive step to commit the cell to a particular program. The potential lethality from inappropriate activation of proteases necessitates their tight regulation at different levels such as their transcription, translation, compartmentalization, processing, and by interaction with inhibitors.

Among the processes regulated by proteolysis are: activation of zymogens, processing and activation of hormones, metabolic control, progression of the cell cycle, development and programmed cell death (PCD). PCD is a genetically determined process present in all multicellular organisms by which cells activate their own demise (Vaux and Korsmeyer, 1999; Lam *et al.*, 1999). PCD takes place in plants during development, under stress conditions, in the senescence process and in response to pathogen infection (Mittler and Lam, 1996; Lam *et al.*, 1999). When a pathogen invades a plant, two types of cell death responses are mounted on the plant side. If the plant is resistant to the pathogen, a rapid cell death is frequently triggered at the primary site of infection, which constitutes the hypersensitive response (HR) and is accompanied by activation of local defense responses (Heath, this issue). If the plant is susceptible, disease develops, and slower cell death develops as local and systemic infections progress. At present, very little is known about the execution processes that lead to PCD in plants. This is in contrast to animal PCD where key participants have been characterized in the past decade. Among the large number of cell death regulators known in animal systems, a family of cysteine proteases specific for target sites containing aspartate residues has been shown to play a central role in signaling and executing the form of PCD known as apoptosis (Cryns and Yuan, 1998). Apoptosis is characterized by a number of morphological and biochemical features such as nuclear condensation, cellular fragmentation into 'apoptotic bodies', appearance of nucleosomal fragments of nuclear DNA, TUNEL-positive staining of nuclei in the dying cells, and their subsequent engulfment by neighboring cells. Although some of these features appear to be conserved in plant PCD, fundamental differences such as the lack of evidence for apoptotic bodies and engulfment also exist.

In this review, we will first discuss current knowledge of the roles and regulation of caspases in animal systems. Since many excellent reviews on this subject can be found elsewhere (Cryns and Yuan, 1998; Thornberry and Lazebnik, 1998; Wolf and Green, 1999), we will focus on describing the general principles that have been revealed instead of being exhaustive. We will then examine current evidence, mostly based on inhibitor studies that point to the conservation of caspase-like proteases (CLPs) as PCD regulators of higher plants, with specific emphasis on the HR as a model system.

Caspases as a core execution switch for PCD in higher eukaryotes

In animals, the central core of cell death effectors was initially characterized in the nematode *Caenorhabditis elegans* using genetic screens (Ellis and Horvitz, 1986). Loss-of-function mutations in two genes, *Ced-3* and *Ced-4*, result in complete abrogation of the cell death program leading to accumulation of extra cells in mutants while mutations in another gene, *Ced-9*, result in inappropriate cell death (Hengartner *et al.*, 1992; Yuan and Horvitz, 1992; Yuan *et al.*, 1993). From genetic analysis, the following functional relationship among these genes has emerged: *Ced-3* induction of cell death is epistatic from *Ced-4* while cell death induction by *Ced-4* is dependent on *Ced-3*. On the other hand, *Ced-9* requires a functional *Ced-4* protein to suppress *Ced-3* (Cryns and Yuan, 1998). *Ced-3* was found to encode a cysteine protease that shares homology with mammalian ICE (interleukin-1β-converting enzyme), which was first known to be responsible for interleukin processing during inflammation (Yuan *et al.*, 1993; Nicholson and Thornberry, 1997). These proteases were later renamed as caspases (cysteine aspases) in reference to their requirement for an aspartate residue at the P1 position of their substrate recognition sites (Alnemri *et al.*, 1996). The subsequent identification of functional mammalian counterparts for the other cell death regulators *Ced-4* and *Ced-9* demonstrated that the basic cell death machinery is conserved within the animal kingdom. A possible mammalian homologue of CED-4 is Apaf-1, which acts as an adaptor protein to activate caspase-9. CED-9 is homologous to Bcl-2, a cell death suppressor originally identified in a chromosomal translocation associated with a B-cell lymphoma (Korsmeyer, 1995). Mammalian cell death is articulated in a more complex network compared to the nematode with 14 different caspases described to date, and Bcl-2-like proteins (BLPs) also constitute a family with a growing list of members in mammals, with some BLPs having pro-cell death activity, such as Bax and Bak, while others having anti-apoptotic functions, such as Bcl-2 and Bcl-X_L (reviewed in Tsujimoto and Shimizu, 2000). The genetic relationship between *Ced-3*, *Ced-4* and *Ced-9* was later found to reflect the direct physical interaction between the protein products encoded by these proteins. Thus, interaction with CED-4 is required to activate CED-3 while CED-9 can interact with both CED-3 and CED-4 to suppress activation of cell death

(Chinnaiyan *et al.*, 1997; Spector *et al.*, 1997). To activate cell death in the nematode, the pro-apoptosis factor EGL-1 binds to and thereby competes away CED-9 (Conradt and Horvitz, 1998). Interestingly, EGL-1 turns out to have structural similarities to BLPs and thus the theme of deploying BLPs to act as positive and negative regulators of the caspase switch is strikingly conserved in metazoans from nematodes to man. However, apparent variations on the theme established by the study of the CED-3/-4/-9 switch have also been observed. A striking example is that for the activation of caspase-9 by Apaf-1, which requires the binding of cytochrome *c* and dATP/ATP to Apaf-1 (Hu *et al.*, 1999). No such requirement for CED-3 activation by CED-4 has been observed. More recently, it has been shown that multiple CED-4-like gene products likely exist in *Drosophila* and mammals to regulate activation of distinct procaspases (Kanuka *et al.*, 1999). Thus, multiple types of mechanisms involving specific protein-protein interactions exist to regulate procaspase activation.

In many model systems for apoptosis in animals, caspases are the first deterministic executioners of PCD, presumably by cleaving key intracellular substrates. A conserved feature within this family of proteases is the QACXG pentapeptide surrounding the active-site cysteine residue as well as other regions of the protein involved in catalysis (Nicholson and Thornberry, 1997; Cryns and Yuan, 1998). However, variations of this conserved sequence have been found in two *Drosophila* caspases: DcP-2 (also called DREDD) contains a glutamic acid instead of the glycine residue (Chen *et al.*, 1998) and DRONC, an ecdysone-inducible caspase, contains PFCRG at its presumed active site (Dorstyn *et al.*, 1999). A more recently discovered caspase from *Caenorhabditis elegans*, CSP-1B (Shaham, 1998), has the sequence SACRG surrounding its active site. These variations might determine novel interactions with different substrates. Despite a requirement for aspartate (Asp) residues at the P1 position of their target site, individual caspases differ in their substrate specificities (Talanian *et al.*, 1997). These different specificities are determined by the amino acids surrounding the P1 site especially at the amino terminus and, most importantly, the P4 residue. Caspases are expressed as inactive zymogens and upon cell death signal reception, they are recruited to cell death signaling complex(es) by protein-protein interactions where oligomerization, self-cleavage and activation take place (Wolf and Green, 1998).

Based on sequence comparison of their primary structure, caspases have been designated into two subgroups: initiators (or signalling) and effectors (or executioner). As Figure 1 shows diagrammatically, the initiator caspases typically consist of a long, variable N-terminal pro-domain that likely functions as an interface for protein-protein interaction. This is exemplified by caspase-8, -9 and -12. Two classes of pro-domain that are known to function in this way are called CARDs (caspase recruitment domains) and DEDs (death effector domains); both function as protein-binding surfaces via homophilic interactions (reviewed in Wolf and Green, 1999; Aravind *et al.*, 1999). This N-terminal region thus plays an important signal-transducing function by coupling initiator caspase activation to receptor proteins that can receive intra- or extracellular signals. Once activated, the receptors are thought to recruit, either through direct interaction or via adaptor molecules, the target procaspase through its cytosolic domain that ultimately mediates the oligomerization of the initiator procaspase. Once bound, the juxtaposition of the initiator caspase molecules allows for autocatalysis and proteolytic cleavage to produce a large subunit containing the active site and a small subunit derived from the C-terminal region of the procaspase precursor. The pro-domain is also removed via inter- or intra molecular cleavage. As expected, both the cleavage site for the pro-domain as well as that between the large and small subunits contains aspartate at the P1 position (see Figure 1). This portion of the coding sequence is more conserved in structure between different caspases than the variable N-terminus, although the actual amino acid sequence remains rather variable with the highest conservation found in the region surrounding the active site. Once activated by cleavage, the large and small subunits assemble into a heterotetramer to form the active caspase and initiates a protease cascade by cleaving and activating downstream proteases such as the effector caspases. This class of caspases, exemplified by caspase-3, -6, -7 and -14, typically have very short pro-domains and do not autoactivate.

With mice strains that have been engineered to 'knockout' specific caspases and other regulators of cell death, at least three distinct pathways of apoptosis activation have now been demonstrated. A striking observation is that the three initiator caspases (caspase-8, -9 and -12) are apparently dedicated to distinct types of regulatory pathways. Caspase-8 is essential for mediating death-inducing signals from transmembrane receptors for ligands such as Fas or TNF. This has been

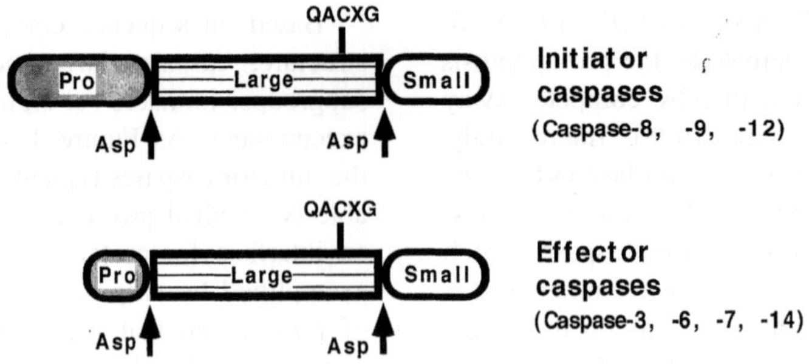

Figure 1. Domain structure and location of intramolecular cleavage sites for caspases. The structural organization of the three different domains found in the two types of procaspases are shown schematically. The initiator caspases are shown to have a longer N-terminal variable domain (Pro) as compared to the effector caspases. The different caspases which fit the functional and structural descriptions of these two types of caspases are listed in parenthesis. The situation for the remaining caspases is not so clear and thus they are not listed. The approximate position of the conserved active site sequence, QACXG, is shown and the cleavage sites for the activation of procaspases and removal of the pro-domain are indicated by arrows. The presence of aspartate residues (Asp) at the P1 position of these cleavage sites are also shown. The relative locations for the 'large' and 'small' subunits of the activated caspases are indicated.

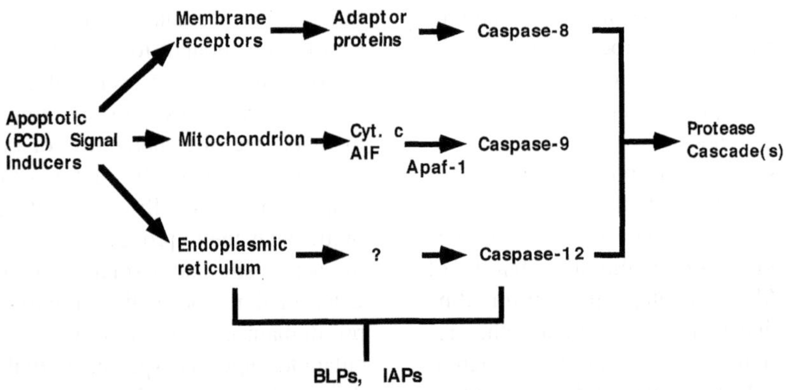

Figure 2. Involvement of initiator caspases in three separate cellular pathways for cell death activation in mammalian systems. The specific pathways that three characterized initiator caspases are known to mediate are compared diagrammatically. The components and order of activation for the protease cascade(s) downstream from the three initiator caspases are not well-defined in most cases and could be cell-type-dependent. Thus, the steps after the procaspase activation step may or may not be common between the three pathways shown.

shown to be mediated by interaction between its DED domain at the N-terminus and the DED domain of adaptor proteins such as TRADD, RAIDD and FADD (reviewed in Aravind *et al.*, 1999). For mitochondrion-derived death signals, cytochrome *c* and dATP/ATP cooperate to activate caspase-9 specifically through Apaf-1 (Slee *et al.*, 1999). On the other hand, for cell death signals from cellular stresses related to the endoplasmic reticulum (ER), caspase-12 has recently been shown to be the key regulator that is dedicated to this pathway (Nakagawa *et al.*, 2000). In contrast to the specificity that these initiator caspases displayed, effector caspases such as caspase-3 have been found to be activated by multiple signals, thus supporting the cascade model in which successive cleavages of a network of proteases ultimately generate the final cell death phenomenon (Cryns and Yuan, 1998; Wolf and Green, 1999).

To regulate caspases tightly, multiple classes of proteins exist to create a balance that determines the fate of the cell. In addition to the BLPs, the study of proliferation factors in animal viruses has also uncovered additional classes of apoptosis regulators for animal cells. Thus, in addition to viral homologues to BLPs that regulate activation of procaspases, such as the E1B19 protein of adenoviruses (White, 1996), studies with baculovirus have uncovered two additional apoptosis suppressors that can directly inhibit activated caspases: IAP (inhibitor of apoptosis) and p35 (reviewed in Clem *et al.*, 1996). IAP homologues have been discovered in both mammalian and *Drosophila* systems and their role as caspase inhibitors has recently been demonstrated by two separate studies (Deveraux and Reed, 1997; Wang *et al.*, 1999). However, IAPs may also function to repress cell death through direct interaction with and sequestering of pro-apoptotic factors such as the *Drosophila* death inducers REAPER, GRIM and HID. Thus, like the BLPs, these novel proteins may also have different modes of action.

The anti-apoptotic protein p35 has been shown to be a broad-range caspase inhibitor that is extremely useful for the study of cell death in animal systems. It has been found to efficiently inhibit all of the caspases examined so far without any significant inhibitory effects on numerous unrelated proteases that have been tested (Zhou *et al.*, 1998). However, no cellular homologue of this protein has been uncovered as yet. The mechanism of p35 as a caspase inhibitor has been well characterized: it first binds to caspases through a flexible, surface-exposed loop that contains multiple aspartate residues (Fisher *et al.*, 1999). Subsequent cleavage at this loop produces a large and small subunits of p35, both of which are required together to form a stable complex with the caspase and to prevent its further catalysis (Bump *et al.*, 1995). Mutations that suppress the cleavage of p35 by caspases inhibit its activity as an apoptosis repressor (Xue and Horvitz, 1995), thus demonstrating that its biological activity is related to cleavage by and subsequent inhibition of caspases.

Caspase-dependent and -independent pathways defining apoptotic and necrotic types of cell death in animal models: the mitochondrion connection

Although the role of caspases as the critical switch for the activation of apoptosis has largely been substantiated by numerous studies in the past decade, several new studies have suggested that there exist additional cell death pathways which may not involve these proteases as key regulators. In one study, nitric oxide (NO) was found to induce cell death in PC12 and HeLa cells with both apoptotic (TUNEL-positive and nuclear condensation) and necrotic (mitochondrial disruption and propidium iodide staining) phenotypes. Cell death in this case was not inhibited by either peptide inhibitors for caspases or p35 expression, but is suppressed by Bcl-2 and enhanced by Bax (Okuno *et al.*, 1998). No activation of CLP activity can be detected upon NO treatment and in fact, NO was shown to inhibit activity of purified caspase-3 and -6.

Studying cell death during development of digits in the mouse, Chautan *et al.* (1999) found that broad-range peptide inhibitors against caspases, as well as mutations in Apaf-1, can suppress the apoptotic changes normally observed in interdigital cells. Death of these cells nevertheless still occurs during development, but with morphologies more resembling necrosis with a 'mottled' rather than condensed nuclear phenotype and disruption of mitochondria. These observations raised the intriguing possibility that a necrotic type of cell death independent of caspases may be a default pathway that can mediate the proper developmental signals to induce PCD in the interdigital cells.

Lastly, it has been shown that expression of the pro-apoptotic BLP Bax activates cell death in yeast and bacteria, although caspase and Apaf-1 are absent in these organisms (Zha *et al.*, 1996; Jurgensmeier *et al.*, 1997; Asoh *et al.*, 1998). In both cases, co-expression of Bcl-2 or Bcl-X_L can effectively suppress the growth arrest and cell killing activity and the requirement of the BH3 domain in Bax, a region necessary for interaction with Bcl-2 and Bcl-X_L, suggests that the effects of the anti-apoptotic BLPs are likely through direct binding to the expressed Bax protein. Interestingly, co-expression of p35 failed to block Bax-induced cell death in yeast, consistent with the absence of caspases in yeast, and no CLP activity can be detected biochemically (Jurgensmeier *et al.*, 1997). The principal site of intracellular localization for Bax expressed in yeasts appears to be the mitochondrial membrane and the targeting of Bax to this compartment is vital to its death-inducing activity. Removal of the transmembrane domain near its C-terminus abolished its localization to the mitochondria and also concomitantly abrogated its induction of cell death in yeast. In contrast, this deletion does not alter mitochondria localization of Bax and it retains death-inducing activity when expressed in animal cells (Zha *et al.*, 1996). This differential dependence on the C-terminus for Bax localization and cell death induction is likely due to the heterodimerization of the truncated Bax with other pro-apoptotic BLPs that are present in animal cells but are absent in yeast. Elucidation of the mechanism for Bax-induced cell killing in yeast and bacteria may reveal an evolutionarily conserved cell death pathway distinct from caspase-dependent apoptosis. In both yeast and bacteria, reactive oxygen species (ROS) are increased by Bax expression and anoxygenic conditions can suppress Bax-induced death in yeast (Asoh *et al.*, 1998; Madeo *et al.*, 1999).

The induction of ROS by Bax in eukaryotes is likely due to its effect on the integrity of the mitochondrial membrane system that is mediated by its dimerization and pore formation activities, and interaction with the adenine nucleotide transporter (Marzo *et al.*, 1998; Vander Heiden *et al.*, 1999; Nouraini *et al.*, 2000). *In vitro* studies have also demonstrated the ability of Bax to form transmembrane pores for ions and cytochrome *c* by interaction with the voltage-dependent anion channel (VDAC), although cytochrome *c* release in yeast may not be functionally relevant to cell killing in Bax-expressing yeast (reviewed in Tsujimoto and Shimizu, 2000; Gross *et al.*, 2000; Kluck *et al.*, 2000). From genetic approaches, it has been demonstrated that the β-subunit of mitochondrial F_1F_0 ATPase is required for cell killing in yeast (Matsuyama *et al.*, 1998; Gross *et al.*, 2000). The biological relevance of this finding is supported by the observation that oligomycin, a specific inhibitor of the F_0 portion of the mitochondrial ATPase, can indeed suppress Bax-induced cell death in yeast as well as mammalian cells (Matsuyama *et al.*, 1998). Taken together, the mitochondrion is now considered to be a likely candidate as an evolutionarily conserved player in orchestrating cell death via its management of ROS while the elaboration of caspase-dependent apoptotic events may be a later event in evolution that provides signal amplification as well as a more rapid and orderly mode of PCD (Green and Kroemer, 1998; Blackstone and Green, 1999; Vander Heiden *et al.*, 1999). However, the exact mechanism(s) of ROS generation and the signaling pathways from ROS that ultimately leads to the observed death phenotypes remain unclear. In addition to ROS and cytochrome *c*, there may be other death factors that originate from the mitochondrion upon cell death induction (Susin *et al.*, 1999). The full extent to which this organelle is involved in driving the death engine of eukaryotic cells remains to be elucidated.

Evidence for caspase involvement in PCD control in plants

Results with peptide inhibitors

Are CLPs present in plants and do they play any role in plant cell death in an analogous fashion to that in animal apoptosis? To date, the evidence for the existence of CLPs in plants is still indirect and largely based on the inhibitory effects of caspase-specific inhibitors *in vitro* or *in vivo* with isolated plant cells or tissues. Nevertheless, it is intriguing that the circumstantial evidence is mounting in favor of the existence of cell death-related proteases which should have functional, if not structural, properties that are strikingly similar to caspases.

One of the first effects of caspase-specific inhibitors on plant cell death was reported in tobacco (del Pozo and Lam, 1998). Coinfiltration of either the caspase-specific inhibitor Ac-DEVD-CHO (more specific for caspase-3) or Ac-YVAD-CMK (more specific for caspase-1) with *Pseudomonas syringae* pv. *phaseolicola* (NPS3121), a bean pathogen that induces non-host resistance in tobacco, abolished HR cell death. Specific HR cell death molecular markers *hsr203J* and *hin1* (Pontier *et al.*, 1994; Gopalan *et al.*, 1996), which are genes induced early during incompatible interactions, were not expressed in the presence of caspase inhibitors. This observation suggests that the step where the relevant CLPs act in tobacco might be an early one in the cell death process. To test the specificity of action for these inhibitors, other protease inhibitors with different target preferences were tested in tobacco. PMSF (phenylmethylsulfonyl fluoride) or TPCK (N-tosyl-L-phenylalanine-chloromethyl ketone) slowed the HR cell death although they did not abolish it; leupeptin (a broad-range cysteine/serine protease inhibitor), TLCK (N-tosyl-lysine-chloromethyl ketone), and E-64 (a cysteine protease inhibitor) had no effect on HR cell death development in this system. TPCK and TLCK contain the CMK group and were used at higher concentrations than used for the CMK caspase inhibitors, providing the appropriate control for the potent and reactive CMK group. Antipain, a broad-range cysteine/serine protease inhibitor, blocks cell death in this system although it does not inhibit caspases. This result suggests that there may be additional proteases involved in cell death activation in this system. However, due to the broad target range of this inhibitor we cannot rule out that it is inhibiting the same caspase-like activity. Interestingly, although the induction of cell death and early HR-specific molecular markers can be completely inhibited by coinfiltration with caspase-specific peptide inhibitors, the induction of PR-1 and PR-2 gene expression was only retarded slightly and eventually expressed to similar levels. In contrast, much lower levels of PR gene expression were observed with an isogenic *Pseudomonas* strain (NPS4000) that is unable to induce the HR in tobacco, in the absence or presence of caspase inhibitors. These results together suggest that the signaling between the pathogen and the host can proceed in the apparent absence of cell death and that the induction of PR gene expression, which is a common marker for resistance induction, can be uncoupled from cell death. This conclusion is consistent with our previous observation in which anoxygenic conditions also suppress cell death without abolishing PR gene induction (Mittler *et al.*, 1996).

Caspase-specific proteolytic activity has been characterized in TMV (tobacco mosaic virus)-infected tobacco tissues that were synchronized to develop HR cell death by a temperature shift protocol (del Pozo and Lam, 1998). The N gene/TMV pathosystem allows systemic spreading of TMV at 30 °C; upon shifting to restrictive temperature (23 °C), massive cell death can be induced synchronously. With a synthetic fluorogenic substrate specific for animal caspase-1 (Villa *et al.*, 1997), increases in caspase-like cleavage activity could be detected in leaf extracts before the appearance of visible lesions and increases in ion leakage. Interestingly, no proteolytic activity could be detected using a caspase-3 fluorogenic substrate (Ac-DEVD-AMC). This CLP activity could be inhibited by caspase-specific inhibitors (Ac-YVAD-CMK, Ac-DEVD-CHO and Ac-DEVD-FMK) but was not significantly affected by the addition of other unrelated protease inhibitors. CLP activity has also been detected in pea plants that are induced to undergo senescence in caspase-specific peptide substrates (long and short substrates). The activity was sensitive to inhibition by caspase-specific inhibitors, suggesting plant CLPs may also be involve in the senescence process (A. Levine, personal communication).

Induction of proteolytic activity that may be relevant to cell death during the HR has also been studied in the cowpea rust fungus/cowpea pathosystem using bovine poly(ADP-ribose) polymerase (PARP) as substrate (D'Silva *et al.*, 1998). The cowpea rust fungus is an obligate biotroph in susceptible cowpea plants and elicits a HR in resistant cultivars. PARP is a well-characterized substrate for caspase-3 (Lazebnik *et al*, 1993) and was found to be endoproteolytically cleaved when added to extracts prepared from fungus-infected cowpea plants that were developing an HR. No PARP cleavage could be detected in the presence of extracts from uninfected cowpea plants or from plants that were undergoing a susceptible interaction. This cleavage activity was inhibited by different cysteine protease inhibitors to various degrees: E-64 (full inhibition), IAAmide (nearly total), NEM (partial), Ac-DEVD-CHO (partial inhibition) and GDEVDGIDEV (peptide spanning the cleavage site of PARP by caspase 3, partial inhibition). Interestingly, replacement of Asp by Ala at the P1 position in the long peptide (GDEVAGIDEV) resulted in a loss of PARP cleavage inhibition, thus demonstrating the specificity. Ac-YVAD-CHO, an inhibitor more specific for caspase-1, showed no effect on PARP cleavage by the cowpea extracts. Other types of protease inhibitor examined in this study include those targeting serine proteases, metalloproteases, calpain, and aspartate proteases, with little effects observed. Although the cleavage of an exogenously added substrate is physiologically and mechanistically irrelevant in the context of HR cell death, this study provides interesting evidence for a CLP in cowpea the activity of which is correlated with the induction of HR cell death.

Cytosolic extracts from carrot cell suspension cultures, upon addition of cytochrome *c*, were able to trigger apoptosis-like changes in purified liver nuclei (Zhao *et al.*, 1999a). The observed changes included chromatin condensation, DNA laddering and cleavage of nuclear lamins to specific fragments. None of these changes could be observed in liver nuclei in the absence of added cytocrome *c*. Interestingly, DNA laddering induced by carrot cell extracts was inhibited in the presence of caspase-1 inhibitor (Ac-YVAD-CHO) and also with higher concentrations of Ac-DEVD-CHO, suggesting that a CLP is involved in this phenomenon. Similar cytochrome *c*-dependent induction of PCD in purified carrot nuclei has also been observed upon addition of the carrot cytosol fraction, CS-100 (Zhao *et al.*, 1999b), although the effect of caspase inhibitors in this more homologous system was not reported. In any case, these studies provided the first evidence that a cytochrome *c*-dependent CLP may be present in the carrot cell suspension culture and factors for cell-free apoptosis in mammalian nuclei are present in the cytosol of plant cells. In another study, Sun *et al.* (1999) found evidence that menadione-induced cell death of tobacco protoplasts in culture resembles apoptosis in animal cells. In addition to the appearance of nucleosomal ladders within 4 to 6 h, these authors detected evidence for menadione-induced cleavage of tobacco PARP using antisera against mammalian PARP and leakage of cytochrome *c* into the cytosol fraction. Nuclear DNA laddering and PARP cleavage, but not cytochrome *c* release, are inhibited by caspase inhibitors Ac-DEVD-CHO and Ac-YVAD-CHO. In addition, in a cell-free system, it was shown that cytochrome *c* addition or purified mitochondria from tobacco protoplasts are sufficient to activate apoptosis-like phenotypes in mouse liver nuclei in combination with tobacco cytosol extracts. This work suggests that menadione-induced death of plant cells may involve a cytochrome *c* repartitioning step that triggers cell death-inducing mechanisms via activation of CLPs. However, the identity of the target for the cytochrome *c*-dependent activity remains unclear. Nevertheless, the parallels between these results and those observed in animal cell models are quite striking and would argue for the conservation of a cytochrome *c*/Apaf-1-like switch for plant CLPs. In addition to this study with menadione treatment, D-mannose addition and heat stress have also been shown to induce cytochrome *c* leakage from the mitochondria into the cytosol with concomitant DNA fragmentation and cell death in plant cells (Stein and Hansen, 1999; Balk *et al.*, 1999). These studies thus indicate that cytochrome *c* release into the cytosol may be a common phenomenon for different types of PCD pathways in plants.

Observations with baculovirus p35 expression in transgenic plants

The broad range and specific inhibition of caspases, as well as its capacity to block cell death when ectopically expressed in multiple systems, have made p35 protein a popular tool. It is often used to implicate caspases as a participant in activating different forms of cell death and in different systems. In order to test the involvement of CLPs in plant cell death *in vivo*, we generated transgenic tobacco plants that express p35 under the control of the constitutive 35S promotor from cauliflower mosaic virus (CaMV) (del Pozo and Lam, submitted). This approach would allow us to confirm the initial results obtained with synthetic peptide inhibitors on HR cell death. Possible non-specific effects of the synthetic peptide inhibitors should be avoided by the use of transgenic expression of this physiological inhibitor protein. We found that p35-expressing transgenic tobacco plants were phenotypically normal and were fertile as in wild-type tobacco. When challenged with incompatible bacteria (*P. syringae* pv. *phaseolicola* strain NPS3121) or virus (TMV) pathogens, these plants exhibited partial inhibition of HR cell death as measured by a delay in electrolyte leakage. Expression of p35 did not affect virus replication and did not delay the accumulation of PR proteins, thus suggesting that only the cell death component of the defense response was significantly affected. High p35 expressors infected with TMV showed a striking phenotype: systemic symptoms of TMV infection (vein clearing, mosaic patterns of chlorosis in leaves and stunting of plant growth) and HR-like lesions in upper uninfected tissues including stems, leaves and floral organs. In fact, these symptoms correlated with virus presence, thus showing that timely induction of cell death is necessary for TMV restriction within the primary infection site. To determine whether the cell death inhibitory effects of p35-expressing plants were related to the ability of p35 to inhibit caspases, different mutant versions of the p35 protein were generated that are impaired in caspase inhibition. Two p35 mutants were altered by site-directed mutagenesis in the caspase target site and the other one was a C-terminal deletion which has previously been shown to be defective in caspase inhibition (Clem *et al.*, 1996; Xue and Horvitz, 1995). Transgenic tobacco plants that were expressing these mutant p35 genes under control of the CaMV 35S promoter did not show cell death inhibition or systemic TMV spread. This result confirms that the effect of p35 in HR cell death is related to its ability to inhibit putative CLPs in tobacco. In addition, this transgenic system provides a unique system to study the dynamics of host-pathogen interaction within a resistant background.

Consistent with our results with HR cell death as a model system, leaves of transgenic tobacco plants expressing p35 were also found to resist cell death activation by UV-B irradiation (Y. Ohashi, personal communication). In this case, however, mutational analyses have not been carried out as yet to ascertain the correlation between the ability of p35 to inhibit caspases and its effect on cell death.

Correlative evidence for the conservation of mitochondrial participation in plant PCD

Expression and targeting of Bax to plant mitochondria resulted in HR-like cell death

If caspases are indeed conserved between plant and animal systems as cell death regulators, we may expect some other components in the tripartite regulatory switch that we discussed above to be conserved as well. Although no clear evidence for the presence of a functional homologue of BLPs has been reported in plants to date, there are observations that may be viewed as consistent with this expectation.

Recently, it has been reported that expression of Bax in tobacco using a TMV-derived vector induces cell death and PR expression that is independent of the N-resistant gene (Lacomme and Santa Cruz, 1999). Bax-induced cell death was inhibited by okadaic acid, a phosphatase inhibitor that blocks cell death induced by TMV in N tobacco plants, which suggests that Bax could be activating downstream elements of the same pathway. Mutational studies were performed by disruption of the Bcl-2 homology (BH) domains in tobacco. TMV expressing wild-type Bax protein produced smaller lesions than those activated by TMV-expressing Bax mutants. BH2 deletion induced slightly larger lesions while BH1 and BH3 deletion mutants of Bax activated much bigger lesions than wild-type Bax, likely as a consequence of a delay in the activation of resistance responses in the case of Bax mutants which leads to more extensive spread of the virus. This is in contrast to Bax deletion and expression studies in yeast and mammals, where the BH3 domain was found to be necessary for Bax cytotoxicity (Zha *et al.*, 1996). Truncating Bax of its transmembrane domain at the C-terminus only gave rise to mild chlorotic symptoms in tobacco, thus showing that this region is essential for cell death activation and mitochondria localization, as it was found in yeast. In mammalian cells, Bax expression can kill cells in a caspase-independent manner (Xiang *et al.*, 1996) although the physiological relevance is still uncertain. The results of Lacomme and Santa Cruz (1999) nevertheless provided evidence that Bax is likely to trigger HR-like cell death in tobacco by affecting mitochondria homeostasis. Whether this involves plant CLPs or not is unclear at the moment. It still remains to be shown if the TMV is arrested at the boundary of the HR-like lesions in Bax-expressing plants and if cytochrome *c* is released from the mitochondria to the cytosol as in yeast and mammals. It should also be interesting to test if coexpression of Bcl-2 or Bcl-X_L can suppress cell death promoted by Bax in plant cells.

Expression of Bcl- X_L *and CED-9 in tobacco confers resistance to UV and pathogen-induced cell death*

Using transgenic tobacco plants expressing anti-death BLPs, Bcl-X_L and CED-9, Mitsuhara *et al.* (1999) found suppression of cell death induced by UV-B, paraquat or TMV/N interaction. The level of cell death inhibition correlated with Bcl-X_L or CED-9 protein expression quantitatively. TMV lesions in Bcl-X_L-expressing transgenic plants were slightly larger than in WT plants two days after infection and HR cell death activation was delayed. Induction of synchronous cell death in Bcl-X_L-expressing plants by keeping the infected plants at non-permissive temperature (30 °C) and shifting them to permissive temperature (20 °C) revealed a 2–4 h delay in the onset of HR and reduced cell death as measured by conductivity compared with WT plants. In addition, PR-1 expression was reduced and delayed after TMV infection. In this system, mutational studies were not provided so it is not clear if mitochondria localization of BLPs is necessary for the delay in cell death or if HR cell death was blocked by interacting with cytoplasmic factors (like CED-4/Apaf-1) to inhibit plant CLPs. Although HR cell death is delayed by BLP expression in the TMV/N system, it is interesting to note that similar levels of cell death nevertheless do take place eventually. This is consistent with previous observations with Bcl-X_L-expressing transgenic tobacco plants where cell death was reported not to be significantly different after TMV challenge of N tobacco without temperature shift to synchronize HR induction (Mittler *et al.*, 1996). Similarly, the effects of BLP expression on death induction by paraquat treatment are quantitative rather than a complete suppression (Mitsuhara *et al.*, 1999). One interpretation of the inability of BLPs to maintain cell death inhibition in plants may be that multiple pathways of cell death could be activated by the resistance pathway and only a subset of these may be repressed by BLP expression. In contrast to HR- and paraquat-induced cell death, the protective effects of BLPs on UV-B treatment are much more dramatic, as quantitated by ion leakage measurements (Mitsuhara *et al.*, 1999). This observation suggests that cell death induced by UV photo-damage is more specifically regulated by BLP-sensitive mediators.

Conserved and plant-specific regulators for plant cell death?

With yeasts that were expressing Bax, suppressors of its cytotoxic effect were found by co-expression of a mammalian cDNA library. Three independent clones encoding the same novel protein were characterized and the gene was named Bax inhibitor-1 (BI-1). When overexpressed, BI-1 was able to suppress apoptosis in mammalian cells triggered by different stimuli such as growth factor withdrawal, etoposide and staurosporine (Xu and Reed, 1998). Interestingly, Fas ligand-induced cell death was not affected by BI-1, suggesting that BI-1 acts on a subset of caspase-dependent apoptotic pathways. Although structural predictions based on the deduced BI-1 amino acid sequence suggested a membrane protein, it is not clear how BI-1 acts to suppress apoptosis, since BI-1 appears not to bind Bax in cells but is able to interact with Bcl-2 and Bcl-X_L. Interestingly, putative homologues of BI-1 were found in the *Arabidopsis* and *C. elegans* databases and, recently, a rice homologue of BI-1 (with 45% sequence identity) was isolated and shown to suppress Bax-induced cell death in yeast (Kawai *et al.*, 1999). The fact that a BI-1 homologue has not been detected in the yeast genome database suggests that this protein may be acquired in a progenitor eukaryote that ultimately gives rise to animals and plants but is distinct from the yeast lineage. Alternatively, the BI-1 homologue in yeast may have been lost during evolution. It remains to be demonstrated whether animal and plant BI-1 homologues do in fact play an essential role in the regulation of PCD by loss-of-function mutation approaches. Nevertheless, BI-1 is at present an interesting candidate for a common conserved cell death-specific regulator in plants and animals.

A plant-specific regulator of PCD could be the alternative oxidase (AOX) that resides in plant mitochondria. This enzyme could function to control the generation of ROS from the mitochondrial electron transport chain when oxidative phosphorylation is inhibited. For example, when Complex III of the mitochondria is inhibited by antimycin A treatment, AOX expression is induced and ROS generation is kept to a minimum so that little cell death is activated. However, when AOX expression is suppressed by antisense strategies, antimycin A treatment leads to high levels of ROS generation with concomitant cell death (Vanlerberghe *et al.*, 1995; Maxwell *et al.*, 1999). Induction of AOX expression by treatment with antimycin A and KCN was also found to correlate with suppression of cell death lesions during TMV infection of tobacco plants when salicylic acid accumulation was inhibited (Chivasa and Carr, 1998). These results suggest that AOX may be involved in the control of ROS generation in plant mitochondria and thereby regulate cell death induction in plants. In addition to ROS, however, it is clear from the work of Glazener *et al.* (1996) that there are likely additional signal mediators that are necessary for HR cell death activation.

Future perspectives

Although the morphological and cytological changes which accompany plant PCD in reponse to pathogens have proven to be diverse and may vary depending on the pathogen and the host (reviewed in Lam *et al.*, 1999), some common events emerged from comparative analysis. Changes in microtubule organization as well as nuclear condensation are among the common features of plant cell death. DNA fragmentation has also been observed during the HR, disease cell death, senescence and cell death under different developmental contexts. Internucleosomal cleavage that leads to DNA laddering, one of the hallmarks of animal apoptosis, has also been shown to occur in some cases of plant cell death (Navarre and Wolpert, 1999; Lam *et al.*, 1999; Heath, this issue; Sugiyama *et al.*, this issue). Interestingly, these changes are shared with animal PCD where caspases or caspase-induced factors are responsible for initiating these cellular changes. In contrast, caspase-independent PCD does not show apoptotic phenotypes such as TUNEL staining of nuclei in dying cells (Chautan *et al.*, 1999).

Using iterative database searches, Aravind *et al.* (1999) found a *C. elegans* open reading frame (encoded in the genomic clone F22D3.6) which is distantly related to caspases. It contains the catalytic residues as well as the critical structural elements of the caspases, which indicates that the three-dimensional fold of the protein is likely conserved. Using this new member of the caspase family to search in the database, distant homologues of putative caspase-like proteases from *Dictyostelium discoideum*, *Streptomyces coelicolor* and *Rhizobium* were identified. The putative caspase found in bacteria is linked to a kinase domain, suggesting that these proteins may have other functions in these systems. The completion of the *Arabidopsis* and rice genomes in the near future should allow a more detailed and complete search for caspase-related proteins that may be expressed in plants. Whether there is structural conservation between plant CLPs and metazoan caspases will be likely resolved by the cloning of the plant enzymes that are responsible for the CLP activities detected as we have discussed in this review. The answer will help to resolve whether the CLPs of plants and animals are evolutionarily related or if it is a case of convergent evolution in which unrelated proteases with similar substrate specificities have been independently selected for the control of cell death in plants and animals.

Acknowledgements

We would like to thank A. Levine, Y. Ohashi for sharing of unpublished observations, and N. Kato for critical reading of this manuscript. Work on cell death in the Lam laboratory is supported in part by a competitive grant from the USDA and the New Jersey Commission of Science and Technology.

References

Alnemri, E.S., Livingston, D.J., Nicholson, D.W., Salvesen, G., Thornberry, N.A., Wong, W.W. and Yuan, J. 1996. Human ICE/CED-3 protease nomenclature. Cell 87: 171.

Aravind, L., Dixit, V.M. and Koonin, E. 1999. The domains of death: evolution of the apoptosis machinery. Trends Biochem. Sci. 24: 47–53.

Asoh, S., Nishimaki, K., Nanbu-Wakao, R. and Ohta, S. 1998. A trace amount of the human pro-apoptotic factor Bax induces bacterial death accompanied by damage of DNA. J. Biol. Chem. 273: 11384–11391.

Balk, J., Leaver, C.J. and McCabe, P. 1999. Translocation of cytochrome *c* from the mitochondria to the cytosol occurs during heat-induced programmed cell death in cucumber plants. FEBS Lett. 463: 151–154.

Blackstone, N.W. and Green, D.R. 1999. The evolution of a mechanism of cell suicide. BioEssays 21: 84–88.

Bump, N.J., Hackett, M., Hugunin, M., Seshagiri, S., Brady, K., Chen, P., Ferenz, C., Franklin, S., Ghayur, T., Li, P., Licari, P., Mankovich, J., Shi, L., Greenberg, A.H., Miller, L.K. and Wong, W.W. 1995. Inhibition of ICE family proteases by baculovirus antiapoptotic protein p35. Science 269: 1885–1888.

Chautan, M., Chazal, G., Cecconi, F., Gruss, P. and Golstein, P. 1999. Interdigital cell death can occur through a necrotic and caspase-independent pathway. Curr. Biol. 9: 967–970.

Chen, P., Rodriguez, A., Erskine, R., Thach, T., and Abrams, J.M. 1998. Dredd, a novel effector of the apoptosis activators reaper, grim, and hid in *Drosophila*. Dev. Biol. 201: 202–216.

Chinnaiyan, A.M., O'Rourke, K., Lane, B.R. and Dixit, V.M. 1997. Interaction of CED-4 with CED-3 and CED-9: a molecular framework for cell death. Science 275: 1122–1126.

Chivasa, S. and Carr, J.P. 1998. Cyanide restores N gene-mediated resistance to tobacco mosaic virus in transgenic tobacco expressing salicylic acid hydroxylase. Plant Cell 10: 1489–1498.

Clem, R.J., Hardwick, J.M. and Miller, L.K. 1996. Anti-apoptotic genes of baculoviruses. Cell Death Differ. 3: 9–16.

Conradt, B. and Horvitz, R.H. 1998. The *C. elegans* protein EGL-1 is required for programmed cell death and interacts with the Bcl-2-like protein CED-9. Cell 93: 519–529.

Cryns, V. and Yuan, J. 1998. Proteases to die for. Genes and Development 12: 1551-1570.

D'Silva, I., Pirier, G.G. and Heath, M.C. 1998. Activation of cysteine proteases in cowpea plants during the hypersensitive response, a form of programmed cell death. Exp. Cell Res. 245: 389–399.

del Pozo, O. and Lam, E. 1998. Caspases and programmed cell death in the hypersensitive response of plants to pathogens. Curr. Biol. 8: 1129–1132.

Deveraux, Q.L. and Reed, J.C. 1997. X-linked IAP is a direct inhibitor of cell-death proteases. Nature 388: 300–304.

Dorstyn, L., Colussi, P.A., Quinn, L.M., Richardson, H., and Kumar, S. 1999. DRONC, an ecdysone-inducible *Drosophila* caspase. Proc. Natl. Acad. Sci. USA 96: 4307–4312.

Ellis, H.M. and Horvitz, R.H. 1986. Genetic control of programmed cell death in the nematode *C. elegans*. Cell 44: 817–829.

Fisher, A.J., de la Cruz, W., Zoog, S.J., Schneider, C.L. and Friesen, P.D. 1999. Crystal structure of baculovirus P35: role of a novel reactive site loop in apoptotic caspase inhibition. EMBO J. 18: 2031–2039.

Glazener, J.A., Orlandi, E.W. and Baker, J.C. 1996. The active oxygen response of cell suspensions to incompatible bacteria is not sufficient to cause hypersensitive cell death. Plant Physiol. 110: 759–763.

Gopalan, S., Wei, W. and He, S.Y. 1996. *hrp* gene-dependent induction of *hin1*: a plant gene activated rapidly by both harpins and the *avrPto* gene-mediated signal. Plant J. 10: 591–600.

Green, D. and Kroemer, G. 1998. The central executioners of apoptosis: caspases or mitochondria? Trends Cell Biol. 8: 267–271.

Gross, A., Pilcher, K., Blachly-Dyson, E., Basso, E., Jockel, J., Bassik, M.C., Korsmeyer, S.J. and Forte, M. 2000. Biochemical and genetic analysis of the mitochondrial response of yeast to BAX and BCL-X_L. Mol. Cell. Biol. 20: 3125–3136.

Hengartner, M.O., Ellis, R.E., and Horvitz, R.H. 1992. *Caenorhabditis elegans* gene *ced-9* protects cells from programmed cell death. Nature 356: 494–499.

Hu, Y., Benedict, M.A., Ding, L. and Nunez, G. 1999. Role of cytochrome c and ATP/ATP hydrolysis in Apaf-1-mediated caspase-9 activation and apoptosis. EMBO J. 18: 3586–3595.

Jurgensmeier, J.M., Krajewski, S., Armstrong, R.C., Wilson, G.M., Oltersdorf, T., Fritz, L.C., Reed, J.C., and Ottilie, S. 1997. Bax- and Bak-induced cell death in the fission yeast *Schizosaccharomyces pombe*. Mol. Biol. Cell. 8: 325–339.

Kanuka, H., Sawamoto, K., Inohara, N., Matsuno, K., Okano, H. and Miura, M. 1999. Control of the cell death pathway by Dapaf-1, a *Drosophila* Apaf-1/CED-4-related caspase activator. Mol. Cell 4: 757–769.

Kawai, M., Pan, L., Reed, J.C. and Uchimiya, H. 1999. Evolutionarily conserved plant homologue of the Bax Inhibitor-1 (BI-1) gene capable of suppressing Bax-induced cell death in yeast. FEBS Lett. 464: 143–147.

Kluck, R.M., Ellerby, L.M., Ellerby, M.H., Naiem, S., Yaffe, M.P., Margoliash, E., Bredesen, D., Mauk, G.A., Sherman, F. and Newmeyer, D.D. 2000. Determinants of cytochrome *c* pro-apoptotic activity. J. Biol. Chem. 275: 16127–16133.

Korsmeyer, S.J. 1995. Regulators of cell death. Trends Genet. 11:101–105.

Lacomme, C., and Santa Cruz, S. 1999. Bax-induced cell death in tobacco is similar to the hypersensitive response. Proc. Natl. Acad. Sci. USA 96: 7956–7961.

Lam, E., Pontier, D., and del Pozo, O. 1999. Die and let live: programmed cell death in plants. Curr. Opin. Plant Biol. 2: 502–507.

Lazebnik, Y.A., Kaufman, S.H., Desnoyers, S., Poirier, G.G. and Earnshaw, W.C. 1993. Cleavage of poly(ADP-ribose) polymerase by a proteinase with properties like ICE. Nature 371: 346–347.

Maedo, F., Frohlich, E., Ligr, M., Grey, M., Sigrist, S.J., Wolf, D.H. and Frohlich, K.-U. 1999. Oxygen stress: a regulator of apoptosis in yeast. J. Cell Biol. 145: 757–767.

Marzo, I., Brenner, C., Zamzami, N., Jurgensmeier, J.M., Susin, S.A., Vieira, H.L.A., Prevost, M.-C., Xie, Z., Matsuyama, S., Reed, J.C., and Koremer, G. 1998. Bax and adenine nucleotide translocator cooperate in the mitochondrial control of apoptosis. Science 281: 2027–2031.

Matsuyama, S., Xu, Q., Velours, J. and Reed, J.C. 1998. The mitochondrial F_0F_1-ATPase proton pump is required for function of the proapoptotic protein Bax in yeast and mammalian cells. Mol. Cell 1: 327–336.

Maxwell, D.P., Wang, Y. and McIntosh, L. 1999. The alternative oxidase lowers mitochondria reactive oxygen production in plant cells. Proc. Natl. Acad. Sci. USA 96: 8271–8276.

Mitsuhara, I., Malik, K.A., Miura, M. and Ohashi, Y. 1999. Animal cell-death suppressors Bcl-x_L and Ced-9 inhibit cell death in tobacco plants. Curr. Biol. 9: 775–778.

Mittler, R. and Lam, E. 1996. Sacrifice in the face of foes: pathogen-induced programmed cell death in plants. Trends Microbiol. 4: 10–15.

Mittler, R., Shulaev, V., Seskar, M. and Lam, E. 1996. Inhibition of programmed cell death in tobacco plants during a pathogen-induced hypersensitive response at low oxygen pressure. Plant Cell 8: 1991–2001.

Nakagawa, T., Zhu, H., Morishima, N., Li, E., Xu, J., Yankner, B.A. and Yuan, J. 2000. Caspase-12 mediates endoplasmic-reticulum-specific apoptosis and cytotoxicity by amyloid-β. Nature 403: 98–103.

Navarre, D.A. and Wolpert, T.J. 1999. Victorin induction of an apoptotic/senescence-like response in oats. Plant Cell 11: 237–249.

Nicholson, D.W. and Thornberry, N.A. 1997. Caspases: killer proteases. Trends Biochem. Sci. 22: 299–306.

Nouraini, S., Six, E., Matsuyama, S., Krajewski, S. and Reed, J.C. 2000. The putative pore-forming domain of Bax regulates mitochondrial localization and interaction with Bcl-XL. Mol. Cell. Biol. 20: 1604–1615.

Okuno, S.-I., Shimizuk, S., Ito, T., Nomura, M., Hamada, E., Tsujimoto, Y. and Matsuda, H. 1998. *bcl-2* prevents caspase-independent cell death. J. Biol. Chem. 273: 34272–34277.

Pontier, D., Godiard, L., Marco, Y. and Roby, D. 1994. *hsr203J*, a tobacco gene whose activation is rapid, highly localized and specific for incompatible plant pathogen interactions. Plant J. 5: 507–521.

Shaham, S. 1998. Identification of multiple *Caenorhabditis elegans* caspases and their potential roles in proteolytic cascades. J. Biol. Chem. 273: 35109–35117.

Slee, E.A., Harte, M.T., Kluck, R.M., Wolf, B.B., Casiano, C.A., Newmeyer, D.D., Wang, H.-G., Reed, J.C., Nicholson, D.W., Alnemri, E.S., Green, D.R. and Martin, S.J. 1999. Ordering the cytochrome *c*-initiated caspase cascade: hierarchical activation of caspases-2, -3, -6, -7, -8, and -10 in a caspase-9-dependent manner. J. Cell Biol. 144: 281–292.

Spector, M.S., Desnoyers, S., Hoeppner, D.J. and Hengartner, M.O. 1997. Interaction between the *C. elegans* cell-death regulators CED-9 and CED-4. Nature 385: 653–656.

Stein, J.C. and Hansen, G. 1999. Mannose induces an endonuclease responsible for DNA laddering in plants cells. Plant Physiol. 121: 71–79.

Sun, Y.L., Zhao, Y., Hong, X. and Zhai, Z.H. 1999. Cytochrome c release and caspase activation during menadione-induced apoptosis in plants. FEBS Lett. 462: 317–321.

Susin, S., Lorenzo, H.K., Zamzami, N., Marzo, I., Snow, B.E., Brothers, G.M., Mangion, J., Jacotot, E., Costantini, P., Loeffler, M., Larochette, N., Goodlett, D.R., Aebersold, R., Siderovski, D.P., Penninger, J.M. and Groemer, G. 1999. Molecular characterization of mitochondrial apoptosis-inducing factor. Nature 397: 441–445.

Talanian, R.V., Quinlan, C., Trautz, S., Hackett, M.C., Mankovich, J.A., Banach, D., Ghayur, T., Brady, K.D. and Wong, W.W. 1997. Substrate specificities of caspase family proteases. J. Biol. Chem. 272: 9677–9682.

Thornberry, N.A. and Lazebnik, Y. 1998. Caspases: enemies within. Science 281: 1312–1316.

Tsujimoto, Y. and Shimizu, S. 2000. Bcl-2 family: life-or-death switch. FEBS Lett. 466: 6–10.

Vander Heiden, M.G., Chandel, N.S., Schumacker, P.T. and Thompson, C.B. 1999. Bcl-XL prevents cell death following growth factor withdrawal by facilitating mitochondrial ATP/ADP exchange. Mol. Cell 3: 159–167.

Vanlerberghe, G.C., Day, D.A., Wiskich, J.T., Vanlerberghe, A.E., and McIntosh, L. 1995. Alternative oxidase activity in tobacco leaf mitochondria. Plant Physiol. 109: 353–361.

Vaux, D.L. and Korsmeyer, S.J. 1999. Cell death in development. Cell 96: 245–254.

Vierstra, R.D. 1996. Proteolysis in plants: mechanisms and functions. Plant Mol. Biol. 32: 275–302.

Villa, P., Kaufmann, S.H., and Earnshaw, W.C. 1997. Caspases and caspase inhibitors. Trends Biochem. Sci. 22: 388–393.

Wang, S.L., Hawkins, C., Yoo, S.J., Muller, H.J. and Hay, B.A. 1999. The *Drosophila* caspase inhibitor DIAP1 is essential for cell survival and is negatively regulated by HID. Cell 98: 453–463.

White, E. 1996. Life, death, and the pursuit of apoptosis. Genes Dev. 10: 1–15.

Wolf, B.B. and Green, D.R. 1999. Suicidal tendencies: apoptotic cell death by caspase family proteinases. J. Biol. Chem. 274: 20049–20052.

Xiang, J., Chao, D.T. and Korsmeyer, S.J. 1996. Bax-induced cell death may not require interleukin 1β-converting enzyme-like proteases. Proc. Natl. Acad. Sci. USA 93: 14559–14563.

Xu, Q. and Reed, J.C. 1998. Bax inhibitor-1, a mammalian apoptosis suppressor identified by functional screening in yeast. Mol. Cell 1: 337–346.

Xue, D. and Horvitz, R.H. 1995. Inhibition of the *Caenorhabditis elegans* cell death protease CED-3 by a CED-3 cleavage site in baculovirus p35 protein. Nature 377: 248–251.

Yuan, J.Y. and Horvitz, R.H. 1992. The *Caenorhabditis elegans* cell death gene *ced-4* encodes a novel protein and is expressed during the period of extensive programmed cell death. Development 116: 309–320.

Yuan, J.Y., Shaham, S., Ledoux, S., Ellis, H.M., and Horvitz, R.H. 1993. The *C. elegans* cell death gene *ced-3* encodes a protein similar to mammalian interleukin-1β-converting enzyme. Cell 75: 641–652.

Zha, H., Fisk, H.A., Yaffe, M.P., Mahajan, N., Herman, B. and Reed, J.C. 1996. Structure-function comparisons of the proapoptotic protein Bax in yeast and mammalian cells. Mol. Cell. Biol. 16: 6494–6508.

Zhao, Y., Jiang, Z.F., Sun, Y. and Zhai, Z.-H. 1999a. Apoptosis of mouse liver nuclei induced in the cytosol of carrot cells. FEBS Lett. 448: 197–200.

Zhao, Y., Sun, Y., Jiang, Z. and Zhai, Z. 1999b. Apoptosis of carrot nuclei in *in vivo* system induced by cytochrome *c*. Chin. Sci. Bull. 44: 1497–1501.

Zhou, Q., Krebs, J.F., Snipas, S.J., Price, A., Alnemri, E.S., Tomaselli, K.J. and Salvesen, G.S. 1998. Interaction of the baculovirus anti-apoptotic protein p35 with caspases. Specificity, kinetics and characterization of the caspase/p35 complex. Biochemistry 37: 10757–10765.

Plant Molecular Biology **44:** 429–442, 2000.
E. Lam, H. Fukuda and J. Greenberg (Eds.), Programmed Cell Death in Higher Plants.

Salicylic acid in the machinery of hypersensitive cell death and disease resistance

María Elena Alvarez
Departamento de Química Biológica, CIQUIBIC-CONICET, Facultad de Ciencias Químicas, Universidad Nacional de Córdoba, Ciudad Universitaria 5000, Córdoba, Argentina (e-mail: malena@dqbfcq.uncor.edu.ar)

Key words: disease resistance, hypersensitive cell death, lesion formation, oxidative burst, pro- and anti-death signals, salicylic acid

Abstract

Although extensive data has described the key role of salicylic acid (SA) in signaling pathogen-induced disease resistance, its function in physiological processes related to cell death is still poorly understood. Recent studies have explored the requirement of SA for mounting the hypersensitive response (HR) against an invading pathogen, where a particular cell death process is activated at the site of attempted infection causing a confined lesion. Biochemical data suggest that SA potentiates the signal pathway for HR by affecting an early phosphorylation-sensitive step preceding the generation of pro-death signals, including those derived from the oxidative burst. Accordingly, the epistatic relationship between cell death and SA accumulation, analyzed in crosses between lesion-mimic mutants (spontaneous lesion formation) and the transgenic *nahG* line (depleted in SA) places the SA activity in a feedback loop downstream and upstream of cell death. Exciting advances have been made in the identification of cellular protective functions and cell death suppressors that might operate in HR. Moreover, the spatio-temporal patterns of the SA accumulation (non-homogeneous distribution, biphasic kinetics) described in some HR lesions, may also reveal important clues for unraveling the complex cellular network that tightly balances pro- and anti-death functions in the hypersensitive cell death.

Abbreviations: AA, arachidonic acid; *avr* gene, pathogen avirulence gene; BA, benzoic acid; BA2H, benzoic acid 2-hydroxylase; COX, cyclooxygenase; HR, hypersensitive response; H_2O_2, hydrogen peroxide; IKK, IκB kinase complex; INA, 2,6-dichloroisonicotinic acid; MAP kinase, mitogen-activated protein kinase; NO, nitric oxide; O_3, ozone; $^{\bullet}O_2^-$, superoxide anion; PAL, phenylalanine ammonia-lyase; PR protein, pathogenesis-related protein; *R* gene, plant resistance gene; ROI, reactive oxygen intermediates; SA, salicylic acid; SAG, SA 2-*O*-β-D-glucoside; SAR, systemic acquired resistance; tCA, *trans*-cinnamic acid; TMV, tobacco mosaic virus

Introduction

The first physiological effects of salicylic acid (orthohydroxybenzoic acid) in living organisms were described in animals as unrelated to cell death. Ancient medicine ascribed relevant therapeutic successes to the willow bark, which for centuries had been used to reduce pain, heat and swelling. Clinical trials of willow bark were reported since 1763 (Stone, 1763) but it was not until 1829 that a glucoside precursor of SA exerting antipyretic effects was purified from willow (reviewed by Vane and Botting, 1998). Pharmacological use of SA took off after its chemical synthesis in 1860 and was largely replaced by a better-tolerated derivative, acetylsalicylic acid (aspirin) since 1899. A century later, when more than 30 thousand tons of aspirin are consumed per year, much remains to be learned about the mechanism of action of salicylates. A main anti-inflammatory effect of these non-steroidal drugs is mediated by the inhibition of cyclooxygenase (COX) (Vane, 1971), the rate-limiting enzyme in the biosynthesis of prostaglandins from arachidonic acid. The two COX isoforms in mammals, COX1 (constitutive) and COX2 (induced by cytokines and hormones), have different susceptibility to salicylates (Mitchell *et al.*, 1994) but are unlikely the only targets of SA in the control of inflammation. Salicylates also affect protein phosphorylation steps regulating cell growth, differentiation and inflammation (Pillinger *et al.*, 1998; Yin *et al.*, 1998) and they inhibit transcription factors such as NF-κB (Kopp and Ghosh, 1994) and AP-1 (Dong, 1997), leading to apoptotic programs in particular cases (Van Antwerp *et al.*, 1996; Beg and Baltimore, 1996).

Unlike animals, plants can synthesize SA and activate SA-dependent physiological programs (Klessig and Malamy, 1994). A key role of SA in plants is the activation of disease resistance that is frequently associated with a special type of cell death in the hypersensitive response (HR) (Dangl *et al.*, 1996; Greenberg, 1997). HR is locally triggered upon pathogen attack to induce defense responses leading to the collapse of infected cells (Hammond-Kosak and Jones, 1996). This local response is often related to the establishment of systemic acquired resistance (SAR) (Ross, 1961) that immunizes the entire plant against further infections which normally cause disease (Ryals *et al.*, 1996). The spatiotemporal pattern of SA accumulation in HR suggests that SA contributes to controlling the timing and extent of cell death. Moreover, biochemical and genetic data demonstrate that SA potentiates defense responses promoting lesion formation. Thus, although it is known that SA per se does not produce hypersensitive cell death, it may play an indirect but essential role in HR, by regulating both pro- and anti-death functions. This review summarizes the extensive data analyzing the role of SA in genetically defined resistance associated with hypersensitive cell death. The inter-relationship among the SA-dependent defense pathway and other programs leading to nonspecific resistance, such as those activated by jasmonic acid or ethylene, has been recently updated (Dong, 1998) and is not discussed here. The basis of inducible plant disease resistance (Ryals *et al.*, 1996; Baker *et al.*, 1997), SA biosynthetic pathways in plants (Klessig and Malamy, 1994; Lee *et al.*, 1995), and features of plant cell death in defense, development and transgenic models are discussed in other papers in this issue as well as in recent reviews (Dangl *et al.*, 1996; Greenberg, 1997; Glazebrook *et al.*, 1997; Pennell and Lamb, 1997).

Physiological involvement of SA in plants

Our knowledge of the physiological processes responding to SA in plants is still limited. Exogenously applied SA modifies ion uptake, stomatal closure, growth rate, photosynthesis and flower development among other processes. However, under physiological conditions, these events are not necessarily related to increases in endogenous SA levels (Raskin, 1992; Pancheva *et al.*, 1996). SA may participate in the induction of thermotolerance in mustard, where its levels rise 4-fold after heat treatment and exogenous SA can reproduce the response in a dose-dependent manner (Dat *et al.*, 1998a, b). A well-characterized role of SA in plants is the control of heat production in the inflorescences of thermogenic species of the genus *Arum.* There, a heating process that precedes volatilization of odor compounds in pollination is mediated by SA increase and is mimicked by exogenous SA (15 μM; Raskin *et al.*, 1987). Energy for heat generation comes from increased cyanide-resistant 'alternative' respiration that drives mitochondrial electron transport from the cytochrome to the alternative oxidase pathway avoiding ATP production (Vanlerberghe and McIntosh, 1997). Interestingly, SA induces the expression of alternative oxidase and activates the alternative respiratory pathway in thermogenic and non-thermogenic tissues (Raskin *et al.*, 1987; Rhoads and McIntosh, 1993; Lennon *et al.*, 1997). In addition, the activity of this oxidase responds to SA accumulated in defense responses against virus (discussed below; Chivasa and Carr, 1998). The best-characterized physiological role of SA in several plant species is the activation of inducible defense programs leading to disease resistance. which is discussed in the following sections.

Accumulation of SA in disease resistance

SA is an inducer of disease resistance in different species (Ryals *et al.*, 1996) and accumulates in infected and uninfected tissues of cucumber (Métraux *et al.*, 1990, Rasmussen *et al.*, 1991; Meuwly *et al.*, 1995), tobacco (Malamy *et al.*, 1990, Yalpani *et al.*, 1991, Enyedi *et al.*, 1992) and *Arabidopsis* (Uknes *et al.*, 1993; Summermatter *et al*; 1995), inducing SAR. In the last two cases, infected tissues accumulate ca. 6 μg SA per gram fresh weight (Enyedi *et al.*, 1992; Uknes *et al.*, 1993) which may represent a cytosolic concentration of ca. 70 μM considering no compartmentalization (Bi *et al.*, 1995). Potato plants contain higher basal levels of SA (5 μg/g in leaves) and induce SAR in response to arachidonic acid (AA) but do not respond to SA (Coquoz *et al.*, 1995). However, potato plants that express a bacterial salicylate hydroxylase which oxidizes SA into catechol (*nahG* plants; Gaffney *et al.*, 1993; Delaney *et al.*, 1994) do not develop SAR in response to AA (Yu *et al.*, 1997). The requirement of SA for SAR in potato does not involve an increase in SA biosynthesis, but probably functions through an enhanced sensitivity to SA. Less information about the function of SA in disease resistance is available for monocots. Rice, which contains the highest SA levels among the plant species tested so far (30 μg/g in shoots), does not significantly accumulate SA in disease resistance (Silverman *et al.*, 1995) and by still unknown mechanisms develops inducible disease resistance.

Feeding experiments and radiolabeling tracer studies revealed some steps of the SA biosynthetic pathways in healthy and infected tissues (Lee *et al.*, 1995). The activation of the phenylpropanoid pathway in disease resistance may generate SA via *trans*-cinnamic acid (tCA) and benzoic acid (BA) at least in tobacco (Yalpani *et al.* 1993), cucumber (Meuwly *et al.*, 1995) and potato (Coquoz *et al.*, 1998). Phenylalanine ammonia-lyase (PAL) catalyzes the first step in the pathway in which phenylalanine is converted into tCA. PAL is transcriptionally activated in response to pathogen attack, mechanical damage, and a variety of abiotic stresses (Dixon and Paiva, 1995). Interestingly, PAL gene activation by an avirulent pathogen can be blocked by inhibiting PAL enzyme activity, suggesting that the pathway can be autoamplified by an intermediate (Mauch-Mani and Slusarenko, 1996) and, as described below, SA may contribute to this kind of amplification. In tobacco, the release of SA from BA involves the activation of benzoic acid 2-hydroxylase (BA2H), a cytochrome P450 monooxygenase. BA2H is synthesized *de novo* upon BA increase or in response to pathogen attack preceding HR development and is induced within 5 min after treatment with hydrogen peroxide (H_2O_2) in a dose-dependent manner (León *et al.*, 1995).

A large part of SA in plants is in conjugates, derived from glucosylation or esterification of the unique hydroxyl or carboxyl groups, or from modifications in other positions of the aromatic ring (Enyedi *et al.*, 1992; Klessig and Malamy, 1994). The main conjugate in tobacco is SA 2-*O*-β-D-glucoside (SAg), which accumulates in the vicinity of HR lesions upon the activation of SA glucosyltransferase. This enzyme is inducible by high concentrations of SA reached in infected tissues (Enyedi *et al.*, 1992). SAg neither exudes from infected leaves nor induces genes encoding PR (pathogenesis-related) proteins effectively, suggesting its poor capacity to signal systemic defenses (Lee *et al.*, 1995). In contrast, methyl salicylate is a proposed airborne defense signal in tobacco (Shulaev *et al.*, 1997). It should be noted that the majority of data concerning the content of SA in disease resistance programs refer to the free or glucosylated forms, while the levels of other derivatives have not been systematically analyzed.

SA and cell death: temporal relationship

A pre-necrotic[1] accumulation of SA has been detected in several tissues developing HR (Métraux *et al.*, 1990; Malamy *et al.*, 1990, 1992; Summermatter *et al.*, 1995, Draper, 1997). Infection of cucumber leaves with tobacco necrosis virus generates macroscopic lesions after 72 h. In the phloem sap of infected cucumber plants, a transient rise of SA (10-fold increase) occurs around 48 h and, 5 days later, a second and higher SA increase parallels the establishment of SAR (Métraux *et al.*, 1990). Likewise, two-phase kinetics of SA accumulation also occurs in tobacco mosaic virus (TMV)-infected tobacco plants (resistant NN genotype) where a transient, small and pre-necrotic peak (10-fold increase) is followed by a sustained, massive and post-necrotic SA rise (50-fold increase) (Malamy *et al.*, 1990). Curiously, biphasic SA accumulation in HR-like lesions is not necessarily induced in response to pathogens. Tobacco transgenic plants deficient in catalase activity (Cat1AS; 90% reduction) respond to high light exposure by accumulating H_2O_2 and activating several HR components including cell death (Chamnongpol *et al.*, 1998). High light treatment induces not only a pre-necrotic rise of SA, but also a later sustained accumulation.

Neither of these studies analyzed the effect of each SA peak in HR development but recent work with tobacco transgenic lines impaired in SA accumulation addresses this question (Mur *et al.*, 1997). The SH-L isoform of salicylate hydroxylase from *P. putida* was placed under the control of two different promoters, AoPR1 (inducible by H_2O_2) or PR1a (inducible

[1] The term 'necrotic' alludes to either apoptotic or necrotic cell death.

by SA), and introduced into tobacco plants harboring the N-resistance gene. After infection with TMV, both lines show a 2-fold reduction in the SA levels in the second phase, although it still accumulates (ca. 3.5 μg/g). Only the AoPR1:SH-L plants strongly suppress SA accumulation in the pre-necrotic phase (ca. 7-fold reduction, ca. 60 ng/g). Lesion formation is abnormal only in the AoPR1:SH-L plants where it initiates later and extends faster and longer than in wild-type plants. While the inefficient second phase of SA accumulation may contribute to such effect, the lack of the first phase appears to be crucial. Thus, in wild-type plants, the early enhancement of SA in HR may speed up both initiation and limitation of cell death. The AoPR1-promoter activity is associated with the lesion margin and residual SA may accumulate beyond this margin in healthy tissues of AoPR1:SH-L plants. In this line, lesions in AoPR1:SH-L plants are better confined than those produced in a transgenic line constitutively depleted of SA (35S:SH-L plants; Mur *et al.*, 1997) suggesting that residual SA levels in healthy tissues surrounding dead cells may contribute to the restriction of lesion spread. SA can interfere with viral replication and systemic movement (Murphy *et al.*, 1999) and thus can directly suppress spreading through a process unrelated to cell death. However, SA can also influence the rate of HR-like lesion formation in lesions not generated by viral infection, such as those caused by ozone (O_3) exposure, where the initiation phase is delayed in the absence of SA enhancement (Rao and Davis, 1999).

Cell death in inducible pathogen resistance programs

Multiple preformed anti-microbial compounds contribute to the constitutive defense machinery of plants against pathogenic organisms. In addition, plants can trigger inducible defense programs upon the perception of invaders. The abrupt hypersensitive cell death activated in infected tissues parallels the induction of other local and systemic defenses leading to SAR (Ryals *et al.*, 1996; Baker *et al.*, 1997). HR is triggered by resistant hosts in response to avirulent pathogens (incompatible interactions) but it is absent or delayed in susceptible hosts infected with virulent pathogens (compatible interactions) that develop disease. A 'gene-for-gene' resistance is defined between confronted plant and pathogen upon the recognition of products encoded by a plant resistance gene (*R*) and its corresponding pathogen avirulence gene (*avr*) (de Wit, 1997). Direct physical contact between R and Avr products was demonstrated for Pto-AvrPto in yeast, although not yet proved in plant cells (Scofield *et al.*, 1996 and references therein). *R*-gene-dependent resistance usually leads to cell death that could also be activated by expressing *avr* genes inside plant cells or overexpressing *R* genes (Tang *et al.*, 1999 and references therein). However, *R*-gene-dependent incompatibility also operates in the absence of the HR lesion (Yu *et al.*, 1998) and, conversely, hypersensitive lesions may not necessarily lead to resistance (Century *et al.*, 1995). Thus, although cell death is not a requisite for the induction of defense its occurrence may reinforce defense pathways. Alternatively, it may merely represent a consequence of defense activation.

ROI and NO in the activation of cell death

The early signaling pathways leading to *R*-gene-mediated resistance include ion fluxes, GTP-binding proteins, protein kinases, phosphatases and phospholipases (Hammond-Kosack and Jones, 1996). A burst in the oxidative metabolism generated within minutes of infection produces the accumulation of reactive oxygen intermediates (ROI) such as H_2O_2 and superoxide anion ($^{\bullet}O_2^{-}$), involving an NADPH-dependent oxidase (Doke 1983; Lamb and Dixon, 1997). Incompatible interactions lead to the generation of a nonspecific oxidative burst followed by an *avr*-dependent sustained enhancement of ROI (Lamb and Dixon, 1997). ROI activate cell wall strengthening and expression of antioxidant defense genes (Levine *et al.*, 1994; Jabs *et al.*, 1996, 1997; Lamb and Dixon, 1997) and, in some conditions, high levels of ROI are sufficient to trigger hypersensitive cell death (Levine *et al.*, 1994; Jabs *et al.*, 1996; Chamnongpol *et al.*, 1998). However, in other cases, plant cells remain alive after a normal oxidative burst induced in response to bacterial infection (Glazener *et al.*, 1996). It is conceivable then that ROI play a determinant role, reinforced by other signals, in the generation of hypersensitive lesions. In soybean and tobacco cells, a rise of nitric oxide synthase activity occurs during the onset of HR. Nitric oxide (NO), which does not activate cell death by itself, synergizes with ROI in the induction of cell death and complements the ROI-mediated defense gene activation (Delledone *et al.*, 1998, Durner *et al.*, 1998).

ROI accumulated in HR also function as second messengers of SAR and induce systemic micro-lesions in periveinal cells from uninfected tissues (Alvarez *et al.*, 1998). In *Arabidopsis*, micro-lesion development is activated by an avirulent race of *Pseudomonas syringae* but not induced by the isogenic virulent race. Secondary bursts and defense gene expression precede the activation of systemic cell death in the so called 'micro-HRs'. Both the primary oxidative burst and the systemic microbursts are required for SAR development and occur almost in parallel. The generation and release of the ROI-dependent signals that induce micro-HRs do not require completion of the local cell death program, and subsequent secondary bursts appear to participate in the activation of systemic cell death by ROI.

SA feeding the defense machine including cell death

The capacity of SA to induce defense programs was originally demonstrated by the 95% reduction in the number of TMV-induced lesions in tobacco leaves pretreated with a 0.01% SA solution (White, 1979). This kind of treatment leads to *PR* gene expression and establishment of SAR where the role of SA in signaling has been extensively characterized (Ryals *et al.*, 1996). Subsequently, several studies demonstrated the requirement of SA for the development of local disease resistance. *NahG* plants, depleted in SA, are not only unable to activate SAR but are also more susceptible to localized infections (Gaffney *et al.*, 1993; Delaney *et al.*, 1994; Mur *et al.*, 1997). Likewise, inhibition of PAL-mediated SA accumulation increases the susceptibility of *Arabidopsis* to infections by *Peronospora*, while the capacity to activate resistance is restored by exogenous SA (Mauch-Mani and Sluzarenko, 1996).

SA functions in HR by enhancing defense responses induced upon pathogen attack that may lead to hypersensitive cell death. It was originally proposed that SA induces H_2O_2 accumulation by inhibiting the activity of catalases and peroxidases (Chen *et al.*, 1993; Durner and Klessig, 1995). However, at the concentrations of SA reached in HR (<1 mM) such inhibitory effects might be negligible (Bi *et al.*, 1995; Summermatter *et al.*, 1995; Tenhaken and Rubel, 1997; Kvaratskhelia *et al.*, 1997). More recent models strongly suggest that H_2O_2 functions upstream of SA in the development of SAR and induces SA accumulation (Bi *et al.*, 1995; Neuenschwander *et al.*, 1995; Summermatter *et al.*, 1995). Although SA contributes to the increase of ROI levels in HR, it does not primarily function by inhibiting ROI turnover. Long-term pre-treatment of tissues with exogenous SA primes cells to trigger stronger defenses in the hypersensitive responses upon elicitation (Kauss and Jeblick, 1995; Mur *et al.*, 1996; Fauth *et al.*, 1996; Thulke and Conrath, 1998). Moreover, in soybean cells, SA (50 μM) simultaneously applied with an avirulent pathogen is able to activate H_2O_2 production, defense gene expression as well as enhance and accelerate cell death (Shirasu *et al.*, 1997). These responses are not induced by SA alone and the SA-mediated potentiation effect does not depend on protein synthesis. The co-treatment of cells with SA and cantharidin (a phosphatase inhibitor) in the absence of pathogen generates a massive oxidative burst. These results suggest that SA activates an agonist-dependent control at an early step of the pathway, probably by affecting the phosphorylation status of a regulatory component preceding ROI generation (Shirasu *et al.*, 1997). Activation of PAL gene expression, through SA-mediated potentiation (Shirasu *et al.*, 1997) as well as in response to NO produced during the HR (Delledone *et al.*, 1998; Durner *et al.*, 1998), may provide efficient ways to amplify SA-dependent effects in the local defense response.

As mentioned above, two-phase SA accumulation occurs in tissues developing HR. In tobacco leaves infected with *P. syringae-avr*, each phase of the oxidative burst is followed by SA increases (Draper *et al.*, 1997). The metabolic pathways involved in this biphasic SA accumulation are uncertain. Although PAL activation may participate along the course of the local defense response (Mauch-Mani and Slusarenko, 1996; Shirasu *et al.*, 1997), a rapid and transient release of SA from preexisting conjugates may occur in response to H_2O_2 accumulation (León *et al.*, 1995).

Does SA signal mitochondrial processes in defense?

In addition to promoting *PR* gene expression and potentiating the activation of early steps in the hypersensitive cascade, SA also appears to signal mitochondrial processes in defense. Pharmacological studies suggest that alternative oxidase, the terminal enzyme of the alternative respiratory pathway (Vanlerberghe and McIntosh, 1997), may be involved in the activation of disease resistance against TMV in tobacco

(Chivasa *et al.*, 1997; Lennon *et al.*, 1997; Chivasa and Carr, 1998). The inhibition of alternative oxidase leads to a loss of viral resistance induced by SA and 2,6-dichloroisonicotinic acid (INA; a SA analogue signaling SAR downstream of SA). Cyanide (a respiratory inhibitor), as well as SA and INA, promote the expression of the oxidase gene in wild-type plants. Interestingly, cyanide is the only one among these inducers that restores viral resistance and suppresses lesion spreading in SH-L plants expressing salicylate hydroxylase (Chivasa and Carr, 1998). These results suggest that the alternative respiratory pathway may contribute to the abnormal development of lesions in SH-L plants (Mur *et al.*, 1997). It is still unclear whether the involvement of SA-dependent defense pathways involving the alternative oxidase may exclusively function in resistance against virus. This enzyme is induced in *Arabidopsis* upon the attack of an avirulent race of *P. syringae* and associated with HR development and ethylene signaling, although its induction is not abolished in the *nahG* background and does not respond to application of SA (Simons *et al.*, 1999). It is possible that the alternative respiration contributes to signal disease resistance pathways responding to alterations in the cellular redox state but further studies are required to demonstrate its involvement in defense. In addition, high levels of SA can affect oxidative phosphorylation. In tobacco cells SA (20–500 μM) inhibits ATP production perturbing electron transport, by a mechanism that appears to require ROI. This toxic effect mediated by SA may contribute to defense responses in infected tissues that accumulate high levels of SA (Xie *et al.*, 1999). In animals, apoptotic processes such as those induced by γ-irradiation and ligation of Fas disrupt mitochondrial electron transport and ATP generation, while others affect mitochondrial functions related to such processes are the release of caspase-activating proteins and the alteration of the cellular redox potential by generation of ROI (Green and Reed, 1998).

SA and anti-death functions

The survival of tissues surrounding a focus of dead cells may basically result from two effects: the incapacity to produce optimal levels of pro-death signals and the activation of anti-death factors. As mentioned above, SA accumulated in the proximity of infected tissues contributes to control the HR lesion spread. SA promotes *PR* gene induction in uninfected neighboring tissues, enhancing defense and suppressing pathogen growth (Ryals *et al.*, 1996). In a particular case, SA may counteract oxidative damage leading to cell death. *Arabidopsis* leaves exposed to O_3 activate a defense program partially resembling HR that involves generation of H_2O_2 and SA and induction of antioxidant responses. Similar levels of H_2O_2 are accumulated in wild-type and *nahG* plants in response to O_3. However, only wild-type plants are able to increase the glutathione redox state (GSH/GSSG ratio) in response to O_3, while *nahG* plants reduce this state (Rao and Davis, 1999).

Chromatin condensation precedes the development of programmed cell death in HR and is induced upon H_2O_2 accumulation (Levine *et al.*, 1996; Pennell and Lamb, 1997; Greenberg, 1997; Alvarez *et al.*, 1998). In tobacco protoplasts high levels of H_2O_2, SA or calcium ionophores induce chromatin condensation and programmed cell death. However, lower levels of those compounds (i.e. 70 μM H_2O_2) solely induce a transient effect in chromatin structure that is fully reversed leaving no trace of death (O'Brien *et al.*, 1998). In this system, the transition from life to death does not appear to be achieved in a single irreversible step dependent on the mere activation of a death program, which is consistent with the participation of cell death suppressors. Although several putative cell death suppressors have been identified in plants recently, their function in HR and their sensitivity to SA are unknown.

LSD1 from *Arabidopsis* is a regulatory component of disease resistance that functions as a cell death repressor at least in HR. *Lsd1* mutant activates hypersensitive cell death and defense responses at low doses of SAR inducers including SA, INA and pathogens (Dietrich *et al.*, 1994; see Table 1). In contrast to other *lsd* phenotypes such as *lsd2–lsd7*, the *lsd1* lesions propagate extensively once initiated and generation of $^{\bullet}O_2^-$ is sufficient to activate runaway cell death (Jabs *et al.*, 1996). LSD1 may negatively control regulatory steps in the initiation and propagation of cell death programs activated by $^{\bullet}O_2^-$, while the *lsd1* mutant continuously activates this pathway (Jabs *et al.*, 1996). LSD1 is a zinc-finger protein proposed to function as a transcriptional regulator of cell death effectors (Dietrich *et al.*, 1997). Interestingly, the *R*-gene-mediated cell-autonomous pathway is intact in *lsd1* (Dietrich *et al.*, 1994), but cell-non-autonomous responses to signals produced by infected cells are defective (Jabs *et al.*, 1996; Dietrich *et al.*, 1997). The LSD1 gene is constitutively expressed in healthy tissues. Its sub-

Table 1. SA levels and cell death status in *Arabidopsis* mutants.

Phenotype (ecotype)	Dominance/ recessivity	Spontaneous lesions	SA increase[a,b]	Lesion status × *nahG*[c]	References
Lesion-mimic					
acd2 (Col-0)	R	+	3/3	ND	Greenberg *et al.*, 1994
acd6 (Col-0)	D	+[f]	+	–	Rate *et al.*, 1999
lsd1 (Ws-0)	R	+	+	–[d]	Dietrich *et al.*, 1994; D. Aviv, R. Dietrich and J. Dangl (personal communication)
lsd2 (Col-0)	D	+	+	+	Dietrich *et al.*, 1994; Hunt *et al.*, 1997
lsd3 (Ws-0)	R	+	+	ND	Dietrich *et al.*, 1994
lsd4 (Ws-0)	D	+	+	+	Dietrich *et al.*, 1994; Hunt *et al.*, 1997
lsd5 (Ws-0)	R	+	+	ND	Dietrich *et al.*, 1994
lsd6 (Col-0)	D	+	3/20	–[e]	Weymann *et al.*, 1995
lsd7 (Col-0)	D	+	5/15	–	Weymann *et al.*, 1995
Constitutive SAR					
cpr1 (Col-0)	R	–	5/21	NL	Bowling *et al.*, 1994
cpr5 (Col-0)	R	+	30/30	+	Bowling *et al.*, 1997
cpr6 (Col-0)	D	–	7/9	NL	Clarke *et al.*, 1998
dnd1 (Col-0)	R	–[f]	14/38	NL	Yu *et al.*, 1998

[a]The number of fold that SA (free/conjugated or total) is increased compared to the levels in uninfected wild type plants is indicated.
[b]According to PR gene expression expected enhanced levels of SA are shown as +.
[c]The existence of spontaneous lesion in the *nahG* background is shown. ND, not determined; NL, no spontaneous lesion in the mutant background.
[d]Lesion formation is reduced but not abolished (see text).
[e]Spontaneous lesion formation is restored by INA treatment.
[f]These mutants are unable to activate HR lesions.

cellular distribution, activity and sensitivity to SA in healthy and infected tissues is still unknown but could provide a clue to understanding its mechanism of function in HR.

LLS1 from maize is a cell death suppressor whose putative targets are phenolic compounds, including SA. The lesion-mimic mutant *lls1* (lethal leaf spot) develops chlorotic lesions that progressively enlarge and exhibits resistance to fungal pathogens (Simmons *et al.*, 1998). Revertant healthy sectors in *lls1* are indicative of the cell-autonomous nature of this mutation suggesting that the death effector regulated by LLS1 functions intracellularly. *Lls1* is recessive and the gene is predicted to encode a suppressor of cell death (Johal *et al.*, 1995). LLS1 is a novel protein conserved in monocots and dicots that presents high homology with dioxygenases (Gray *et al.*, 1997). Two consensus motifs present in LLS1 suggest that it may function as an aromatic ring-hydroxylating dioxygenase degrading phenolics. The identification of the LLS1 targets, including its substrates in plant cells should enhance our knowledge of mechanisms suppressing cell death in plants.

Homologues to putative cell death suppressors from animals have also been isolated in plants. The human anti-apoptotic gene *DAD1* (defender against apoptotic death) has its homologue in *Arabidopsis* (Gallois *et al.*, 1997) and rice (Tanaka *et al.*, 1997). DAD1 from *Arabidopsis* is constitutively expressed in many organs including leaves and in cell transformation assays, it complements the apoptotic phenotype of a mutant hamster. The yeast OST2 protein which is 50% homologue to DAD1 from *Arabidopsis* turns out to be a subunit of the oligosaccharyl-transferase enzyme (Gallois *et al.*, 1997). This simple result suggests, for the first time, a functional homology between cell death control in plant and animal kingdoms.

What do mutants tell us about SA and cell death?

After the identification of several components of HR, much effort has been directed to define functional links between them. It is still an unresolved question whether cell death and gene-for-gene resistance respond to a common genetic program. In a similar sense, the causal relationship between SA and cell death is still unclear.

Mutants that activate cell death programs resembling those developed in disease resistance have been isolated in different plant species (Walbot, 1991; Johal *et al.*, 1995; Dangl *et al.*, 1996; Glazebrook *et al.*, 1997). In addition, a variety of transgenic models that produce spontaneous lesion formation have been used for the study of disease resistance (Dangl *et al.*, 1996; Mittler and Rizhsky, 2000). Two families of *Arabidopsis* mutants proved to be particularly useful for the analysis of the functional relationships between SA accumulation and cell death in disease resistance. The first corresponds to the 'lesion mimic mutants', which develop HR-like lesions in the absence of pathogens, and include the *lsd* class (lesion-simulating disease) and *acd2* (accelerated cell death) (Table 1; see references in the table). The second refers to the 'constitutive SAR mutants' which contain high basal levels of SA, constitutive expression of *PR* genes and enhanced resistance to virulent pathogens. The latter family is represented by the *cpr* (constitutive expression of *PR* genes), *cim* (constitutive immunity) and *dnd1* (defense with no HR cell death) mutants. Other *Arabidopsis* disease resistance mutants such as *ndr1*, *npr1/nim1/sai1*, *edr1* and *eds* are less informative for this particular point and are not discussed here (see Glazebrook *et al.*, 1997).

In the *lsd* and *acd* mutants, lesion phenotype correlates with features of SA accumulation although SA enhancement does not necessarily determine lesion formation (Table 1). *dnd1* activates gene-for-gene resistance in response to several R products and, like *cpr* mutants, constitutively accumulates SA and exhibits SAR. However, unlike *cpr* mutants, *dnd1* does not develop HR lesions in response to avirulent pathogens (Yu *et al.*, 1998). On the other hand, wild-type plants do not develop lesions in response to SA treatments. Taken together these data suggest that SA accumulation occurs as a consequence of cell death and that cell death development is not influenced by SA levels. However, the epistatic relationship between SA accumulation and cell death-related mutants revealed a different pattern.

Lesion formation was analyzed in the *nahG* background for *cpr5*, *acd6* and several *lsd* mutants. *cpr5*, *lsd2* and *lsd4* retain the spontaneous lesion phenotype in the *nahG* background (Hunt *et al.*, 1997; Bowling *et al.*, 1997). In contrast, *nahG* abolishes lesion formation in *lsd6*, *lsd7* and *acd6* (Weymann *et al.*, 1995; Rate *et al.*, 1999) and delays and reduces lesions in *lsd1* (D. Aviv, R. Dietrich and J. Dangl, personal communication). The second group of results place SA accumulation upstream of lesion formation in these mutant backgrounds. A probable explanation for these apparently contradictory data is the existence of a feedback regulatory pathway involving a rheostat amplification loop between SA and cell death (Weymann *et al.*, 1995). In agreement with this hypothesis, treatment with SA or INA restored the lesion phenotype in the *lsd6 nahG* background (Weymann *et al.*, 1995).

Protein kinases, SA, and phosphorylation-dependent gene expression

As mentioned above, it is likely that SA affects a phosphorylation-dependent step early in the HR signaling pathway. Plant disease resistance involves general signaling mechanisms mediated by protein phosphorylation. *R* genes such as *Pto* and *Xa21* encode serine-threonine protein kinases, as well as *Pti* that functions downstream of *Pto* in the local defense pathway (see Hammond-Kossack, 1996; Baker *et al.*, 1997).

Recently, the induction of mitogen-activated protein (MAP) kinases by elicitors or Avr products activating disease resistance was analyzed by using artificial substrates. In parsley cells, fungal elicitors activate the 45 kDa ERM (elicitor-responsive MAP) kinase that functions downstream of ion channels and upstream or independently of the oxidative burst in the defense pathway. Interestingly, ERM translocates into the nucleus within minutes of elicitation (Ligterink *et al.*, 1997). Transgenic tobacco plants expressing the tomato *R* gene *Cf-9* activate two MAP kinases of 46 and 48 kDa within 5 min upon elicitation with Avr9, responding to a gene-for-gene-dependent recognition process. The activation of these kinase pathways is not required for induction of the Avr9-dependent oxidative burst (Romeis *et al.*, 1999). The recent finding that 46 and 48 kDa kinases are immunologically related to N-terminal domains of the earlier characterized WIPK (wounding-induced protein kinase; Seo *et al.*, 1995) and SIPK (SA-induced protein kinase; Zhang and Klessig, 1997, 1998; Zhang *et al.*, 1998) MAP kinases from tobacco (Romeis *et al.*, 1999) helped to clarify previous data concerning kinase involvement in defense. WIPK and SIPK, like other MAP kinases from tobacco (Suzuki *et al.*, 1995, Adám *et al.*, 1997), are activated by nonspecific elicitors. In addition, WIPK and SIPK activation by TMV infection occurs in an N-gene-dependent manner (Seo *et al.*, 1995; Zhang and Klessig, 1998; Zhang *et al.*, 1998). In tobacco cells SIPK is rapidly induced by treatment with SA, at the physiological levels reached in HR (ca. 100 μM; Zhang and Klessig, 1997). However, nonspecific elicitation or mechanical damage also activates SIPK suggesting that diverse biotic and abiotic signals stimulate common MAP kinase-dependent signaling pathways (Romeis *et al.*, 1999).

Protein phosphorylation could also regulate SA-dependent defense gene expression (Stange *et al.*, 1997). In tobacco, phosphorylation-dependent transcriptional activation mediated by SA regulates the activity of the *as-1 cis* element. The tobacco SARP protein (salicylic acid response protein) binds to the *as-1* element of the cauliflower mosaic virus 35S promoter activating gene expression in response to SA (Jupin and Chua, 1996). In the absence of SA no binding occurs unless dissociating agents were included in the system, suggesting that inactive forms of SARP are associated with inhibitors. Moreover, phosphatase treatment abolished the SARP DNA-binding capacity in SA-treated cells indicating that phosphorylation regulates the SA-dependent gene expression of this promoter (Jupin and Chua, 1996). The recently characterized genes that are rapidly induced by SA in the absence of protein synthesis (Horvath *et al.*, 1998) could facilitate the dissection of phosphorylation-dependent mechanisms regulating gene expression in response to SA.

In animals, salicylates modulate the NF-κB pathway that triggers inflammation and immune responses (Koop and Ghosh, 1994). NF-κB is maintained in the cytoplasm by binding to the IκB proteins (α and β). The IκB kinase complex (IKK) phosphorylates both IκBs promoting their degradation (DiDonato *et al.*, 1997) and the function of both IKK subunits (α and β) is required for IκB phosphorylation (Zandi *et al.*, 1997). It was recently suggested that salicylates behave as competitive inhibitors of ATP binding to IκB kinase-β through a noncovalent but probably irreversible binding to the enzyme (Yin *et al.*, 1998). In this way, salicylates can indirectly inhibit the nuclear translocation of NF-κB thus avoiding inflammatory effects.

NPR1 from *Arabidopsis* shows partial sequence homology to IκBα (Ryals *et al.*, 1997). The *Arabidopsis npr1/nim1/sai1* allelic mutations abolish *PR* gene expression and SAR development in response to SA (Cao *et al.*, 1994; Delaney *et al.*, 1995; Shah *et al.*, 1997). These mutants accumulate SA after infection (Delaney *et al.*, 1995) and it is likely that NPR1/NIM1/SAI (hereafter referred to as NPR1) functions downstream of SA and upstream of *PR* gene expression. It is expected that interaction of NPR1with other proteins is required for SAR signaling. NPR1 contains several ankyrin-rich repeats (involved in protein-protein interactions) where a point mutation characterized for the *npr1-1* mutant allele is localized (Cao *et al.*, 1997). Due to the recessive nature of the *npr1/nim1/sai1* mutations, the wild-type product may directly work as a positive regulator of *PR* gene expression although NPR1 has not a relevant homology to known transcription factors (Cao *et al.*, 1997; Ryals *et al.*, 1997). Alternatively, NPR1 may function by repressing a negative transcriptional regulator of *PR* genes. This makes a difference with the regulatory function of IκB whose null mutation is predicted to constitutively activate gene expression.

Interestingly, the region and residues harboring the two phosphorylation sites of IκBα are conserved in NPR1 (Ryals *et al.*, 1997) and may represent targets for phosphorylation events in the latter. The mechanism by which NPR1 responds to SA accumulation is unknown and could either involve its function as an intracellular SA-receptor or as a SA-sensitive transcriptional regulator. Recently, the physical interaction in yeast and *in vitro* between NPR1 and a subclass of basic leucine zipper (bZIP) protein transcription factors (TGA6 and AHBP-1b) from *Arabidopsis* has been described (Zhang *et al.*, 1999). Truncated versions of NPR1 retaining the ankyrin-repeat domain are able to interact with AHBP-1b in the yeast two-hybrid system and this interaction appears to be favored by the NPR1 N-terminal domain. Point mutations affecting the NPR1 ankyrin-repeat domain (*npr1-1*) or the N-terminal domain (*npr1-2*) abolish the NPR1 interactions with TGA6 or AHBP-1b. In addition,

AHBP-1b specifically binds to the *as-1*-like element of the *Arabidopsis* PR1 gene promoter required for the SA-dependent gene activation. It is conceivable that the SA-dependent PR1 gene activation involves the AHBP-1b activity and responds to its association with NPR1 (Zhang *et al.*, 1999).

Integrated model for SA functions in the HR

A distinctive characteristic of the hypersensitive lesion is the sharp limit between dead and live tissues that is established subsequent to the first appearance of death symptoms. Although the activation of *R*-mediated resistance is cell-autonomous (Bennetzen *et al.*, 1988), extracellular or diffusible signals generated in the active infected cells may trigger responses in the neighboring uninfected cells. A pivot in the HR network may be the non-homogeneous distribution of SA within lesions. In TMV-infected tobacco tissues that accumulate SA, a gradient of SA is established early and maintained over 6 days along with lesion formation (Enyedi *et al.*, 1992). The highest levels of SA are found in the center of the lesions, decreasing towards the borders, while in healthy surrounding tissues the SA levels are slightly higher than in uninfected plants (Enyedi *et al.*, 1992). The maintenance of this SA gradient may suggest that structural features are preserved at some levels in the 'dead' tissue that apparently do not redistribute SA from dead cells into healthy parts of the plant.

A model for the involvement of SA in the development of the HR lesion is provided in Figure 1. The model is based on the non-homogeneous SA distribution in lesions, that is, lower levels in healthy tissues and higher ones in dead tissues. It is assumed that diffusible signals generated in infected cells activate non-cell-autonomous responses. H_2O_2 can diffuse across membranes and reach lower concentrations in the lesion margins. H_2O_2 may play a dual role in HR, activating cell death at high concentrations and inducing the expression of antioxidant genes at low concentrations (Levine *et al.*, 1994; Jabs *et al.*, 1996; Jabs *et al.*, 1997). Superoxide is not a diffusible molecule but its generation precedes lesion formation in the *Arabidopsis* mutant *lsd1* and it is accumulated in the margins of the formed lesions (Jabs *et al.*, 1996). In addition, it is unknown whether or not NO functions as a diffusible signal in HR and may induce, at low concentrations, protective functions in uninfected cells (Beligni and Lamattina, 1999).

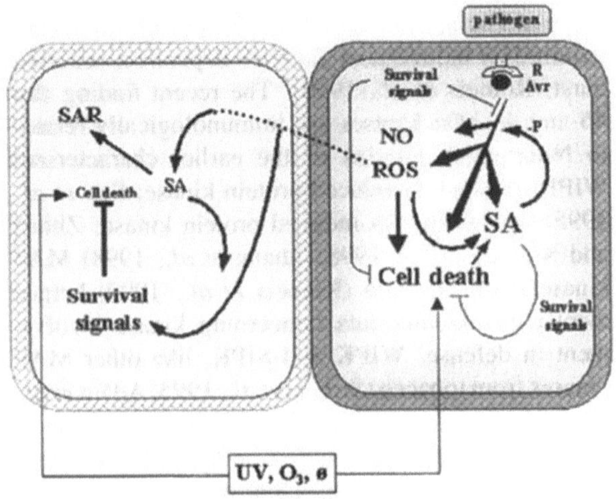

Figure 1. Model of SA function in the hypersensitive cell death context. The scheme represents two cells from a HR lesion at a final stage. These cells belong to infected (right) and uninfected neighbor (left) tissues. Constitutive suppressors of cell death programs (survival signals) counteract the effect of biotic and abiotic injuries promoting cell death. The perception of a pathogen activates the HR pathway. Diffusible signals transiently generated at infected cells (i.e. ROI) induce a direct (León *et al.*, 1995) or indirect modest increase of the SA levels in the area (Malamy *et al.*, 1990; Métraux *et al.*, 1990; Mur *et al.*, 1997; Chamnongpol *et al.*, 1998; Draper *et al.*, 1997). At this stage SA and ROI levels are sub-optimal for the induction of cell death but sufficient to activate survival signals (antioxidant responses, cell death suppressors such as LSD1) allowing cells to reach a 'primed state'. Once primed infected cells perceive optimal levels of the Avr factor, the *R*-gene-dependent specific defenses are activated. Accumulated SA amplifies the pathway by up-regulating a phosphorylation-dependent step (−P), preceding the branch between oxidative burst and phenylpropanoid biosynthesis (Kauss and Jeblick, 1995; Shirasu *et al.*, 1997). The function of a SA-dependent feedback loop promoting lesion formation is also predicted by genetic studies (Weymann *et al.*, 1995). In infected cells, protectant functions are unable to shut down the massive accumulation of ROI and SA and other pro-death signals (i.e. NO) (Levine *et al.*, 1994; Delledone *et al.*, 1997; Durner *et al.*, 1997; Kvaratskhelia *et al.*, 1997; Allan and Fluhr, 1997) thus leading to cell death. In uninfected primed cells pro-death signals are maintained below the threshold for activation of cell death. Alternatively, amplification of signals leading to cell death in uninfected cells could be performed under continuous generation of pro-death factors (i.e. H_2O_2; Chamnongpol *et al.*, 1998).

Perspectives

Many recent studies suggest the function of SA in genetically determined disease resistance beyond signaling SAR. SA is a rheostat sensor of the HR network that executes a tight control in localized disease resistance and HR lesion formation. It involves, among other unknown features, two-phase generation kinetics and a spatial gradient distribution. A better understanding of its function in HR may require the elucidation of which SA forms (free/conjugated) are effective in the different roles it has in HR. Isolation of mutants impaired in SA biosynthetic pathways and engineering strategies targeting the rate-limiting enzymes in SA synthesis/conjugation should help us to achieve this goal. Biochemical approaches to identify the sources of the two-phase SA accumulation would also contribute to this matter. Cross-talks between SA and other HR components, including cell death and regulatory steps dependent on protein phosphorylation, are beginning to be elucidated. The identification of substrates for MAP kinases recently characterized in plant defense and the analysis of their hierarchical relationship could help to position the predicted SA-mediated regulatory steps in the hypersensitive cell death pathway. Finally, further studies on the function of LSD1 and NIM1 in defense may allow us to gain access to the nuclear events signaled by SA in disease resistance.

Acknowledgements

I am grateful to E. Lam, J. Dangl and E. Taleisnik for discussions and comments on this work. I would like to thank A.L. Rosa and G. Boriolli for critical reading of the manuscript. This work is supported by grants from ANPCyT (Agencia Nacional de Promoción Científica y Tecnológica), CONICET (Consejo Nacional de Investigaciones Científicas y Tecnológicas) and Fundación Antorchas.

References

Adám, A.L., Pike, S., Hoyos, M.E., Stone, J.M., Walker, J.C. and Novacky, A. 1997. Rapid and transient activation of myelin basic protein kinase in tobacco leaves treated with hairpin from *Erwinia amylovora*. Plant Physiol. 115: 853–861.

Alvarez, M.E., Pennell, R.I., Meijer, P-J., Ishikawa, A., Dixon, R.A. and Lamb, C. 1998. Reactive oxygen intermediates mediate a systemic signal network in the establishment of plant immunity. Cell 92: 773–783.

Baker, B., Zambryski, P., Staskswicz, B. and Dinesh-Kumar, S.P. 1997. Signaling in plant-microbe interactions. Science 276: 726–733.

Beg, A.A. and Baltimore, D. 1996. An essential role for NF-κB in preventing TNF-α-induced cell death. Science 274: 782–784.

Bennetzen, J.L., Blevins, W.E. and Ellingboe, A.H. 1988. Cell-autonomous recognition of the rust pathogen determines *Rp1*-specified resistance in maize. Science 241: 208–210.

Beligni, M.V and Lamattina, L. 1999. Nitric oxide counteracts cytotoxic processes mediated by reactive oxygen species in plant tissues. Planta 208: 337–344.

Bi, Y.-M., Kenton, P., Mur, L., Darby, R. and Draper, J. 1995. Hydrogen peroxide does not function downstream of salicylic acid in the induction of PR protein expression. Plant J. 8: 235–245.

Bowling, S.A., Guo, A., Cao, H., Gordon, A.S., Klessig, D.F. and Dong, X. 1994. A mutation in *Arabidopsis* that leads to constitutive expression of systemic acquired resistance. Plant Cell 6: 1845–1857.

Bowling, S.A., Clarke, J.D., Liu, Y., Klessig, D.F. and Dong, X. 1997. The *cpr 5* mutant of *Arabidopsis* expresses both *NPR*1-dependent and *NPR*1-independent resistance. Plant Cell 9: 1573–1584.

Cao, H., Glazebrook, J., Clarke, J.D., Volko, S. and Dong, X. 1997. The *Arabidopsis NPR1* gene that controls systemic acquired resistance encodes a novel protein containing ankyrin repeats. Cell 88: 57–63.

Century, K.S., Holub, E.B. and Staskawicz, B.J. 1995. *NDR1*, a locus of *Arabidopsis thaliana* that is required for disease resistance to both a bacterial and fungal pathogen. Proc. Natl. Acad. Sci. USA 92: 6597–6601.

Chamnongpol, S., Willekens, H., Moeder, W., Langebartels, C., Sandermann, H., Van Montagu, M., Inzé, D. and Van Camp, W. 1998. Defense activation and enhanced pathogen tolerance induced by H_2O_2 in transgenic tobacco. Proc. Natl. Acad. Sci. USA 95: 5818–5823.

Chen, Z., Silva, H. and Klessig, D.F. 1993. Active oxygen species in the induction of plant systemic acquired resistance by salicylic acid. Science 263: 1883–1886.

Chivasa, S. and Carr, J. 1998. Cyanide restores N-gene mediated resistance to tobacco mosaic virus in transgenic tobacco expressing salicylic acid hydroxylase. Plant Cell 10: 1489–1498.

Chivasa, S., Murphy, A.M., Naylor, M. and Carr, J.P. 1997. Salicylic acid interferes with tobacco mosaic virus via a novel salicylhydroxamic acid sensitive mechanism. Plant Cell 9: 547–557.

Coquoz, J.L., Buchala, A., Meuwly, P. and Métraux, J.P. 1995. Arachidonic acid induces local but not systemic synthesis of salicylic acid and confers systemic resistance in potato plants to *Phytophthora infestans* and *Alternaria solani*. Phytopathology 85: 1219–1224.

Coquoz, J.L., Buchala, A. and Métraux, J.P. 1998. The biosynthesis of salicylic acid in potato plants. Plant Physiol. 117: 1095–1101.

Dangl, J.L., Dietrich, R.A. and Richberg, M.H. 1996. Death dont't have no mercy: cell death programs in plant-microbe interactions. Plant Cell 8: 1793–1807.

Dat, J.F., Lopez-Delgado, H., Foyer, C.H. and Scott, I.M. 1998a. Parallel changes in H_2O_2 and catalase during thermotolerance induced by salicylic acid or heat acclimation in mustard seedlings. Plant Physiol. 116: 1351–1357.

Dat, J.F., Foyer, C.H. and Scott, I.M. 1998b. Changes in salicylic acid and antioxidants during induced thermotolerance in mustard seedlings. Plant Physiol. 118: 1455–1461.

Delaney, T.P., Uknes, S., Vernooij, B., Friedrich, L., Weymann, K., Negroto, D., Gaffney, T., Gut-Rella, M., Kessmann, H., Ward, E. and Ryals, J. 1994. A central role of salicylic acid in plant disease resistance. Science 266: 1247–1250.

Delaney, T.P., Friedrich, L. and Ryals, J.A. 1995. *Arabidopsis* signal transduction mutant defective in chemically and biologically induced disease resistance. Proc. Natl. Acad. Sci. USA 92: 6602–6606.

Delledonne, M., Xia, Y., Dixon, R.A. and Lamb, C. 1998. Nitric oxide functions as a signal in plant disease resistance. Nature 394: 585–588.

de Wit, P.J.G.M. 1997. Pathogen avirulence and plant resistance: a key role for recognition. Trends Plant Sci. 2: 452–458.

DiDonatto, J.A., Hayakawa, M., Rothwarf, D.M., Zandi, E. and Karin, M.A. 1997. Cytokine responsive Iκ-B kinase that activates transcription factor NF-κB. Nature 388: 548–554.

Dietrich, R.A., Delaney, T.P., Uknes, S.K., Ward, E.R., Ryals, J.A. and Dangl, J.L. 1994. *Arabidopsis* mutants simulating disease resistance response. Cell 77: 565–577.

Dietrich, R.A., Richberg, M.H., Schmidt, R., Dean, C. and Dangl, J.L. 1997. A novel zinc finger protein is encoded by the *Arabidopsis LSD1* gene and functions as a negative regulator of plant cell death. Cell 88: 685–694.

Dixon, R. and Paiva, N.L. 1995. Stress-induced phenylpropanoid metabolism. Plant Cell 7: 1085–1097.

Doke, N. 1983. Generation of superoxide anion by potato tuber protoplasts during the hypersensitive response to hyphal wall components of *Phytophthora infestans* and specific inhibition of the reaction by suppressors of the hypersensitive response. Physiol. Plant Path. 23: 359–367.

Dong, X. 1998. SA, JA, ethylene and disease resistance in plants. Curr. Biol. 1: 316–323.

Dong, Z. 1997. Inhibition of ultraviolet B-induced activator protein-1 (AP-1) activity by aspirin in AP-1-luciferase transgenic mice. J. Biol. Chem. 272: 26325–26329.

Draper, J. 1997. Salicylate, superoxide synthesis and cell suicide in plant defense. Trends Plant Sci. 2: 162–165.

Durner, J. and Klessig, D.F. 1995. Inhibition of ascorbate peroxidase by salicylic acid and 2,6-dichloroisonicotinic acid, two inducers of plant defense. Proc. Natl. Acad. Sci. USA 92: 11312–11316.

Durner, J., Wendehenne, D. and Klessig, D.F. 1998. Defense gene induction in tobacco by nitric oxide, cylic GMP, and cylic ADP ribose. Proc. Natl. Acad. Sci. USA 95: 10328–10333.

Enyedi, A.J., Yalpani, N., Silverman, P. and Raskin, I. 1992. Localization, conjugation, and function of salicylic acid in tobacco during the hypersensitive reaction to tobacco mosaic virus. Proc. Natl. Acad. Sci. USA 89: 2480–2484.

Fauth, M., Merten, A., Hang, M.G. Jeblick, W and Kauss, H. 1996. Competence for elicitation of H_2O_2 in hypocotyls of cucumber is induced by breaching the cuticule and is enhanced by salicylic acid. Plant Physiol. 110: 347–354.

Gaffney, T., Friedrich, L., Vernooij, B., Negrotto, D., Nye, G., Uknes, S., Ward, E., Kessmann, H. and Ryals, J. 1993. Requirement of salicylic acid for the induction of systemic acquired resistance. Science 261: 754–756.

Gallois, P., Makishima, T., Hecht, V., Despres, B., Laudié, M., Nishimoto, T. and Cooke, R. 1997. An *Arabidopsis thaliana* cDNA complementing a hamster apoptosis suppressor mutant. Plant J. 11: 1325–1331.

Glazebrook, J., Rogers, E.E. and Ausubel, F.M. 1997. Use of *Arabidopsis* for genetic dissection of plant defense responses. Annu. Rev. Genet. 31: 547–569.

Gray, J., Close, P.S., Briggs, S.P. and Johal, G.S. 1997. A novel suppressor of cell death in plants encoded by the *Lls1* gene of maize. Cell 89: 25–31.

Green, D.R. and Reed, J.C. 1998. Mitochondria and apoptosis. Sicience 281: 1309–1312.

Greenberg, J.T. 1997. Programmed cell death in plant-pathogen interactions. Annu. Rev. Plant Physiol. Plant Mol. Biol. 48: 525–545.

Greenberg, J.T., Guo, A. Klessig, D.F. and Ausubel, F.M. 1994. Programmed cell death in plants: a pathogen-triggered response activated coordinately with multiple defense functions. Cell 77: 551–563.

Hammond-Kosack, K.E. and Jones, J.D.G. 1996. Resistance gene-dependent plant defense responses. Plant Cell 8: 1773–1791.

Horvath, D.M., Huang, D.J. and Chua, N-H. 1998. Four classes of salicylate induced tobacco genes. Mol. Plant-Microbe Interact. 11: 895–905.

Hunt, M., Delaney T.P., Dietrich, R.A., Weymann, K.B., Dangl, J.L. and Ryals, J.A. 1997. Salicylate-independent lesion formation in *Arabidopsis lsd* mutants. Mol. Plant-Microbe Interact. 10: 531–536.

Jabs, T., Dietrich, R.A. and Dangl, J.L. 1996. Initiation of runaway cell death in an *Arabidopsis* mutant by extracellular superoxide. Science 273: 1853–1856.

Jabs, T., Tschöpe, M., Colling, C., Hahlbrock, K. and Scheel, D. 1997. Elicitor-stimulated ion fluxes and O_2 from the oxidative burst are essential components in triggering defense gene activation and phytoalexin synthesis in parsley. Proc. Natl. Acad. Sci. USA 29: 4800–4805.

Johal, G.S., Hulbert, S.H. and Briggs, S.P. 1995. Disease lesion mimic in maize: a model for cell death in plants. BioEssays 17: 685–692.

Jupin, I. and Chua, N-H. 1996. Activation of the CaMV *as-1 cis*-element by salicylic acid: differential DNA-binding of a factor related to TGA1a. EMBO J. 15: 5679–5689.

Kauss, H. and Jeblick, W. 1995. Pretreatment of parsley suspension cultures with salicylic acid enhances spontaneous and elicited production of H_2O_2. Plant Physiol. 108: 1171–1178.

Klessig, D.F. and Malamy, J. 1994. The salicylic acid signal in plants. Plant Mol. Biol. 26: 1439–1458.

Kopp, E and Ghosh, S. 1994. Inhibition of NF-κB by sodium salicylate and aspirin. Science 265: 956–959.

Kvaratskhelia, M., George, S.J. and Thorneley, R.N. 1997. Salicylic acid is a reducing substrate and not an effective inhibitor of ascorbate peroxidase. J. Biol. Chem. 272: 20998–21001.

Lamb, C.J. and Dixon, R.A. 1997. The oxidative burst in plant disease resistance. Annu. Rev. Plant Physiol. Plant Mol. Biol. 76: 419–422.

Lee, H-I., León, J. and Raskin, I. 1995. Biosynthesis and metabolism of salicylic acid. Proc. Natl. Acad. Sci. USA 92: 4076–4079.

Lennon, A.M., Neuenschwander, U.H., Ribas-Carbo, M., Giles, L., Ryals, J.A. and Siedow, J.N. 1997. The effects of salicylic acid and tobacco mosaic virus infection on the alternative oxidase of tobacco. Plant Physiol. 115: 783–791.

León, J., Lawton, M.A. and Raskin, I. 1995. Hydrogen peroxide stimulates salicylic acid biosynthesis in tobacco. Plant Physiol. 108: 1673–1678.

Levine, A., Tenhaken, R., Dixon, R. and Lamb, C. 1994. H_2O_2 from the oxidative burst orchestrates the plant hypersensitive disease resistance response. Cell 79: 583–593.

Levine, A., Penell, R.I., Alvarez, M.E., Palmer, R. and Lamb, C. 1996. Calcium-mediated apoptosis in plant hypersensitive disease resistance response. Curr. Biol. 6: 427–437.

Ligterink, W., Kroj, T., zur Nieden, U., Hirt, H. and Scheel, D. 1997. Receptor-mediated activation of a MAP kinase in pathogen defense in plants. Science 276: 2054–2057.

Malamy, J., Carr, J.P., Klessig, D.F. and Raskin, I. 1990. Salicylic acid: a likely endogenous signal in the resistance response of tobacco to viral infection. Science 250: 1002–1004.

Malamy, J., Henning, J. and Klessig, D.F. 1992. Temperature-dependent induction of salicylic acid and its conjugates during the resistance response to tobacco mosaic virus infection. Plant Cell 4: 359–366.

Mauch-Mani, B. and Slusarenko, A. 1996. Production of salicylic acid precursors is a mayor function of phenylalanine ammonia-lyase in the resistance of *Arabidopsis* to *Peronospora parasitica*. Plant Cell 8: 203–212.

Métraux, J.P., Signer, H., Ryals, J., Ward, E., Wyss-Benz, M., Gaudin, J., Raschdorf, K., Schmid, E., Blum, W. and Inverardi, B. 1990. Increase in salicylic acid at the onset of systemic acquired resistance in cucumber. Science 250: 1004–1006.

Meuwly, P., Mölders, W., Buchala, A. and Métraux, J.-P. 1995. Local and systemic biosynthesis of salicylic acid in infected cucumber plants. Plant Physiol. 109: 1107–1114.

Mitchell, J.A. Akarasereenont, P., Thiemermann, C. Flower, R.J. and Vane, R. 1994. Selectivity of non-steroideal anti-inflammatory drugs as inhibitors of constitutive and inducible cyclooxygenase. Proc. Natl. Acad. Sci. USA 90: 11693–11697.

Mittler, R. and Rizhsky, L. 2000. Trangenic-induced lesion mimic. Plant Mol. Biol., this issue.

Mur, L.A.J., Naylor, G., Warner, S.A.J., Sugars, J.M., White, R.F. and Draper, J. 1996. Salicylic acid potentiates defence gene expression in tissues exhibiting acquired resistance to pathogen attack. Plant J. 9: 559–571.

Mur, L.A.J., Bi, Y.-M., Darby, R.M., Firek, S. and Draper, J. 1997. Compromising early salicylic acid accumulation delays the hypersensitive response and increases viral dispersal during lesion establishment in TMV infected tobacco. Plant J. 12: 1113–1126.

Murphy, A.M., Chivasa, S., Singh, D.P. and Carr, J.P. 1999. Salicylic acid-induced resistance to viruses and other pathogens: a parting of the ways? Trends Plant Sci. 4: 155–160.

Neuenschwander, U., Vernooij, B., Friedrich, L., Uknes, S., Kessmann, H. and Ryals, J. 1995. Is hydrogen peroxide a second messenger of salicylic acid in systemic acquired resistance? Plant J. 8: 227–233.

O'Brien, I.E.W., Baguley, B.C., Murray, B.G., Morris, B.A.M. and Ferguson, I.B. 1998. Early stages of the apoptotic pathway in plant cells are reversible. Plant J. 13: 803–814.

Pancheva, T.V., Popova, L.P. and Uzunova, A.N. 1996. Effects of salicylic acid on growth and photosynthesis in barely plants. Plant Physiol. 149: 57–63.

Pennell, R.I. and Lamb, C.J. 1997. Programmed cell death in plants. Plant Cell 9: 1157–1168.

Pillinger, M.H., Capodici, C., Risenthal, P., Kheterpal, N., Hanft, S., Philips, M.R. and Weissmann G. 1998. Modes of action of aspirin-like drugs: salicylates inhibit Erk activation and integrin-dependent neutrophil adhesion. Proc. Natl. Acad. Sci. USA 95: 14540–14545.

Rao, M.V. and Davis, K.R. 1999. Ozone-induced cell death occurs via two distinct mechanisms in *Arabidopsis*: the role of salicylic acid. Plant J. 17: 603–614.

Raskin, I., Ehman, A., Melander, W.R. and Meusse, B.D.J. 1987. Salicylic acid: a natural inducer of heat production in *Arum* lilies. Science 237: 1601–1602.

Raskin, I. 1992. Role of salicylic acid in plants. Annu. Rev. Plant Physiol. Plant Mol. Biol. 43: 439–463.

Rasmussen, J.B., Hammerschmidt, R. and Zook, M.N. 1991. Systemic induction of salicylic acid accumulation in cucumber after inoculation with *Pseudomonas syringae* pv. *syringae*. Plant Physiol. 97: 1324–1347.

Rate, D.N., Cuenca, J.V., Bowman, G.R., Guttman, D.S. and Greenberg, J.T. 1999. The gain-of-function *Arabidopsis acd6* mutant reveals novel regulation and function of the salicylic acid signaling pathway in controlling cell death, defenses and cell growth. Plant Cell 11: 1695–1708.

Rhoads, D.M. and McIntosh, L. 1993. The salicylic acid-inducible alternative oxidase gene *aox1* and genes encoding pathogenesis-related proteins share regions of sequence similarity in their promoters. Plant Mol. Biol. 21: 615–624.

Romeis, T., Piedras, P., Zhang, S., Klessig, D.F., Hirt, H. and Jones, J.D.G. 1999. Rapid Avr-9 and Cf-9-dependent activation of MAP kinases in tobacco cell cultures and leaves: convergence of resistance genes, elicitor, wound and salicylate responses. Plant Cell 11: 273–287.

Ross, A.F. 1961. Systemic Acquired Resistance induced by localized virus infections in plants. Virology 14: 340–358.

Ryals, J., Weymann, K., Lawton, K., Friedrich, L., Ellis, D., Steiner, H.-Y., Johnson, J., Delaney, T.P., Jesse, T., Vos, P. and Uknes, S. 1997. The *Arabidopsis NIM1* protein shows homology to the mammalian transcription factor inhibitor IκB. Plant Cell 9: 425–439.

Ryals, J. Neuenschwander, U.H., Willits, M.G., Molina, A., Steiner, H.-Y and Hunt, M. 1996. Systemic acquired resistance. Plant Cell 8: 1809–1819.

Scofield, S.R., Tobias, C.M., Rathgen, J.P., Chang, J.H. Lavelle, D.T., Michelmore, R.W and Staskawicz, B.J. 1996. Molecular basis of gene-for-gene specificity in bacterial speck disease of tomato. Science 274: 2063–2065.

Seo, S., Okamoto, M., Seto, H., Ishizuka, K., Sano, H. and Ohashi, Y. 1995. Tobacco MAP kinase: a possible mediator in wound signal transduction pathways. Science 270: 1988–1992.

Shah, J. Tsui, F. and Klessig, D. 1997. Characterization of salicylic acid insensitive mutant (*sai1*) of *Arabidopsis thaliana*, identified in a selective screen utilizing the SA-inducible expression of the *tms2* gene. Mol. Plant-Microbe Interact. 10: 69–78.

Shirasu, K., Nakajima, H., Krishnamachari Rajasekhar, V., Dixon, R.A. and Lamb, C. 1997. Salicylic acid potentiates an agonist-dependent gain control that amplifies pathogen signals in the activation of defense mechanisms. Plant Cell 9: 261–270.

Shulaev, V., Silverman P. and Raskin I. 1997. Airborne signaling by methyl salicylate in plant pathogen resistance. Nature 385: 718–721.

Silverman, P., Seskar, M., Kanter, D., Schweizer, P., Métraux, J-P. and Raskin, I. 1995. Salicylic acid in rice. Plant Physiol. 108: 633–639.

Simmons, C., Hantke, S., Grant, S., Johal, G.S. and Briggs, S.P. 1998. The maize lethal leaf spot 1 mutant has elevated resistance to fungal infection at the leaf epidermis. Mol. Plant-Microbe Interact. 11: 1110–1118.

Simons, B.H., Millenaar, F.F. Mulder. L. van Loon, L.C. and Lambers H. 1999. Enhanced expression and activation of the alternative oxidase during the infection of *Arabidopsis* with *Pseudomonas syringae* pv. *tomato*. Plant Physiol. 120: 529–538.

Stange, C., Ramirez, I., Gómez, I., Jordana, X. and Holuigue, L. 1997. Phosphorylation of nuclear proteins directs binding to salicylic acid-responsive elements. Plant J. 11: 1315–1324.

Stone, E. 1763. An account of the success of the bark of the willow in the cure of agues. Phil. Trans. R. Soc. 53: 195–200.

Summermatter, K., Sticher, L. and Métraux, J.P. 1995. Systemic responses in *Arabidopsis thaliana* infected and challenged with *Pseudomonas syringae* pv. *syringae*. Plant Physiol. 108: 1379–1385.

Suzuki, K and Shinshi, H. 1995. Transient activation and tyrosine phosphorylation of a protein kinase in tobacco cells treated with fungal elicitors. Plant Cell 7: 639–647.

Tanaka, Y., Makishima, T., Ichinose, Y., Shiraishi, T., Nishimoto, T. and Yamada, T. 1997. *dad-1*, a putative programmed cell death suppressor gene in rice. Plant Cell Physiol. 38: 379–383.

Tang, X., Xie, M., Kim, Y.J., Zhou, J.; Klessig, D.F. and Martin, G.B. 1999. Overexpression of *Pto* activates defense responses and confers broad resistance. Plant Cell 11: 15–29.

Tenhaken, R. and Rubel, C. 1997. Salicylic acid is needed in hypersensitive cell death in soybean but does not act as a catalase inhibitor. Plant Physiol. 115: 291–298.

Thukle, O. and Conrath, U. 1998. Salicylic acid has a dual role in the activation of defense-related genes in parsley. Plant J. 14: 35–42.

Uknes, S., Winter, A.M., Delaney, T., Vernooij, B., Morse, A., Friedrich, L., Potter, S., Slusarenko A., Ward, E. and Ryals, J. 1993. Biological induction of systemic acquired resistance in *Arabidopsis*. Mol. Plant-Microbe Interact. 6: 680–685.

Van Antwerp, D.J., Martin, S.J., Kafri, T., Green, D.R. and Verma, I.M. 1996. Suppression of TNF-α-induced apoptosis by NF-κB. Science 274: 784–787.

Vane, J.R. 1971. Inhibition of prostaglandin synthesis as a mechanisms of action of the aspirin-like drugs. Nature New Biol. 231: 232–235.

Vane, J.R. and Botting, R.M. 1998. Anti-inflammatory drugs and their mechanism of action. Inflamm. Res. 47: 78–87.

Vanlerberghe, G.C. and McIntosh, L. 1997. Alternative oxidase: from gene to function. Annu. Rev. Plant Physiol. Plant Mol. Biol. 48: 703–734.

Vernooij, B., Friedrich, L., Morse, A., Reist, R., Kolditz-Jawhar, R., Ward, E., Uknes, S., Kessmann, H. and Ryals, J. 1994. Salicylic acid is not the translocated signal responsible for inducing systemic acquired resistance but is required in signal transduction. Plant Cell 6: 959–965.

Walbot, V. 1991. Maize mutants for the 21st century. Plant Cell 3: 857–866.

Weymann, K., Hunt, M., Uknes, S., Neuenschwander, U., Lawton, K., Steiner, H.Y. and Ryals, J. 1995. Suppression and restoration of lesion formation in *Arabidopsis Lsd* mutants. Plant Cell 7: 2013–2022.

White, R.F. 1979. Acetylsalicylic acid (aspirin) induces resistance to tobacco mosaic virus in tobacco. Virology 99: 410–412.

Xie, Z. and Chen, Z. 1999. Salicylic acid induces rapid inhibition of mitochondrial electron transport and oxidative phosphorylation in tobacco cells. Plant Physiol. 120: 217–226.

Yalpani, N., Silverman, P., Wilson, T.M.A., Kleier, D.A. and Raskin, I. 1991. Salicylic acid is a systemic signal and an inducer of pathogenesis-related proteins in virus-infected tobacco. Plant Cell 3: 809–818.

Yalpani, N., León, J., Lawton, M. and Raskin, I. 1993. Pathway of salicylic acid biosynthesis in healthy and virus-inoculated tobacco. Plant Physiol. 103: 315–321.

Yin, M.-J., Yamamoto, Y. and Gaynor, B. 1998. The anti-inflammatory agents aspirin and salicylates inhibit the activity of IκB kinase-β. Nature 396: 77–80.

Yu, D., Liu, Y., Fan, B., Klessig, D. and Chen, Z. 1997. Is the high basal level of salicylic acid important for disease resistance in potato? Plant Physiol. 11: 343–349.

Yu, I.-C., Parker, J. and Bent, A. 1998. Gene-for-gene disease resistance without the hypersensitive response in *Arabidopsis dnd1* mutant. Proc. Natl. Acad. Sci. USA 95: 7819–7824.

Zandi, E., Rothwarf, D.M., Delhase, M., Hayakawa, M. and Karin, M. 1997. The IκB kinase complex (IKK) contains two kinase subunits IKKα and IKKβ necessary for IκB phosphorylation and NF-κB activation. Cell 91: 243–252.

Zhang, S. and Klessig, D.F. 1998. Resistance gene N-mediated the novo synthesis and activation of a tobacco mitogen-activated protein kinase by tobacco mosaic virus infection. Proc. Natl. Acad. Sci. USA 95: 7433–7438.

Zhang, S. and Klessig, D.F. 1997. Salicylic acid activates a 48-KD MAP kinase in tobacco. Plant Cell 9: 809–824.

Zhang, S., Du, H. and Klessig, D.F. 1998. Activation of the tobacco SIP kinase by both a cell wall-derived carbohydrate elicitor and purified proteinaceous elicitins from *Phytophthora* spp. Plant Cell 10: 435–449.

Zhang, Y., Fan, W., Kinkema, M., Li, X. and Dong, X. 1999. Interaction of NPR1 with basic leucine zipper protein transcription factors that bind sequences required for salicylic acid induction of the *PR-1* gene. Proc. Natl. Acad. Sci. USA 96: 6523–6528.

Plant Molecular Biology **44:** 443–444, 2000.
E. Lam, H. Fukuda and J. Greenberg (Eds.), Programmed Cell Death in Higher Plants.

Index, Vol. 44 No. 3 (2000)

MIX
Papier aus verantwortungsvollen Quellen
Paper from responsible sources
FSC® C105338

If you have any concerns about our products,
you can contact us on
ProductSafety@springernature.com

In case Publisher is established outside the EU,
the EU authorized representative is:
Springer Nature Customer Service Center GmbH
Europaplatz 3, 69115 Heidelberg, Germany

Printed by Libri Plureos GmbH
in Hamburg, Germany